ARNOLD J. TOYNBEE

ARNOLD J. TOYNBEE

A Life

William H. McNeill

OXFORD UNIVERSITY PRESS
New York   Oxford

Oxford University Press

Oxford   New York   Toronto
Delhi   Bombay   Calcutta   Madras   Karachi
Petaling Jaya   Singapore   Hong Kong   Tokyo
Nairobi   Dar es Salaam   Cape Town
Melbourne   Auckland

and associated companies in

Berlin   Ibadan

Copyright © 1989 by William H. McNeill

First published in 1989 by Oxford University Press, Inc.,
200 Madison Avenue, New York, New York 10016

First issued as an Oxford University Press paperback, 1990

Oxford is a registered trademark of Oxford University Press

Library of Congress Cataloging-in-Publication Data

McNeill, William Hardy, 1917–
Arnold J. Toynbee : a life
Includes bibliographical references and index.
1. Toynbee, Arnold Joseph, 1889–1975.   2. Historians—
Great Britain—Biography.   I. Title.
DA3.T68M37   1989   907'.2024   88–23188
ISBN 0-19-505863-1
ISBN 0-19-506335-X (pbk.)

2   4   6   8   10   9   7   5   3   1

Printed in the United States of America

To My Wife

Acknowledgments

I entered upon the task of writing this biography in 1986 at the invitation of Arnold Toynbee's sole surviving son, Lawrence. He, together with his wife and other members of the immediate family, have been unfailingly helpful and cooperative throughout the book's gestation. Those to whom I am particularly indebted, in addition to Lawrence and Jean Toynbee, include Mrs. Philip Toynbee, Margaret Toynbee, Polly Toynbee, Rosalind Pennybacker, Anne Powell Wollheim, and Alexander Murray.

Judgments and opinions are my own, and inferences and guesses I have ventured about some important aspects of Toynbee's life are no more than hypotheses invented to make the story make sense. I have found that writing biography required greater leaps of the imagination than I remember making when writing world or any other kind of history. The reason is obvious: some things are hidden, systematically and deliberately, because they seem shameful or embarrassing. Hence the documentary record of anyone's life will have gaps—yawning gaps—which only an act of the imagination can fill in. I certainly found this to be true for Toynbee, despite the abundance of his letters and writings.

This biography is also, in a small way, an essay in world history, since the contagion of Toynbee's ideas was something of a global phenomenon. The country I needed most help in trying to understand was Japan. A long list of correspondents and consultants helped me with that task, and with the practical question of getting access to some of Toynbee's most important writings that are lodged at Nihon University Library in Tokyo. Those to whom I am indebted include Mikio Sato, Muneaki Naganuma, and Tokuharu Shoji of Nihon University; Keisuke Kawakubo of Reitaku University; Shigeru Nakayama, whose affiliation I do not know; Daisaku Ikeda and Fred M. Nakabayashi of Soka Gakkai; Masao Nishikawa of the National Committee of Japanese Historians; Kei Wakaizumi of Kyoto Technical University; Mike Mansfield and Eugene A. Nojek of the U.S. Embassy, Japan; Donald Keene of Columbia University; my colleagues Akira Iriye and Tetsuo Najita of the University of Chicago and R.E. Schecter of American Family Foundation. Yet as in the case of my efforts to understand the dynamic of Toynbee's personality, my efforts to describe the circumstances in Japan that led to Toynbee's sudden rise to fame in the 1970s is a work of my imagination, feeding on the materials provided by those who helped me. They ought not to be held responsible for my conclusions any more than the family should be supposed to concur with all I have said about Toynbee's private life.

A fellowship from the Guggenheim Foundation helped meet the expenses of my time in England, and for that I am doubly grateful since it was a second award to me and an unusual honor.

Another set of obligations is to the libraries that allowed me to use their resources for preparing this book. Pride of place must go to the Bodleian Library, Oxford, and particularly to Colin Harris and other members of the staff of the Western Manuscripts division, where most of Toynbee's papers are stored. The backbone of my biography comes from the array of materials under their control. The vague style of annotation I have used when citing these holdings conforms to the advice of the Curator of Western Manuscripts, who explained that references ought not to be made to the cartons in which the Toynbee papers are temporarily (and confusedly) stored, since in future, when they have been properly catalogued, new carton and sheet numbers will be asssigned. Other libraries used in the preparation of this book are the Library of Congress, Yale University Library, the University of Chicago Library, and the interlibrary loan system of the Beardsley Memorial Library of Winsted, Connecticut. The archivists of Time, Inc., New York also were very helpful.

Finally, a number of individuals went out of their way to share information of one sort or another. These include Christian Peper, Imogen Seger-Coulborn, Christopher Collins, Richard Stern, Jane Caplan, Spiros Doxiadis, David Astor, Richard Clogg, Osman Okyar, George Curry, Father Columba of Ampleforth Abbey, Father Sweeney of Boston College, Dorothy Hamerton of the Royal Institute of International Affairs, P.-F. Moreau and Mme. Laffite-Larnaudie of the Institut de France, and Robin Denniston and Sheldon Meyer of Oxford University Press. Last but not least I owe a special debt of gratitude to Elizabeth Meyer of Yale University, who translated Toynbee's Greek and Latin poems for me, and to members of my own family who read the manuscript in its formative stage and asked for explanations of matters that were self-evident to one of my generation but not to theirs.

Colebrook, Conn.   W. H. M.
21 January 1988

Contents

ARNOLD J. TOYNBEE

I

Great Expectations

Arnold Joseph Toynbee was born on Palm Sunday, 14 April 1889, in his great uncle's house near Paddington station, London. His name was a burden in itself, commemorating, as it did, the infant's famous uncle, Arnold, and his spectacularly successful grandfather, Joseph. The burden became the greater as his father's career faltered, focusing his mother's hopes and aspirations more and more on her son. Only heroic achievement could live up to such expectations, and Arnold Joseph Toynbee responded from early childhood by cultivating an exceptional memory and extraordinary verbal skills so industriously that he soon excelled his contemporaries in schoolwork and went on in later life to become world famous as a historian and publicist.

The Toynbee family came from Lincolnshire, where Arnold's great grandfather, George Toynbee (1783–1865), was a prosperous farmer. George's third son, Joseph (1815–1866), became a fashionable London doctor, specializing in problems of the ear and throat. He treated Queen Victoria for deafness—successfully—and moved familiarly amid the intellectual elite of his day, counting John Stuart Mill, John Ruskin, Michael Faraday, Benjamin Jowett, and Giuseppi Mazzini among his friends and acquaintances.[1] Joseph Toynbee was active in medical research, published abundantly, and met his death prematurely when he exposed himself, experimentally, to an overdose of chloroform.

This catastrophe had a chilling effect on the family. Doctor Toynbee, to be sure, had charged unprecedentedly high fees for his services and accumulated sufficient capital at the time of his death to maintain his widow in the style to which she was accustomed. Eventually, after her death in 1897, her youngest son, Harry Valpy, father of Arnold Joseph Toynbee, the historian, received as his share of the inheritance (divided among eight children), a sum of capital yielding between £200 and £300 a year, which was exactly half his salary at the time.[2] Nevertheless, the expansive, adventurous spirit that prevailed during

Joseph Toynbee's prime disappeared; careful conservation of resources and something of a struggle to retain upper middle class status took its place.

As far as Joseph's second son, Arnold (1852–1883), was concerned, this meant that going to university required him to win a scholarship. This he was able to do, matriculating at Oxford in 1873 and receiving a B.A. degree in 1877. As an undergraduate, the young man made a remarkable impression on his contemporaries and teachers. Upon completion of his undergraduate studies at Balliol College, the master, Benjamin Jowett, first found him a post as special tutor to the second son of the Duke of Bedford, and then, when a place opened, made him a regular tutor at Balliol. He continued there until his death, of "brain fever," in 1883, when he was just thirty years of age.

In retrospect, it is hard to appreciate Arnold Toynbee's fame, for he made his mark by talking rather than by writing; and the surviving summaries of his public speeches and lectures, together with the few letters his sister published after his death, do not seem particularly noteworthy. But, as Benjamin Jowett wrote after his death: "The really interesting and striking thing in his life was not what he actually produced, but Himself, that is to say . . . his unlikeness to everyone else."[3] And in a letter to Arnold's sister, Jowett declared: "I am sure that he was one of the best persons I have ever known."[4]

Two things are clear. First, Arnold Toynbee became the animating spirit of a group of young Balliol men who aspired to reform church and society in such a way as to bridge the gap between themselves and the working class. This involved living among the poor in London during university vacations, and discussing "the laws of nature and of God" with working men under the aegis of an energetic Anglican clergyman. Eventually Arnold Toynbee's effort to communicate with the poor led a group of his admirers to establish Toynbee Hall in the east end of London. This, the first settlement house, was intended to replicate within its walls something of the life of an Oxford college, thus propagating at least some of the benefits of university education among people hitherto excluded.

In the second place, Arnold Toynbee set out to revise and update the dismal science of economics, thus bridging the gap between rich and poor in theory as well as in practice. "Morality must be united with economics as a practical science," he declared in his lectures on the industrial revolution in England. "The historical method has revolutionized Political Economy, not by showing its laws to be false, but by proving that they are relative for the most part to a particular stage of civilisation."[5] Just what remedies he might have found for economic inequality, had he lived, cannot be known. He was not a socialist, perhaps had never even heard of Marx, and his published fragments, none of them actually finished works, give almost no clue. Instead, his early death aroused something of a cult among his family and in a broader circle of those he had influenced through private conversation. His name became one to invoke in any discussion of "the social question"; discussion which became endemic among serious-minded Oxford undergraduates and dons in the decades before World War I.

Arnold Toynbee's career had telling impact on his youngest brother, Harry Valpy Toynbee (1861–1941). Harry, the baby of the family, was only five years old when his father Joseph died so unexpectedly. After finishing school, he somehow blundered into the tea business, though young Harry soon found the routine of the office and the company of his fellow clerks thoroughly unattractive. He aspired to a university career like his brother; but Arnold assured him, in a very patronizing letter, that university men were really no different from men of business. "Don't think of throwing up your present work until you see quite clearly what other work there is you can do which will suit you better and enable you to earn a livelihood. . . . The tea business is surely not hopeless."[6]

In 1883, the year of Arnold's death, Harry Toynbee did find another job, becoming District Secretary in Hempstead for The Charity Organisation Society. This was a private association of aristocratic and well-to-do middle class people who hoped to help the poor to help themselves. They aimed at healing the breach between the lower classes and themselves, just as Arnold Toynbee had wished, but in a more immediately practical fashion. Between 1875 and 1913 the COS was headed by another Balliol man, Charles Stuart Loch; and under his leadership the Charity Organisation Society pioneered the case work method of helping the poor. Funds came from private donations; case work, too, was on a volunteer basis. Only a skeleton staff of district secretaries received salaries, which remained at a modest, not to say skinflint, level. Even when Harry Toynbee became one of two Organizing Secretaries in 1898, with several district secretaries under him, his salary was only £400 a year.[7]

Thus in acting upon his brother Arnold's example, and devoting his life to bridging the gap between rich and poor, Harry Toynbee created an intractable problem for himself. His income was simply insufficient to sustain the upper middle class style of life he had known as a child, and the fact that most of his siblings were better off than he made the burden of his ideals even harder to bear. The contrast was particularly sharp in the case of his brother Paget, three years younger than Arnold, who became a sort of unofficial head of Joseph Toynbee's family after Arnold's death. Paget married well enough to become a country gentleman and was able to devote his life to Dante scholarship. The gap between his comfortable way of life and Harry's cramped circumstances strained family solidarity. Paget's efforts to be helpful somehow smacked of condescension to poor relations. As a result, Harry and Paget never got on smoothly, and the prickliness was on both sides.[8]

Harry's marriage, in July 1887, to Sarah Edith Marshall (she was known always by her second name) did nothing to relieve the pecuniary pinch, for his wife's father, a Birmingham industrialist, had lost his factory, and everything else, when bankers refused to give him loans he needed to shift from wood to metal construction of the railroad rolling stock built by his firm.[9] Instead of setting up a home of their own, the newlyweds therefore moved in with Joseph's youngest brother, the elderly and opinionated "Uncle Harry" (1819–1909). He was a retired sea captain and pioneer meterologist, whose wife's death had left him rattling around in a modest but respectable house at 12 Upper Westbourne Terrace, London. Here it was, twenty-one months later, that Arnold Joseph Toynbee was born.

Almost eight years passed before a second child, Jocelyn, arrived in 1897; Margaret followed in 1900, completing the family. His sisters were thus too young to be playmates for Arnold. His important relations in early life were with adults: his mother first and foremost, a well-loved Nanny and Uncle Harry competing for second place. His father worked too hard and came home too tired and with emotions too tense to establish easy and warm relations with his son, however much he loved, admired, and wished him well.

Edith Marshall Toynbee (1859–1939) became the dominating figure in the life of her son, and retained that role until his marriage in 1913. She was a strongminded woman, utterly firm in her Anglican faith, her English patriotism, her sense of duty, and her attachment to her son. As a young woman, she had defied convention to the extent of pursuing higher education in an age when women were not admitted to English universities. But at Cambridge an arrangement had been made whereby young women could study university subjects and take the same exams as the men without, however, receiving degrees. Edith Marshall, accordingly, enrolled in what subsequently became Newnham College, Cambridge, followed a course in modern history, and won a First in the examinations. She and another woman were, in fact, the only Firsts in history that year: a feat long remembered, and remarkable enough in an antifeminist age. She later taught school but gave it up when she married, since convention required married women of the middle class to remain at home, manage the servants, and keep house.

Surviving records do not allow us to know how Edith Marshall Toynbee felt about being squeezed into the mold expected of her. Nothing suggests that she did not heartily endorse the propriety and necessity of surrendering her own career as a teacher to take on her new role as wife and mother. Yet she certainly faced unusual difficulties in adapting to her new circumstances. First and foremost, she was not really mistress of the household, for Uncle Harry owned the house and had already lived there for years when Edith and Harry moved in. Long since retired, he usually remained on the premises all day long and, being accustomed to his own ways, was not in the least backward about insisting on them.[10] Thus Edith's sovereignty within the household was sharply cut back from what a Victorian lady expected. Her temperament, naturally imperious, must have chafed at such constriction.

In addition, there was the problem of income. Her husband's salary did not permit them even to contemplate a separate household. Real respectability, as defined by the pattern of prosperous middle class life both she and her husband had known in their own childhoods, remained beyond their reach. Yet both Harry and Edith had their full share of pride; and if the outward graces of life were too expensive for them to enjoy, the role of poor relation was much too humiliating to be accepted. A solution in practice seems to have been to turn inward: minimizing contacts with other, more prosperous members of the Toynbee family, for example; entertaining little; and maintaining an appropriate facade of gentility, culture, and independence by dint of the utmost economy of daily expenditure. Yet there were servants: a cook as well as a nanny living in, and part-time cleaning help as well. Anything less would have meant irreparable surrender of middle class status.

Something of the difficulties Edith Toynbee confronted in the first years of her marriage can be glimpsed from the story her son tells of how she came to write the only book that carries her name, a child's history of Scotland. When the young Arnold was turning four, his father decided that they could no longer afford to pay the nanny who had been hired to help care for him. She had become an important figure in the household, and Edith persuaded her husband to relent on condition that she used the time freed from baby care to write a book, in the hope of bringing in enough to pay the nanny's wages. The arrangement worked: Edith's book was duly written, and the £20 fee she received from the publisher sufficed to pay the nanny's wages for an additional year. As a result, the distinctive smell of printers ink, coming from the galley proofs his mother had to correct, were among Arnold's earliest childhood memories.[11]

Edith tested the chapters of her book on her son, reading each aloud to him at bedtime. This was only a part of her persistent and successful effort to share her view of history and of the world with the young child. Stories from the past, improvised from her store of memories of what she had studied at Cambridge, or read aloud from others' books, were regular bedtime rituals. The young Arnold's interests in history and his youthful determination to become a historian arose directly from his mother's teaching and example.[12]

Until he started school, playmates must have been few to nonexistent. Both his mother and Uncle Harry encouraged a precocious bookishness and cultivated the child's skill with language. Uncle Harry, for example, gave him pennies for memorizing Bible passages, with the result that in adult years Arnold Toynbee could quote accurately and from memory a substantial part of both the Old and New Testament. Uncle Harry was, indeed, something of a religious fanatic. Heir and exemplar of the puritanical tradition, he viewed popery as an ever-lurking danger to the Anglican establishment and could find only a single clergyman in London whose Low Church views met his own exacting standards. His frequent diatribes against Catholicism and High Church rituals that smacked too much of popery rather embarrassed Arnold's parents, since their Anglicanism was middle of the road and far more tolerant than the old man's puritanical version of the national faith.

Just what the young boy absorbed from his extended exposure to biblical texts and to his great uncle's theological rancor is hard to tell. There was no unquestioning commitment or concurrence; his parents' reservations prevented that. Yet it is interesting to know that when Arnold was still in his cradle Captain Harry Toynbee published a religious tract, entitled *The Basest Thing in the World,* in which he found "man's greatest failing to be idolatry of self."[13] His great nephew echoed that indictment exactly, some forty-five years afterward, with the difference that he made collective self-idolatry, that is, nationalism, the central sin of humankind, rather than concentrating, as Uncle Harry had done, solely on personal and clerical forms of "self-worship." Suffice it

to say, therefore, that his extraordinary exposure to an aggressive, puritanical
form of protestant piety seeped into the young boy's consciousness, and, though
he held himself aloof, never accepting his great uncle's viewpoint uncritically,
nevertheless a residue remained, and it constituted an important background
and context for all of Arnold J. Toynbee's later thinking and writing.

Much the same might be said for his mother's patriotic version of English
history, which also pervaded his earliest conscious years. In that history wars
and battles figured largely, though the advance of liberty and its slow perfection
through the growth of the British constitution provided the organizing principle
of her vision of the past. But the stories that lodged in the young boy's mind
were mainly military: deeds of valor, critical battles, skillful generalship. Oddly,
naval derring-do seems, from the scattered juvenilia that survive, not to have
attracted his attention.

Arnold's earliest surviving composition, lavishly illustrated by childish drawings, describes a battle fought by rival armies of animals. The victorious general, Pug (probably one of his nursery toys), then prophesies: "Your generation
will grow slovenly and cowardly and shall be beaten down by the nation called
men. . . . But soon men will begin to fight each other and get disorderly, and
then will Peppo's children rule all men. It will take possetion [sic!] of all India
and America and part of this land that we live in, called Africa, and they will
live in a distant land called England, and call themselves Englishmen. The
most famous of their reigns will be under a Queen called Victoria, and all your
other children, your generation, Peppo will kill and nothing will be left." The
tale concludes with a final battle in which "Peppomights massacred all the
Pugwrights. The End." [14]

The story actually combines biblical echoes and traces of fairy tale with a
childish version of his mother's stalwart patriotism, for it begins "Pug was a
cloud, but he tried to blot out the sun," and as a punishment he was turned
into an animal, in which guise he "made all the trees and eatables," and, as
we saw, then played the role both of Judge and Prophet to the rest of the animal
kingdom. Thus it appears that even before his formal schooling had begun, the
future historian was already capable of an imaginative and imaginary world
history. But the most striking thing about this childish composition is the odd
way in which his precociously adult vocabulary contrasts with the naive blood-
thirstiness of the sentiments expressed in the story itself.

Those sentiments show, perhaps, that the special constraints and stimuli of
his childhood did not prevent the young Arnold from enjoying a reasonably
normal boyishness. Summer vacations, often by the shore, were a regular feature of the family life. Then Arnold could bicycle, swim, or play a game called
"Fives" with his father; but at other times of year Harry was too busy to spend
much time with his son. As a result, Arnold's games were mostly solitary. In
particular, playing with toy soldiers engrossed many hours and continued to
absorb him throughout his schoolboy years. His sister Margaret long remembered how, on getting back from school, he hastened to take his various regiments from their storage boxes and put them on parade—thus reenacting one
of his important preschool pastimes, and exercising, in imagination, the de-

lights of command! For his sisters, Arnold's toy soldiers were tantalizingly forbidden; but once, when he was sixteen years old, he gave Margaret a much coveted goat mascot, belonging to his detachment of Welsh Fusiliers. This was meant as consolation for the impending departure of her nanny, decreed, on financial grounds, by their father. She added the goat to her Noah's ark collection, and vividly remembered her brother's generosity more than eighty years later.[15]

Arnold was physically vigorous, and as a small child his nanny or his mother took him to Kensington Gardens almost every day. That park lay conveniently close to where he lived and provided open space where he could run about and play. But he never became adept at games. This was not due to muscular deficiency. Both his parents were athletic. His father won several cups at school for sports, and his mother was a formidable walker all her life. Consequently, Arnold inherited a healthy body and muscularity that made him indefatigable at cross-country walking and a good long-distance runner. What was missing, presumably, was sufficient early practice at cricket and the like to allow him to keep up with his fellows when he started school. And being accustomed to excel at his books, he found it galling to fall short on the playing field.

His response was to distance himself from the noisy fellowship of games, minimizing participation in school sports and affecting disdain for all such activities. Solitary occupations, especially the literary and historical studies in which he delighted, were far more attractive; but this, of course, enhanced a desolate loneliness, which pervaded his school years. Even at the beginning of his last term at school, when his intellectual triumphs had already assured him a leading place in the school's pecking order, he wrote to his mother: "I always feel lonely here: I can never settle down or really identify myself with the School, and that is one reason why I don't want to be Prefect of hall."[16]

His mother inaugurated her son's education, teaching him to read and encouraging him to draw. Mutual admiration blossomed between them, for Arnold's quick mind responded precociously to his mother's guidance; and his promise soon became for her a counterweight to the disappointments that increasingly haunted her husband's career. These disappointments were not simply pecuniary—though financial stringency, and consequent worry about how to pay for their children's education, was the most obvious aspect of Harry's difficulty. Still, there was more bothering him than just money.

For one thing, the assumptions upon which the Charity Organisation Society was based were becoming antiquated by the last decade of the nineteenth century. The condescension implicit in asking well-to-do persons to volunteer to help the poor by giving advice, approving small payments when their morals, on inspection, proved deserving, and, more generally, trying to establish commonality between rich and poor by the exercise of noblesse oblige in an urban environment smacked of what Harry himself found so unsatisfactory in his relations with his brother Paget. Attitudes and class relationships which were still viable in rural England could not be replicated in London; yet that, in essence, was what the Charity Organisation Society was trying to do.

Moreover, as a paid member of the staff, Harry found himself in an ambig-

uous and awkward position vis-à-vis the leisured volunteers who actually did
the work of the society. In technical matters he was their superior in the sense
that he knew more of what to expect in dealings with the poor; but in face to
face encounters, he was their inferior because his actual social and economic
status was so far below theirs. Thus, in practice, Harry's chosen life of service
to the poor compelled him to fawn upon the rich in order to elicit a condescending benevolence that embraced him as well as the ostensible objects of their
charity. This, presumably, was not what he had bargained for in abandoning
the tea business.

Finally, from the 1890s, socialist ideas had begun to percolate into British
society, undermining the intellectual assumptions upon which the activities of
the society were based. Out and out socialists remained few before World War
I, but their voices were loud enough to challenge the virtue and effectiveness
of the Charity Organisation Society. As a result, its approach to the "social
question" was no longer an unquestioned good. Yet Harry Toynbee's career,
giving up personal gain for the good of society, could be justified only if the
Charity Organisation's activities were both right and effectual. Insofar as doubts
crept in—and no surviving documents show whether they did or not—Harry's
career lost its moral justification. What alone is sure is that he took leave from
the Charity Organisation in 1907, accepting a temporary position with a Royal
Commission on the Poor Laws instead—an appointment that took him on prolonged visits to the north of England.[17] When that job terminated, Harry Toynbee began to suffer from acute depression and, in 1909, entered a mental
hospital. He never again emerged into normal life, though he lived until
1941.

It was against this family background that young Arnold's formal education
proceeded. When his mother decided that he needed more instruction than she
herself could provide, he was sent to study under the aegis of a governess in a
friend's house nearby. Here he was thrown with two older children, a boy and
a girl, who teased Arnold by deciding that when they grew up he would be
their servant. After a year or so of this, Arnold became a day student at War-
wick House in Maida Vale, London. Then at age ten a small legacy (and reduced fees) allowed his mother to send him off to Wootton Court, Kent, where
Greek as well as Latin was available.[18] The curriculum at Wootton Court made
entry into one of the prestigious public schools conceivable: and this was indeed the hope and intention Arnold's parents held for him. But such schools
cost far too much for Harry Toynbee to pay the fees. The only path to the top
open to his son therefore lay through winning a scholarship. Arnold buckled
down to the challenge, endured all the pangs of loneliness his banishment from
home implied, and disciplined himself to unceasing application to schoolwork—all in the hope of qualifying for a scholarship to one of England's top
schools.

His first years away at school were therefore critically important in forming
the habits that sustained Arnold throughout his life. He did not meet with complete and universal success. Mathematics was difficult for him, and he once
had to repeat a term of Euclid. On the other hand, he easily excelled in lan-

guage studies, where his extraordinary inherent abilities had already been sharpened by the adult company his mother and great uncle provided.

Letters from home pricked him on. Three weeks after bidding her ten year old son a tearful farewell at Victoria Station, Edith Toynbee inquired: "Do you have marks and places in class? I hope not, but if you do let me know what your place is each week." A month later: "I am longing for your letter on Monday to hear what your prizes are." [19] After his first summer vacation, the pressure relaxed a little. "So you are moved up into the First. Well, you won't need to worry about the prize at Xmas or even Midsummer. We shan't expect to see another prize for two years. . . . Work on quietly and steadily and we shall be pleased. You'll soon get on with football." And again, "Of course you will find the work difficult, and do some of it badly: you needn't be discouraged." [20]

Arnold's letters to his mother are not preserved, but it takes little imagination to picture him working away at his books, faltering on the playing field, and feeling miserably lonely without the loving attention to which he had become addicted at home. His mother had a Spartan streak in her, to be sure, and tried to brace her son against shyness. "It is a pity to fancy the boys don't like you. If you are jolly and friendly with them they will be with you. R.L. Stevenson, all through his ill-health, made it a duty to be happy, a lesson you and I sometimes forget, eh laddie?" [21]

A "Nature Notebook" from his time at Wootton Court contains water colors of flowers, animals, and butterflies, some of which attain a considerable realism. Yet on the last page Arnold drew a monkey in a dunce cap, with spectacles, cane, and book. The balloon issuing from its mouth reads: "I am Toynbee and an ass." [22] Obviously, school, despite (or even because of) his triumphs in the classroom, was not an unmitigated delight. He found some consolation in the company of a dormouse, which the gardener at Wootton Court gave him. He built a cage for it, and took "Dordy" home with him in 1901. Thereafter, regular reports on Dordy's appetite and behavior became a staple of his mother's letters, along with fiercely patriotic comments on the treacheries of the Boers and the slowness of English victories in South Africa.

But the real agendum for his parents was to get their son into one of the famous public schools of England. They set their sights on the very top by enrolling Arnold in the scholarship examinations for Winchester in the summer of 1901. He fell short of success by a single place. Harry Toynbee wrote: "It is no doubt very disappointing to you to have missed the scholarship, but you have every reason to be encouraged. . . . I am delighted with the way you behaved all through the exam, and am very proud to think you have done so well." [23] On the next day his mother wrote: "Father and I are determined that if we can by any means accomplish it you shall go to a good public school, out of London, scholarship or no scholarship, and have a start in life. . . . Instead of fretting over the result of this exam, it is an occasion to take fresh heart and feel very cheerful and encouraged. 'Rest in the Lord: wait patiently for Him, and he shall give thy heart's desire'. That will be your reading tonight and a message straight from God." [24]

The next summer Arnold tried again, and this time he cleared the hurdle triumphantly, ranking third among candidates for the eleven vacant scholarships. As a result, in the autumn of 1902 he entered Winchester as a member of the College. This was the corps of scholarship holders, seventy in number, who enjoyed the benefit of an endowment made in the 1380s by the school's founder, William of Wykeham, Bishop of Winchester. Bishop William was a medieval churchman who had risen from obscure origins to become Chancellor of England under Edward III, and in endowing the school he intended to make it easier for others to imitate his career. From its foundation, therefore, the College was supposed to support the schooling of gifted boys so as to allow them to qualify for Oxford, where the good bishop had already established New College. A special connection between Winchester and New College therefore continued into the twentieth century, since graduates of the school could more or less count on places at New College, Oxford.

But the school Arnold Toynbee entered at the age of thirteen was no longer medieval, having been fundamentally transformed in the sixteenth and again in the eighteenth centuries. In the sixteenth century, classical Latin and Greek had been substituted for medieval Latin as the staple of the curriculum. What happened in the eighteenth century was that well-to-do families, wishing to send their sons to Oxford, came to recognize the advantage of the training Winchester provided. The schoolmasters accommodated them by admitting boys who were off the endowment on condition that they pay suitably high fees for the privilege. This practice expanded throughout the nineteenth century with the result that by 1902 the great majority of the students were such "commoners."

Another important change took form mainly in the nineteenth century when most of the everyday governance of the school was entrusted to senior prefects, chosen from among the older students. Soon the boys developed a complex pattern of rules and taboos, defining what could and could not be done throughout the day. A special language went along with these "notions," and newcomers to the school had to be initiated into the system, partly by formal instruction and partly by beatings administered by the prefects for infractions of the rules.

Relations between Collegians and fee-paying commoners were ambiguous. The scholarship holders were an intellectual elite, and insofar as the school was an intellectual community, their prestige stood correspondingly high. But those whose families were able to pay the school's fees were a different kind of elite—the rich and well born of Great Britain. Their sons were destined by birthright (suitably tempered by education, of course) to become leaders of British society and rulers of an empire which in 1902 encompassed a great deal of the globe.

Thus Winchester, like the other famous public schools of England (Eton, Harrow, and the rest, were all younger than Winchester and in some degree were modeled upon it), mingled talent with birth and wealth. The school became the setting for a remarkable amalgamation of ability and ambition on the one hand with inherited upper class status on the other. Old school boys shared a body of knowledge (primarily a familiarity with Greek and Latin classics) and a community of manners, inculcated perhaps more on the playing fields than anywhere else. Winchester's motto, "Manners makyth Man," was an apt one in the sense that those who had passed through its discipline in youth emerged with a distinct ethos and with a mode of speech that allowed others to recognize and respond to what was universally perceived as membership in the "upper class."

No other country of Europe had such a pattern of recruitment into the ruling elite; only Confucian China could compare. Arnold J. Toynbee's subsequent career, both in government and on its fringes, illustrates the power and flexibility of the system, for it gave him access to the seats of the mighty, even while still a young man, and made his voice one that counted in public affairs from the time he had completed his formal education. The fact that he was a "Wykehamist" and then became a Balliol man at Oxford made him part of an old school network that ramified throughout government, reaching to the very top.

Yet for the newcomer in 1902 the ways of the school were hard to accept. A special vocabulary for places and persons within the school grounds had to be mastered along with all the rules of what was and was not allowed. Living conditions were, if not medieval, at least uncomfortable. Early rising, cold water ablution out of doors, even in freezing weather, and similarly bracing rites were part of the mandatory routine; and the schoolboys were quite literally walled in and cut off from contact with the outside world, save on strictly supervised occasions.

A new boy was assigned to a "father," who was supposed to instruct him in the school's folkways and notions. Toynbee described his initiation, many years later, as follows: "At the end of the [first] fortnight, the new man was examined by the two prefects of his chamber. If he passed, he had to give a meal to his 'father' at School Shop—the 'father' doing the ordering. If he failed (I failed the first time), his 'father' was beaten by the prefects with a wooden bat. . . . I strongly objected [to Winchester notions] at the time."[25] He objected, also, to the food and the chill, judging by an undated letter from his mother that urged him to eat more so as to "feel jollier," and he much disliked the nickname, "Tubbs," that his fellows soon foisted upon him. At the end of his first "half" at Winchester he fell ill with pneumonia—a hazard of exposure to drafts in unheated rooms that provoked mortal illness in at least one of his schoolmates. Arnold, of course, recovered, aided by the gift of a live owl which for a few months replaced the now dead dormouse in his affections, until it, too, died.[26]

Yet there were compensations. First of all, Arnold found himself among boys with abilities comparable to his own, and he made some lifelong friends. Two stood out in later years: "Eddie" Morgan, who became a bishop, and David Davies, who became a judge. Davies was probably Toynbee's closest friend. Late in life they had lunch together nearly every week when both were in London, and Davies signed as witness at Toynbee's first wedding in 1913, and again in 1946.

Within the school, the female sex was conspicuous for its absence. Since

Arnold had grown up enveloped by his mother's love, and with two little sisters who obediently echoed their mother's opinion of his abilities, the absence of admiring females was a severe deprivation for him. But once he had proven his mettle by attaining distinction in schoolwork, Arnold was able to find a pale but valuable substitute for his absent mother in Mary Carter, a sister of one of the schoolmasters, who lived nearby and invited Arnold and other boys to have tea with her on occasions when leaving school grounds was permitted. Like his schoolmasters, she admired Arnold's prowess in Greek and Latin, and after he had gone off to Oxford wrote him several warm letters, signed "Aunt Mary."

But what really mattered, of course, was Arnold's place within the school. His lackadaisical participation in team sports kept him marginal to the rough and tumble side of schoolboy life. But at Winchester, more perhaps than in other public schools, intellectual distinction also commanded real admiration. Accordingly, Arnold did not struggle in vain when he chose to concentrate on winning the respect of his fellows solely by his performance as a student.

Classroom exercises were central of course, but Winchester also sustained a lively extracurricular intellectual life. This took the form of voluntary gatherings with masters in leisure hours at which some special topic was addressed, or more formal clubs, like the XVI Club, whose members read papers in history to one another in imitation of the practice at Oxford. The character and quality of these gatherings obviously depended on the schoolmasters who presided over them. By all odds, the most influential teacher Arnold encountered at Winchester was the "Ostiarius," or Second Master, Montague John Rendall, a parson's son and teacher of Greek. His office also made him housemaster to the College, putting him in daily contact with the seventy privileged, and yet underprivileged, scholarship-holders.

Rendall was, as Toynbee later described him, "both an aesthete and a puritan," and a gifted teacher as well. By using lantern slides to project images of the Greek landscape before his students' eyes, he brought a deadly dull textbook of ancient Greek history to life, at least for Arnold, whose extraordinary sensitivity to historical landscapes must have been either sharpened or provoked by these slides and by Rendall's commentaries on how he had taken the various photographs himself when traveling in Greece. Rendall was also a devotee of Italian Renaissance painting and had a collection of art slides which he showed in his rooms to groups of boys on weekends and after hours. His enthusiasm for painting and architecture was contagious, and it gave Arnold a richer appreciation of visual art than the bookish curriculum, by itself, could have done. Once, after Rendall had taken a group of boys for a picnic and visit to an art gallery, Arnold wrote home. "I should think there are few schoolmasters who would do all that Mr. Rendall does for us; but then one does not think of him as a master but as a friend."[27] Music, by contrast, seems to have been almost absent from his schooling, and he never exhibited the least sensitivity to or interest in that art, so far as surviving evidence shows.

Schoolwork was largely literary and historical, though in his earliest years at Winchester, Arnold still had to labor with mathematics and even encountered chemistry and some other branches of science. Since he failed to excel in these fields, his reaction was to evade exposure to them as soon as the rules of the school allowed him to do so. As a result, he never really understood modern science. The Christian religion, likewise, had surprisingly little influence within the school. It seems to have been completely taken for granted in its moderate Anglican garb, and genteelly neglected. What remained was ancient Greek and Latin, which had been the core and staple of the Winchester curriculum since the sixteenth century, supplemented with a little modern history, language, and literature (English and German principally), taught (oddly) by mathematics masters.

The boys were not only required to read widely in the classics, but also had to learn to compose both prose and verse in the ancient languages, conforming to strict rules of prosody and using only those words and grammatical forms for which Ciceronian or Attic precedents could be found. The school even possessed special dictionaries of acceptable terms, against which Arnold checked his prize essays, lest he fail by inadvertently using a "silver" Latin expression or degenerate New Testament Greek! Reading skill in Greek and Latin was taken for granted; real admiration was reserved for those who could compose faultlessly in the ancient languages, conforming to rules the school had inherited from humanist reformers of the Tudor age who had banished medieval Latinity in favor of the pure, authentic classic tongues. A boy of Arnold's linguistic ability could in fact achieve this pinnacle of literary artificiality. He made composition in Greek and Latin so much a habit that, throughout his life, when he had something emotionally important to say he resorted to Greek and Latin verse. Those in a position to judge agree that his prosody is almost faultness, and some of his verses, for all their artificiality, have the ring of genuine poetry to them.

Years of practice were required to attain such skill, and it was only toward the end of his career at Winchester that Arnold was able to reap the rewards of his application to these schoolroom exercises. By 1906, his penultimate year, prizes began to come his way, and, perhaps not accidentally, some of his letters home have been preserved. On 3 July 1906, for example, he sent a card to his parents: "I have got the German prize! Almost as unexpected as English Lit. Is it not splendid? Latin verse not done yet." And at the end of that same month: "I HAVE GOT THE GODDARD: is not that splendid?" His mother could not forbear exclaiming at the head of the letter he sent her: "What a boy it is!"[28]

The next year he did even better, winning four out of a total of seven prizes reserved for award at the school-leaving ceremony. Arnold's prizes were for Latin prose, on the assigned topic of the then still new Anglo-French entente; Latin verse, for a poem on Garibaldi; Greek verse, for a prescribed translation of a now almost forgotten English play, John Home's *Douglas,* Act I, scene I; and for an essay on satire in English literature—again, a prescribed subject. The prize for English verse escaped him; so did prizes for Latin and English speech, but his friend David Davies won one of the prizes that Arnold missed.[29]

A more brilliant climax to a school career would be hard to imagine, and his schoolmasters were impressed. He was able to inform his parents that the head-master, colloquially known as "the Burge," had "said that it was the highest standard that he had ever seen in English Lit. . . . He said about my essay [on satire] . . . that it was the best he had had both in style and composition since he had been here. You must not shew this letter to anybody else, or people will think I am conceited."[30] And again, "I have got the Latin Essay! The Burge was extraordinarily complimentary."[31] No doubt his fellow students also admired Arnold's extraordinary literary accomplishments, and his feeling about the school, and the exile it meant from his emotionally supportive home environment, mellowed accordingly. In November 1906 he wrote: "I am long-ing to be home again and get all this over." But by the following July, with his glittering assortment of prizes in hand, he wrote to his mother: "Only four more days now till you come; but I am beginning to feel very sad that my time is so near the end: it is a great plunge starting again at Oxford, and one does love this place more than one thinks."[32]

Prolonged saturation in the classical languages and literature certainly shaped Arnold Toynbee's schoolboy mind, but his reaction was not simply to accept the ideals and ideas presented to him. For one thing, he reacted strongly against the narrowness of the literary taste he was compelled to conform to. Exclusive attention to the short period of time when Ciceronian Latin prevailed seemed to him like a set of blinkers on serious study of Roman history. "After all, Diocletian had more influence on the world than Virgil," he remarked, and went on to criticize his classics teachers for taking "absurd attitudes towards sciences and math and everything outside their own particular subject."[33]

The youthful historian also ranged geographically far beyond the school's focus upon Greece (really Athens) and Rome. A "Drawing Book," compiled when he was recovering from his bout of pneumonia, and dated 1903, reflects this fact. Seeking the benefit of country air, Arnold went to stay with his Aunt Grace, who lived near Birmingham, for his recuperation. There he encountered a heady atmosphere of scientific adventure, for both his aunt and uncle were engaged in very active biochemical research and discovery. More important for Arnold, however, was the fact that his uncle bought him a historical atlas to help him while away his time. Long afterward, Toynbee declared "I learnt volumes from it,"[34] and this is no great exaggeration, to judge from the con-tents of his "Drawing Book."

Most drawings are of soldiers—Egyptians, Assyrian, Parthian, Persian, Byz-antine, medieval, and even Cromwell's infantrymen—all with meticulous atten-tion to details of costume and weaponry. This clearly was an extension of his play with toy soldiers, allowing him to familiarize himself with antique armies whose regiments were not available in toy stores to be added to his collection. But the real originality of the "Drawing Book" lies in its maps. One map prefigures a major theme of one of his last books, for it depicts the military provinces of the Byzantine Empire in great detail. Another illustrates the par-titioning of Africa among European powers in the nineteenth century. But most of the maps were traced from a projection designed to show both the Roman Empire and Alexander's empire in a single spread, extending, therefore, from

Britain in the west to the Oxus and Indus in the east. What Arnold did during his convalescence from pneumonia in 1903 was to fill in these maps with the political pattern of different ages, depicting, for example, "The World in the 10th Century" and "The World in the 12th Century," getting his data, presumably, from his new historical atlas and from various history books.

The effect was to pull his mind eastward, making him aware that the Near East had a history, not just at the time of Alexander's conquest, but across all the centuries. One may guess that his play with this map projection was what first led him to transgress the usual geographical limits on historical curiosity inculcated by the prevailing pattern of English education. To know about Georgian rulers of the Caucasus, and to have such exotica as the Il-Khans, Urartu, Tamerlane, and the Chaldaeans ready on his tongue, gave the boy an easy, sure-fire superiority over everyone around him—his mother and his schoolmasters as well as his fellow students. It was an easy way to show off. This, as well as the intrinsic interest of all the strange names and distant places with which he familiarized himself, impelled him to persist in a truly remarkable effort at self-education in Near Eastern history and geography.

From the latter part of his school career, a total of eleven essays, including most of his prize essays, are preserved. They range widely and include a sketch of Venetian history, an account of the Russo-Japanese war, then just concluded, a history of Byzantium under the Macedonian dynasty, together with his essay on satire and another on poetry and oratory. These are in English, and all are impressive when one realizes that they were written in his seventeenth and eighteenth years.

The historical essays are lengthy pastiches of information culled, commonly, from half a dozen learned works, and not really original. Still, traces of ideas later important in Toynbee's thought are clearly recognizable. In particular, the essay on Byzantium under the Macedonian emperors declares: "Two civilisations, which the genius of Rome had momentarily blended, separated again in an altered form, and became more antagonistic than ever." The notion of a plurality of civilisations, and their capacity to survive as recognizable, autonomous entities, so central to his later work, is thus simply assumed in his youthful composition. His exploration of the history of the Near East allowed him to follow Herodotus' lead and think habitually of East versus West; but unlike others who used the same dichotomy, the young Toynbee was not prepared to assume or declare that the West was always right, or the real bearer of civilization. On the contrary, he deplored "the base ingratitude displayed by the Westerners towards the nation which had taught them so much, and which had been for so many centuries their strong bulwark against the ever-encroaching East." Byzantium was thus a civilization different from "the East" as well as from "the West"—caught awkwardly in between. Such usage was not really Toynbee's own; what matters is that he picked up these notions from casual turns of phrase in scholarly books and proceeded to make them central to his own way of thinking. This was largely because he insisted on approaching the past on a grand scale, bridging time and space as specialists habitually refused to do, and he needed the concept of separate civilizations to make sense of it all. His childish one-upmanship as against his mother's and his teacher's parochialism of outlook had become habitual—and, one must say, also immensely fertile, potentially if not yet in actuality.[35]

On the other hand, his essay "Satire in English Literature" seems remarkably mature. He begins with a backward look at the Roman origins of the genre and then discusses the well-known English satirists of the seventeenth and eighteenth centuries. The tone is magistral throughout, and he moves from author to author and age to age by suggesting how their satire reflected the state of society and responded to the existence of an audience receptive to such criticism. This essay matches and surpasses many an M.A. thesis, and on reading it one sees why the headmaster declared it the best he had ever seen. Had Arnold Toynbee not been so firmly committed to history, he might well have become a notable literary scholar and critic.

A second group of his Winchester essays is in Latin, addressing prescribed themes in an arch and unhappily artificial manner. One of them responds to the question: "Whether a state can preserve its freedom at home while holding empire over others?" The relevance of such an essay topic to British affairs at the beginning of the twentieth century was obvious; Toynbee chose to address it by aping the voice of an unknown Roman senator of Augustus' era. His conclusion, incidentally, was: "It is impossible for a free people to bear rule over other peoples without forfeiting its freedom." Such a view was not exactly an endorsement of Britain's imperial grandeur, of which his parents were so proud! But of course the verdict was not young Arnold's, but that of a crusty Roman senator.

His prize-winning Latin essay, "Societas inter Britannos Gallosque conjuncta," is even more arch, purporting to be a dream experienced by the Roman emperor Magnus Clemens Maximus on the eve of his death. (Maximus, proclaimed Emperor by troops stationed in Britain in A.D. 383, won control of Gaul and Italy but then quarreled with Theodosius, Emperor of the eastern provinces, who defeated and killed him at Aquileia in A.D. 388.) In his dream, Maximus is puzzled to find himself in the streets of London just as the English were celebrating their new entente with France in 1903. His effort to understand what was happening allowed young Toynbee to display his knowledge of fourth century Roman affairs as well as of twentieth century international relations, all garbed in ripe Ciceronian rhetoric! Here, obviously, the form was what mattered. Substance there is none, or next to none. Much the same is true of Toynbee's translation into Greek of a speech by Richard Brinsley Sheridan, of his Latin verses (three separate and lengthy poems) on Minoan and Italian history, and of an English poem on the Christianization of Mercia. Skill in prosody may be there, but correct quantities and accurate scanning ranked ahead of intellectual or any other kind of message. A schoolmaster's comment on one of his poems makes that clear: "Very good and free in treatment. Some unfortunate false quantities." He marked it "83."[36]

Religious questions did not assume much importance for Toynbee while at Winchester. In his last year he read the New Testament in Greek and plunged into textual problems and the history of biblical scholarship with gusto and

Toynbee Growing Up

At day school

At Winchester

At Balliol

remarkable thoroughness for one so young.[37] If this study had any impact on his religious faith, it did not show in his letters. Instead he declared that "the history of thought and religion from the time of Euripides . . . to the conversion of St. Augustine proves the truth of Christianity."[38]

On another significant issue he did express disagreement with his parents. Rather perfunctory military training was part of school routines, but his parents, who had reacted to English embarrassments in the Boer War by enthusiastically endorsing universal service, wanted their son to enroll for a week or two of strenuous military training during his school holidays. Drill exercises at school no doubt smacked of the tedium of organized sport as far as Arnold was concerned, and he made this quite clear to his parents, stating that he was "absolutely out of sympathy with the whole thing. . . . But if it would really make you and Father unhappy if I did not go, I will go at once, whatever it is like."[39] No surviving record reveals the upshot, but the fact that Arnold was a complete military novice in 1914 suggests that his will prevailed in this instance, as was so often the case within the bosom of the family.

The young man who entered Oxford in October 1907 was, in fact, used to having his own way—at home, thanks to the enthusiastic support of an imperious and loving mother and his two little sisters, and at school by the more strenuous feat of living for and by his studies, a domain in which he so excelled his contemporaries as to gain their admiration if not their love. His intellectual development was certainly precocious, measured both by his mastery of Greek and Latin and by the breadth and variety of historical information he had crammed into his head. His memory was phenomenal; his diligence no less so. He was shaped, poised, and ready for a distinguished career as scholar and historian. As before, much was expected of him, now not just by his family but also by his schoolmasters, by Oxford dons who remembered his uncle and noticed the names of Winchester prize winners as published in *The Times,* and by his fellow Wykehamists.

Yet at the same time in some respects he was still very childish. Easy relations with his schoolfellows had escaped him. He felt inwardly compelled to excel and be admired; otherwise he retreated into himself and fell silent. And what must have been even more difficult for a vigorous eighteen year old adolescent, relations with girls, other than his sisters, remained totally, or almost totally, unexplored. In short, he was gifted, shy, and gauche, simultaneously sure and unsure of himself—a young man who had already found it hard to live up to his family's great expectations, and who would continue to find it hard to fulfill the great expectations aroused in a wider circle by the extraordinary record he had made at school.

II

Balliol and the Breakup
of Toynbee's Parental Home
1907–1911

Graduates of Winchester were more or less expected to apply to New College, Oxford in conformity with the wishes and expectations of the school's founder. But while still at Winchester, Toynbee met Cyril Bailey, a tutor at Balliol, who came to the school as examiner and was so impressed by the young schoolboy's command of Greek and Latin that he invited Toynbee to apply to Balliol instead.[1]

This was an invitation to aim for the top, since, at the beginning of the twentieth century, Balliol College was the preferred launch pad, especially among Liberals, for a distinguished public career. Balliol's prestige stood near its peak in 1906, the year Toynbee took entrance exams, seeking, as usual, a scholarship to see him through. In that year the Chancellor of the Exchequer (soon to become Prime Minister), Herbert Asquith, was a Balliol man; so was the Foreign Secretary, Sir Edward Grey. Lord Curzon, recently Viceroy of India (1898–1905), and one of his Viceregal predecessors, Lord Landsdowne (1888–1894), were likewise Balliol alumni. In 1907, the college put on a dinner for Asquith, attended by no fewer than thirty other Balliol men who were members of Parliament; and it was on this occasion that Asquith coined a famous phrase, decribing his old college as the seat of "effortless superiority."[2]

Clearly, membership in the college opened a path to the topmost posts of government. Moreover, graduates commonly kept connection with their old college, visiting it at least once a year for special "Gaudies" (from Latin, *gaudeo*, "rejoice"), in remembrance of old times. This allowed each crop of undergraduates to enter into a vivacious old boy network that assured young talent of early opportunities for advancement in whatever branch of public affairs the beginner chose to enter.

The college dons did all they could to cultivate such connections. Ever since Benjamin Jowett (fellow, 1838–1870; master, 1870–1893) had become the leading spirit of the college in the 1860s, Balliol had deliberately sought to combine aristocracy of birth with an aristocracy of talent. Jowett did so by recruiting birth and talent more openly than the genteel, clerical tradition of other colleges permitted. Cyril Bailey's invitation to Toynbee to apply was a case in point: he saw superior abilities and set out to attract their possessor to Balliol. Parallel efforts to attract young men of aristocratic birth were no less successful. In 1906, the year before Toynbee matriculated, nineteen out of fifty-two entering students were Etonians, nearly all of them aristocrats by birth, while only ten were men whose talent was their talisman for entry into Balliol.

The two groups lived and played together, but a yawning social distance remained. The aristocrats were often arrogant and rowdy, and they could be cruel, as when one of them attacked Philip Sassoon with a whip and drove him from the college. (Sassoon's ancestors had once been Jewish bankers in Baghdad, but the family had become fully Anglicized.)[3] Moreover, some of the dons deferred to the well born, giving them privileges and immunities that ordinary students were not allowed. On the other hand, a few young dons had begun to harbor egalitarian principles and resented aristocratic exclusiveness and occasional rudeness. Similar fissures had existed at Winchester too, but they were less glaring. Moreover, the course of public events between 1909 and 1911, which pitted the privileged House of Lords against a graduated income tax proposed by the Liberal government, tended to exacerbate social frictions within Balliol during Toynbee's undergraduate years, 1907–1910.

Yet the incipient erosion of Benjamin Jowett's synthesis of birth and talent was not yet obvious in 1907. Indeed, the newly elected master, James Leigh Strachan-Davidson (fellow, 1866–1907; master, 1907–1916), was a living embodiment of the old regime. In his inaugural speech before what he called "the very flower of the youth of England" he declared: "We are the citizens of no mean city, and its future depends on us. Let us rise to the height of this responsibility. Let us all resolve to live up to the past greatness of our College." His biographer summed up what Strachan-Davidson meant by past greatness as "an imposing row of Balliol names in the Class lists, a high place on the river, and a maintenance in strength of Public School—or as it might be called aristocratic—tradition."[4] Class lists recorded results of the examinations Oxford students took at the end of their studies; colleges competed for Firsts in the exams almost as vigorously as they did for victory in the boat races that took place each spring, and determined the "place on the river" each college occupied until the next year.

An important trait of Oxford's collegiate life was the celibate tradition which lingered on long after dons were legally permitted to keep their fellowships even after marrying and setting up a private household. Strachan-Davidson belonged to the old school: membership in the college meant total immersion for him, and he felt that fellows who married and lived outside college walls were traitors to their calling. Only by being on the spot all the time and sharing in all collegiate activities could a don play his role properly. Strachan-Davidson therefore remained a bachelor, made college routines into a religion, and in his old age did not have time to keep up with current scholarship on Roman constitutional history—his specialty—because his tasks and duties within the college were so all-consuming.

Yet the intensity of collegiality, as embodied by the master and others who felt as he did, was entirely compatible with sharp class distinctions within the college. Students, for example, ate their meals in the great hall, where the dons' high table and its silver vessels advertised different, better, and more variegated food—not to mention the wines, which were something of a religion at most Oxford colleges. At the bottom of the college hierarchy came college servants, who might, nevertheless, achieve a kind of eminence by their longevity and helpfulness to generations of students. A porter named Ezra Hancock, for example, became a central and beloved figure in the life of Balliol College, from Jowett's time until his death in 1916. He was a font of information about everything that happened in the college; he welcomed each new member as he arrived at the gate, delivered everyone's mail, and remembered each face and name long after a man had "gone down." He and the master between them embodied the old regime, sustaining by their example a powerful collective loyalty among Balliol's chosen few.

New winds of doctrine, emphasizing modern philosophy, politics, and economics, had already begun to stir at Balliol; and new ideas of what an Oxford don ought to do with his time—research and writing, for example, on the German model—were also beginning to challenge the preoccupation with tutorials and genteel sociability which nineteenth century reforms had made dominant at the Oxford (and Cambridge) colleges. But such newfangled notions were still subordinate to the humanistic tradition, dating back to the sixteenth century. Latin and Greek remained the preferred staple of study, and when Toynbee entered Balliol, no fewer than five of the thirteen tutors on the Foundation were teachers of classics. Three taught modern history, and single tutors taught Sanskrit, philosophy, law, and chemistry respectively, while two were listed without any subject specialty. The teaching staff was supplemented by five additional tutors, "Off the Foundation." One taught mathematics, one modern languages, and two taught chemistry, while the sole representative of the theological training that had once been the backbone of Oxford's education was a chaplain, who completed the roster of dons.

In 1907–1908, there was a total of 180 undergraduates in residence, and the Balliol old boy network consisted of the 964 graduates "paying dues to the University," of whom 449 held an M.A. and were therefore entitled to vote in university elections. In such a small, tight-knit community, everybody knew everyone else, if not personally, at least by reputation. Getting in was the key: men were counted by their year of matriculation, not of graduation, as in American colleges. And admission depended on the dons' judgment, assisted, in the case of scholarship candidates, by the results of written examinations. They were interested in men who would add to the luster of the college, during their time at Oxford and subsequently. Scholarships per se was not the aim: verbal skill, both written and spoken, together with personal address and physical vigor were what mattered.

The dons were thus faithful to the tradition they had inherited, for sixteenth century humanists had held that study of Greek and Latin was the proper preparation for public life. This ideal was, on the whole, realized, although the generation that entered Balliol with Toynbee fell short of the political prominence that earlier graduates had achieved. When, in 1953, the college compiled a record of its members, the forty-five survivors of the fifty-eight undergraduates who had matriculated with Toynbee in October 1907 reported their occupations as follows: education, twelve; law, nine; government service, eight; church, five; 'noble,' four; business, three; research, two; author, one; social work, one.[5]

This then was the community of gentlemen and scholars to which Cyril Bailey invited the eighteen year old Toynbee to apply. Family circumstances made financial aid essential, so in December 1906 the schoolboy traveled to Oxford for the first time to take the Balliol entrance exams. "Latin verse was quite the hardest paper" he told his parents; and the essay topic "Equality as a Political Principle" allowed him to argue that "absolute political equality was not possible and was not right."[6] Toynbee's command of Greek and Latin was such that he won the scholarship he needed to meet the costs of membership in the college. Accordingly, in October 1907 he matriculated as a student at Balliol.

Toynbee's reaction to Balliol and Oxford paralleled his career at Winchester exactly. At first he was shy, withdrawn, and critical. "They certainly extract your money from you here: fees seem endless," he wrote home. "The dinner with the Master was not such a trouble after all, especially as he is a bachelor." But, "I am disappointed with Oxford so far, and foresee that I shan't have any desire to settle there permanently. . . . As far as I can see, Oxford people are just like other people and are just as afraid of being caught 'talking shop' as a merchant would be of talking about his cotton or hardware. . . . One does not feel that one is in an intellectual atmosphere or a centre of learning in the least: it is all a business affair: it is your tutor's business to get you through the schools, and yours to get through as efficiently as possible with the least amount of work." Worst of all, the chaplain had told him "Rowing was the great thing at Oxford"; drunkenness in the quad further contributed to his disenchantment.[7] His mother responded: "Don't keep aloof: do be human, which is quite compatible with real work."[8]

Yet in due course, Toynbee's attitude changed. He won scholarships and prizes from his first year, adding to his income and prestige within the college with each success. In 1908 he won the Pitt Exhibition, in 1910 the Craven—a more substantial prize—and in 1911: "I have got the Jenks, and am accordingly in high spirits, especially for sordid reasons: for it is a good advertisement from the point of view of getting a job, and gives you plenty of money to go on with until a job comes. Moreover, I really enjoyed doing the papers."[9] Sometimes he failed: "I was very weary after the Ireland (in which I came in fourth, as against second last year—O foolish mod tutors of Balliol who prodded me, sorely unwilling, into doing it)."[10]

But success was more frequent than failure, and toward the end of his undergraduate career Toynbee enjoyed exactly the same intellectual primacy that he had attained at Winchester. Quite simply, he was the best among his contemporaries at what the college most valued in the way of learning, that is, reading, writing, and thinking in Greek and Latin, and knowing in painstaking detail what classical authors of the best period said about things in general and public affairs in particular. Success altered his behavior: "I agree about being less shy—I am a different person through and through from what I was when I came up, though there is a lot too much of that malady still left in me. What a wonderful thing this College is: I love it more and more."[11]

A major reason for this change of view was that the ambitious young historian gradually discovered friends and fellow spirits at Oxford, both within and beyond Balliol's walls. In his book *Acquaintances* (1966), Toynbee sketched his recollections of some of them: the Margoliouths, Alfred Zimmern, Lewis Namier, R.H. Tawney. But he chose to write about these figures, at least partly, because they were acquaintances rather than persons close and important enough to count fully as friends. Zimmern and Namier were occasional colleagues (sometimes also critics) through most of Toynbee's adult career, and from each of them, as well as from other acquaintances made in his undergraduate years, Toynbee learned a great deal, especially about contemporary affairs. His genial recollections of encounters with them, as recorded in *Acquaintances,* gracefully acknowledged this debt.

But Toynbee also made at least two real friends at Balliol: Alexander D. Lindsay and Robert Shelby Darbishire. Lindsay had come to Balliol as a don in 1906, just a year before Toynbee entered. He was a classics tutor, with a specialty in philosophy, but was actually more interested in modern than in ancient thought. He soon became ringleader of those within the college who aspired to a more rigorous and active style of scholarship than anything the master's ideal of collegiality required or expected, for Lindsay was not content simply to teach and attend to other collegial duties; he also aspired to write and publish. While Toynbee was still an undergraduate, Lindsay's ambition took concrete form with the appearance in 1911 of his book on the contemporary French philosopher Henri Bergson.

Lindsay was a Scot, and he despised toadying to the rich and well born, who were then so conspicuous at Balliol. Instead he believed in informal, democratic manners, calling undergraduates by their first names and expecting to be called "Sandy" in return. He was, in short, intent on disrupting the alliance of aristocracy and talent that Jowett had painstakingly created at Balliol, believing that talent alone should prevail. Such attitudes and ideas appealed strongly to the young Toynbee, who remarked in 1912, "Far the biggest man here is Lindsay."[12] He soon became Lindsay's close friend and protégé, even though his prowess at philosophy always fell short of his mastery of the more strictly literary and historical subjects that together comprised Oxford's course of study in *Litterae Humaniores*. Later, as we shall see, this friendship shattered when Toynbee resigned his fellowship at Balliol in 1916. It was never resumed. But from 1907 to 1916, Alexander Lindsay exercised a powerful influence on Toynbee, personally inasmuch as he provided a model for the scholarly, ambitious young don that Toynbee soon decided he, too, wished to become, and

intellectually inasmuch as Lindsay's version of Bergsonian evolutionary thinking helped to wean Toynbee away from his inherited Anglican faith.

Robert Shelby Darbishire was two years senior to Toynbee at Balliol, having
entered in 1905 from Rugby. He was Anglo-American, raised partly in Kentucky and partly in England. His interests were historical and literary, like
Toynbee's, but what brought them together, in all probability, was that they
both had difficulty in breaking childhood bonds to their mothers. Darbishire
was an only child. His father, also an only child, had died when he was three
years old, whereupon the bereaved grandfather, a prominent solicitor in
Manchester, began to compete with his mother, an impoverished Kentucky
aristocrat, for control over the young lad's affection, education, and career.
The mother won, hands down, but was nonetheless humiliatingly dependent on
financial help from Manchester. When her son attended Rugby, she moved into
the town to be near at hand; when he went to Oxford, she refrained from
following. But loosening childhood ties was very difficult for both mother and
son, as was also true for Toynbee and his mother.

No positive evidence proves that their shared Mother problem brought the
two men together, but the unique freedom with which Toynbee expressed his
private ambitions and feelings to Darbishire in letters they exchanged after the
older of the two left Balliol for apprenticeship in his grandfather's law firm
attests the intimacy and warmth of their relationship. No other surviving letters
are so confiding; and, not surprisingly, as soon as the two young men broke
away from their mothers' apron strings by marrying, their correspondence lan-
guished, though it continued—sporadically—as long as Darbishire lived.

Gilbert Murray was a third person who loomed larger and larger in Toynbee's mind and feelings as his undergraduate years at Oxford went by. Murray
was an Australian who had come to England in childhood and, after a brilliant
undergraduate career at Oxford, became professor of Greek at Glasgow at the
age of twenty-three. He resigned his chair in 1899 on grounds of ill health and
spent the next five years as a man of letters, writing a play that was produced
in London with only mediocre success. He then focused his formidable energies on arranging for the professional presentation of his translations of ancient
Greek plays in London. Some of these met with solid success, with the result
that between 1904 and 1939 Murray's translations of Euripides, Sophocles,
Aristophanes, and even Aeschylus flickered across the English stage, in London and in the provinces, sometimes professionally presented and sometimes
entrusted to amateurs.

To treat the ancients as though they were contemporaries with things to say
to twentieth century audiences was both novel and intriguing at the time. It
reflected Murray's conviction that "the scholar's special duty is to turn the
written signs in which old poetry and philosophy is now enshrined back into
living thought and feeling. He must so understand as to relive."[13] In 1905 he
returned to Oxford as a fellow at New College, and then in 1908 became Regius Professor of Greek, a post he held until retirement in 1936.

In his capacity as Regius Professor, Murray did a good deal to set the tone
for Oxford classicists. What he sought, always, was to connect the ancients

with the contemporary world, making their words, thoughts, and lives relevant to his own time. He saw no inconsistency, therefore, in engaging, as a classical expert, in public affairs. He was a stalwart Liberal, married to the eldest daughter of the Earl of Carlisle, whose family maintained the style of eighteenth century Whig grandees, almost unmodified, until World War I. This opened contacts with men like Prime Minister Herbert Asquith, whom he knew well, while his associations with the stage brought him the friendship of George Bernard Shaw, John Galsworthy, and other literary lights.

On top of this, Gilbert Murray was charming—handsome, witty, elegant— an excellent raconteur, discreetly ambitious, personally abstemious, and capable of extraordinary feats of dexterity and balance such as climbing a ladder without holding on. He also experimented with extrasensory perception and, as a sort of parlor game, often exhibited a truly uncanny ability to read the minds of other persons, especially that of his eldest daughter, Rosalind.[14]

Such a man, with such connections, was bound to dazzle young Toynbee, and that was what happened when the Murrays' habitual hospitality reached out to embrace him. Lady Mary and Gilbert Murray held open house for promising undergraduates on Sunday evenings, and Toynbee soon became a frequent visitor. In March 1910 he was even invited to visit the Carlisle family seat at Castle Howard. He reported to his friend Darbishire: "I am going off today week with Gilbert Murray to Castle Howard, a fenced city of his parents-in-law, somewhere in Yorkshire. I wonder if Lady Carlysle (is it so spelt?) will be in residence? Like Lady Mary, I am told, plus temperance, raised to the tenth power. It will be amusing and delightful." And from the spot: "It is a great and marvelous place, early 18th century style on the vast scale, with pictures and lakes and statues and libraries and all manner of things. Lady Carlysle [sic] is obviously a mighty force, but not, perhaps, so formidable. Would though that I were less entirely at sea about politics—domestic matters I mean. . . . However I shall doubtless know plenty about politics before I go away. Do you like Canalettos? They cover all the walls of the room where we eat—I won't call it the dining room, for there are at least twenty like it. There are also some Wattses, and crowds of nice solid books of the eighty years ago kind. Altogether it is more peaceful and less of a 'fearful' joy than I had expected."[15]

Murray and Toynbee were drawn together by another facet of their lives, for Murray could and presently did play a fostering, fatherly role with respect to the young Toynbee which his own father, increasingly, was incapable of doing. Conversely, Toynbee could and eventually did serve as a surrogate for Gilbert Murray's own sons, who turned out to have disappointing careers and took refuge in recurrent bouts of alcoholism. But this bond did not become significant until after Toynbee's marriage to Rosalind Murray in 1913. During Toynbee's undergraduate days, the Murrays remained dazzlingly alien, admirable, and aristocratic. Their friendship was still something of a "fearful joy" to the shy, ambitious young man.

Obviously, Oxford friends, acquaintances, and teachers contributed to Toynbee's intellectual as well as psychological maturation; the most distinctive traits

of his development, however, cannot be attributed to anything but his own
genius, elaborating upon the idiosyncrasies of his schoolboy achievement. Surviving essays and notes allow us to know something of how Toynbee's mind
changed during his undergraduate career.

The most important intellectual landmark was that Toynbee ceased to believe
in the doctrines of the Christian church, even in its broadest and most flexible
Anglican form. This involved a distinct break with his parents, especially with
his mother, whose conventional piety was nearly as firm as her patriotism. But
escaping from his mother's apron strings was attractive in itself, and Toynbee
found himself impelled to reject Christianity both because of philosophical arguments, brought to his attention by Lindsay, and because of literary and historical considerations, brought to his attention by Gilbert Murray and others.

The first of these lines of thought is illustrated by an essay Toynbee wrote
sometime between 1907 and 1911, entitled "The Machine: A Problem of Dual-
ism." This is systematic philosophy in the grand style, beginning: "The whole
universe—my consciousness, my body, my safety razor—is all a machine, and
the paradox of the machine is inherent in the whole of it. It partakes of two
orders of being. . . . Life, which we know because that is what we are, and
that other, which we do not know because we are not it, but which is Life's
environment and object of activity." Toynbee goes on to assert that these two
orders of being "fight" each other, like a pair of wrestlers, and that with every
success that comes to "Life," its opposite, "mechanization," advances as well.
In becoming conscious, "Life" devitalizes itself; language becomes mechanical through grammar; morals harden into law and custom as a result of mechanical application to infinitely varying circumstance.

After thus distilling into about thirty pages a rather jejune version of Bergsonian philosophy, Toynbee concluded his essay by affirming the "complex,
tragic, dualistic world which is the only reality there is."[16] That final phrase
constituted a Parthian shot at his Anglican upbringing, repudiating not just the
Incarnation and personal immortality, but the existence of God as well. He
remained firm in the first of these opinions throughout his life, but soon changed
his mind about God when personal and public distresses multiplied around him,
and Bergsonianism ceased to carry the weight of a new revelation.

A second essay, "What the Historian Does," read to an undergraduate club
at Oxford in the academic year 1910–1911, also attempted the heights of abstract thought, rejecting purposive Hegelian Spirit and affirming all-embracing
change in a vaguely Bergsonian fashion. The essay becomes more interesting
when Toynbee comes down to a more practical level, declaring that the "historian must have second sight that is called intuition," so as to be able to see
the past as though it were present before him. And again: "Art and history
resemble each other in both being activities of the imagination working upon
experience."[17] These views never altered, and he cultivated a (sometimes visionary) capacity for imaginative re-creation of the past throughout his life.

In general, Toynbee was never at home with philosophy. "Don't talk of
philosophy," he wrote to his friend Darbishire," . . . down with Kant."[18]
And again: "I am overwhelmed in Moral Philosophy. . . . It is a river of

death which has to be passed through, if only to discover that they are all talking nonsense—but it is weary wading."[19] The two philosophical essays that survive from his undergraduate days are also weary wading, but they nonetheless attest that systematic philosophy played a part in prying him away from his inherited Christian outlook on the world.

The other intellectual current pulling him in the same direction was literary and historical: the notion that Christianity emerged from pagan and Jewish elements and was a monument to the sensibilities of a particular time and place, with no valid claim to universality and eternal truth. In 1907, J.G. Frazer's famous study of magic and religion, *The Golden Bough* (1890), was still echoing strongly through Oxford, affecting the way classicists looked at Greek and Roman paganism. Frazer's comparative method for making sense of myths and rituals invited them to compare Gospel miracles with stories told by other Graeco-Roman saviors—kings, rebels, gods. From the comparative angle of vision, Christianity became only one of a number of similar "mystery" religions competing for popular support when the Roman Empire was in its prime. This historical, agnostic outlook on early Christian history constituted a new and heady brew in the first decade of the twentieth century, depriving Christian theology of any claim to absolute truth, and making Christianity one among many responses to the special pressures and circumstances of Roman imperial society—no more, no less.

The humanistic tradition of European learning had flourished since the Renaissance by tacitly leaving Christian scriptures and doctrine to theologians, while concentrating attention on narrowly defined eras when language and literature were supposed to have attained peculiar perfection—the Greece of Pericles and Plato, for example, and the Rome of Cicero and Virgil. By the beginning of the twentieth century, however, this compromise, like the compromise Jowett had made within Balliol between aristocracy and talent, was wearing thin at Oxford. Theology was in retreat; biblical texts had long since been torn apart by German scholars using the tools of humanistic philology; now British anthropological study of "primitive" peoples invited classicists of the humanistic tradition to extend their jurisdiction over popular and irrational aspects of ancient thinking as manifested both before and after the "golden ages" upon which their predecessors had centered attention. Gilbert Murray, a lapsed Roman Catholic and aggressive agnostic in private life, was a leader of this assault on domains once left to theologians. His book *Four Stages of Greek Religion* (1912) made that clear. The chapter dealing with the rise of mystery religions was entitled "The Failure of Nerve," and Christianity, of course, was the most important of the new faiths that Murray characterized in this dismissive fashion.[20]

The young Toynbee, who had already extended his curiosity far beyond conventional limits in the course of his schoolboy explorations of Middle Eastern and Byzantine history, took to these currents at Oxford like a duck to water. As a substitute for Christian morality he adopted an agnostic ideal of heroic accomplishment: this-worldly, thoroughly classical, in tone. As he explained to a fellow student from Winchester in a letter of condolence: "If the personality

perishes in Lucretius' fullest sense (and I have taken Lucretius much to heart)
. . . our meaning lies in our work, and our fellowship with one another through
it.''[21] That ideal, and the intellectual judgment on which it rested, was summed
up in F.W. Hasluck's apothegm "God is not the man he was." Toynbee's still
recalled that witty sacrilege thirty-five years after he first heard it from the older
man's lips.[22]

Unlike the experience of agonized souls in the late Victorian age, when they
were first exposed to these currents of thought, abandonment of traditional
Christianity did not involve any noteworthy emotional struggle for Toynbee.
His education had centered so much on classical, pagan writings, that accep-
tance of an Epicurean or Stoic world view seemed simple and straightforward.
His Christian faith had been tepid; giving it up was easy, almost casual.

Socialism ranked with agnosticism among the radical currents of the time.
It, too, had begun to affect Oxford, and Toynbee was inoculated accordingly.
"I have just become a Fabian," he told Darbishire in 1909; but his socialism
was little more than a fad. A few months later, home at vacation time, he
explained that he habitually took a walk in Kensington Gardens "when I feel
depressed or socialistic."[23] What he meant by socialism is hinted at in his
essay, "What the Historian Does," when he remarks, casually, that "economic laws are wholly man made" so that "trades union 'co-ops' may succeed
in mastering economics and bridling it and riding it along whatever road we
choose."[24]

The shade of the first Arnold Toynbee clearly pushed his nephew and namesake toward this kind of socialism, but the young man refused serious personal
commitment to bettering the lot of the poor, with which his father wrestled
professionally, and only skirted the religious question, with which his uncle
had been centrally concerned. Instead, Arnold Joseph Toynbee was intent on
developing and declaring his own intellectual independence, rejecting the model
of his parents, of his teachers and tutors, and of his uncles (for Paget Toynbee,
the Dante scholar, also had firm ideas about how learning should be pursued,
and sought to guide his nephew along the careful paths of specialization).[25]

Well before completing his undergraduate career, Toynbee had begun to nur-
ture the ambition of writing a great book that would comprise ancient and
modern times, as well as the East and Europe, in a single conspectus. He called
it a "philosophy of history" but remained entirely unclear about the guiding
principles he would need to make so much detail intelligible. While still at
Winchester he wrote: "It would be a splendid task to carry on Herodotus' story
[of the wars between Europe and the East] . . . but it would be too vast."[26]
Yet he continued to harbor that ambition, or something like it, throughout his
undergraduate years.

Nothing at Oxford encouraged macrohistorical thinking. Tutorials centered
on specifics; texts were read word by word. Accuracy and detail were all. Truth
was expected to take care of itself as long as the student stayed close to his
sources. Toynbee obediently conformed to these expectations and, indeed, excelled in his mastery of detail.[27] While still an undergraduate, he published a
note "On Herodotus III,90 and VII,75,76,"[28] discussing the meaning of ob-

scure place names in Thrace, Asia Minor, and Cyprus. It was on the strength of such performances that he was invited to confer with the historian H.A.L. Fisher of New College and a fellow of Brasenose named Dr. Bussell about plans for an Oxford edition of Byzantine historians in which the young Toynbee was presumably expected to play the role of editor.[29]

But Toynbee never gave himself unreservedly to detail. He wanted an overview, always, and his curiosity was insatiable. It even embraced distant China, at least in the era of the Mongol empire, as attested by an essay, "The Mongols in East and West," which he read to the Acton Club in November 1910. This is a lengthy political and military narrative, not very well put together. When Darbishire, reacting to an early draft, pointed this out, Toynbee replied: "Thanks for the criticism. You have touched on the weak spot, but I think I could have given the causes: what I meant to do was to present a picture, make them imagine this vast wave of conquest, like the riders on Scorpions in the Apocalypse; but I am afraid the necessary compression left it a mere skeleton of facts."[30] That is, indeed, the impression the essay leaves: ill-digested facts, assembled from the best available sources—which meant modern secondary works—and selected on a rather narrowly constructed military–political axis. Personal quarrels among the members of the ruling family provided the primary focus; military campaigns sank to the level of literary exercises in vivid writing.

Clearly, Toynbee had not yet achieved the insight needed to realize his grand ambition, but the goal was there all the same. "As for Ambition," he wrote, "with a great screaming A, I have got it pretty strong. . . . I want to be a great gigantic historian—not for fame but because there is lots of work in the world to be done, and I am greedy for as big a share of it as I can get." And from the same letter, "I am going to research and become a vast historical Gelehrte."[31]

His use of the German word "Gelehrte" to describe his ideal for himself points to another potent force in Toynbee's intellectual experience at Oxford: exposure to German classical scholarship, and, in particular, to Eduard Mayer's massive and magistral synthesis of Egyptian, Babylonian, Greek, and Roman history. The five volumes of Meyer's *Geschichte des Altertums* (1884–1902) were indeed impressive: a summation and synthesis of centuries of European classical scholarship and of decades of (mostly German) Egyptology and Assyriology. Meyer did for the ancients what Toynbee intended to do for both ancients and moderns; that is, he wove the whole together into a single tapestry, having first mastered more detail than any ordinary mind could cope with. No English historian equalled Meyer's mastery of ancient languages—Oriental as well as European—or aspired to his scope across time and space. More than any other single exemplar, Eduard Meyer therefore provided the young Toynbee with a working model for his own projected magnum opus. But Meyer had done his work so well that Toynbee saw some difficulty in overtaking or surpassing him. "At the moment," he wrote, "I feel that Altertumswissenschaft [classical studies] is a bit of a Moloch, and I am not going to cast myself into the furnace of his belly without thoughtful consideration: all the same, I still rather think that learning, like poetry and preaching of a Religion, demands either all or nothing."[32]

Toynbee already had an inkling of the cyclical pattern upon which, in the 1920s, he founded his mature vision of history. He habitually looked for parallels, or seeming parallels, between ancient and modern events. Thus commentaries on Herodotus, compiled during his undergraduate years, abound in comparisons of the Persian invasion of Greece with Napoleon's Grande Armée and the Ottoman attack on Europe.[33] He saw ancient history as a rise and fall and suspected that modern Europe's historical career might conform to that ancient cycle. Musing on the unfortified sites of Minoan civilization, for example, he compared ancient Crete with England of his own time, remarking how invaders had conquered the Minoans so that "the infusion of northern blood turned Minoans into Greeks. It is the previous cycle to our own Decline and Fall, Dark Ages and Renaissance: though I don't believe that history repeats itself in any accurate or scientific sense. Anyhow it is a curious parallel."[34]

Apprehension of impending decay was very widespread in Edwardian Britain. Troubles in Ireland and threat of class war at home made the Victorian era seem almost serene in retrospect. Rivalry with Germany raised the specter of cataclysmic war, or perhaps of a purging of civil ills through war. Toynbee shared these sentiments fully. As he wrote to his mother, "The world lies at present (that is, till China comes and eats us up) between the English public school man and the German."[35] And again: "War is fascinating, in spite of what people like Murray very reasonably argue to the contrary: it is all waste and destruction, and yet one admires a fine campaign . . . because there is so much human effort and human efficiency blazing out in war, instead of the 'latent heat' of 'Progress and Civilisation'—and it is the human effort and victory . . . that is the real living spirit of history."[36] And again: "A big war with Germany (which is surely coming) may do for us what the Revolution did for France—or it may shatter our nerves altogether, and cripple our industry, and exhaust all Europe into the bargain."[37]

Still all was not seriousness and striving for mastery. In his undergraduate years Toynbee found time to read several of Galsworthy's novels; he found them "both beastly and delightful,"[38] while Walter Pater's *Marius the Epicurean* was "so delicious it must be poison."[39] He remained intensely shy, especially of women. On a train he once found himself in the same compartment with a suffragette, and "when she got out I gallantly lifted my cap and wished her success in her cause. . . . I have been amazed at my temerity ever since."[40] He was also capable of ironic self-caricature, as when he told Darbishire that he was writing "a philosophico-literary-artistic hyperaesthetic masterpiece" on Blake for delivery at an Oxford society of undergraduates.[41] Or again: "I hope to be able to speak something of the [Italian] language—in fact to be in general au fait and superior and intolerable."[42]

Toynbee's intellectual development during his undergraduate years at Balliol played itself out against a tumultuous emotional backdrop provoked by the breakup of his parental home. The critical event was his father's psychological collapse. Surviving records do not permit a precise reconstruction of Harry Valpy Toynbee's descent into incapacitating mental depression. The process was gradual. In a letter written after their mother's death, Margaret Toynbee referred to "terrible nights from 1908 to 1909" when "we knew she [Edith Toynbee] was in terrible trouble, but of course we didn't understand."[43] The crisis came either at the very end of 1909 or early in the following year when Harry Toynbee was adjudged incompetent to manage his own affairs and was committed to St. Ann's, an institution for the mentally deranged.

English law did not permit the wife of a madman to control his property. Instead, a Testimonial Fund was established, under the control of a court, which paid costs incurred for Harry Toynbee's care. At the discretion of the court, the fund could also contribute to Edith Toynbee's household expenses and the education of her two daughters, who were both still of school age. Harry Toynbee's inheritance from his parents presumably went into the Testimonial Fund; so, probably, did special contributions from friends, colleagues, and other members of the family. But there was not enough income both to care for Harry Toynbee and to support his wife and daughters. The court's first priority was Harry's support; Edith had therefore to depend on gifts from other members of the family to keep her household going. For someone as proud as she, this was well-nigh intolerable.[44]

The practical question of where to live presented Mrs. Toynbee with another acute and all but insoluble problem. On the one hand, the house at 12 Upper Welbourne Terrace, London, where she had lived since her marriage, was now steeped in so many horrible memories that staying there was hard for her to bear. She had never been fully mistress of the place, since it belonged to her husband's Uncle Harry. The crusty old sea captain had first become senile[45] and then, after spending some time in a nursing home, died in 1909, the same dreadful year in which her husband went mad. Recollection of these two catastrophes soiled the house for Edith, twice over. Her response was to wish for a new place to live. Yet if her husband should get better, might he not expect, or even need to return to familiar haunts? She also feared breaking off contacts with her London friends, and altogether found herself unable to decide, being oppressed by her immediate environment, crippled by financial dependency, and unsure when or whether her husband would recover. She was both unable to continue and unable to escape.

Toynbee was all but torpedoed by his mother's behavior. Instead of being the tower of strength and ever-ready font of loving support that he had known from his infancy, she became so immersed in her own troubles as to have scant emotional stability herself, and little or no warmth left over for him. As he confided to his friend, "These four years since my Father's illness have taken the life out of her . . . Absolute despair of all things and all men, and hatred of living, is a terrible thing to drive out of anyone. Everything one could catch hold of, care for my small sisters, religion, seems gone, and only bitterness in its stead. . . . but I am writing just the sort of stuff I did not want to."[46]

Toynbee was also haunted by the fear that madness was hereditary. If so, what had happened to his father might also happen to him. Throughout his life, this anxiety continued to lurk in the background of his consciousness, coming to the fore whenever personal difficulties closed in on him. As he confessed to Darbishire in 1946: ''During the past few years, I have been in sight of the madness that overtook my father.''[47] The same specter had bothered him many times before.

Toynbee's immediate response to the news of his father's commitment to an institution for the deranged seems to have been flight. At least this is a possible interpretation of his letter dated only ''Monday evening'' and sent from The Dell, Northfield, Birmingham.[48]

Dear Mother,
    I feel it all through my weakness and folly that I have deserted you. . . . I felt for a time that I was being monstrous in letting myself be absorbed in my work in all these troubles: now I see that it was just that that gives me strength to be at all cheerful and helpful and that the human and Paracelsus side, in me at any rate, can't at all get on without one another. I think the better I work the better I can love, and I will keep my promise Father made me give, that I would go on with my work in any case ''without letting it make any difference'' in that I will consecrate it all to you and him: it shall never cut me off and shut me up in a barren self. So I am at harmony with myself again, and whatever you do, don't box things up in yourself for fear of upsetting me. Let me share everything, and it will be better for both of us. And don't lets [sic] countenance the thought of any choosing between Father and us children: what is best for him is best for us, but also what is best for us must be best for him. . . .
    Remember I am thinking of you now and always: nothing has or will ever come between us. God bless you always.
                                Your v.v.v.l.s. [very, very, very loving son]
                                Arnold

What took him to Birmingham? Presumably when his mother failed him he sought comfort from his Aunt Grace, the biochemist; but nothing in the accessible record makes that sure, and, for that matter, the exact time when this letter was written cannot be fixed for certain. Nothing else survives from the crisis period; correspondence with his mother resumes only when Toynbee started on his grand tour of Italy and Greece in 1911–1912.

Two other letters shed some light on Toynbee's conduct when the wound to his parental family was still fresh. Apparently, when Oxford's summer vacation came round in 1910 he returned to 12 Upper Westbourne Terrace only to collapse himself. The prescribed remedy was massage, but as he confessed to Darbishire: ''the 'cure,' 'treatment' or what you will, which I rashly let myself be persuaded into, was so potent that it fused my nerves (what a whole new field of metaphor electricity gives), and I lay most of the time on a sofa, in the extreme edge of hysteria. I fled from it, however, a week ago today, and went with my Mother to Stockholm.''[49]

News of Toynbee's summer collapse must have triggered the second letter that survives from this time. This remarkable missive, sent by Lady Mary, Gilbert Murray's wife, reveals as much about her as it does about him.

Castle Howard, 15 September 1910

My Dear Mr. Toynbee:

Yes, I know the grief about your father . . . but I did not know how recent the crisis of the trouble had been. . . .

You say it is worse for your mother than for you. I doubt whether you realise how incalculably worse it is for her. . . . You have brilliant gifts, solid learning, public spirit, and power of winning affection and respect, physical strength enough, life and work and love before you—just this one great grief to bear manfully. . . . And life can be such a splendid thing! But your Mother has been made an old woman, her life smashed, all her joy behind her; it seems to me there is almost nothing to help her bear the intimate, close presence of this lasting grief—nothing except you (the little girls are too young and uncertain) and you've been failing her.

Well, there it is , the only thing that makes life worth living is the temper of heroism. . . . What do you matter—you and your pain and suffering—if only you can keep at the helm, be a proper tool in the hands of the wholly unknown thing one may call God. . . . You cannot and must not fail, never again.

This letter should not be answered. I have taken a great liberty in writing it and can only beg your forgiveness.

Yrs always,
Mary Murray[50]

Toynbee did try to play the hero as Lady Mary had commanded him to do, helping his mother to find a new place to live, for example, even though it meant another bitter loss for him. From childhood, he had had an unusually strong attachment to his home. It was his refuge from the exposed loneliness of school, a place where he could command his toy soldiers to his heart's content, and dazzle his little sisters by his accomplishments. To have his mother no longer fully at his beck and call was bad enough; to lose the physical security of home as well made everything worse, especially since it was she who insisted on abandoning it.

Arnold, the child, could not forgive his mother for breaking up his home; Toynbee the man cooperated in the move, but it made him sullen and distraught. As his sister wrote many years later, at the time of Edith Toynbee's death: "If you realised that she was not herself and wasn't, as you say, *deliberately* trying to wind the whole life of the family up, then, however awful it was for you, I can't understand why it made you feel differently to her."[51] But it did alter the young man's feeling for his mother fundamentally, and all his efforts to restore earlier links with her failed. As he confessed to his friend Darbishire, "I have been in Southwald the vac [this vacation], hoping to get my Mother settled into a small house we have rented there. But it is a hard job. . . . When I go home, I don't help my Mother, and get clouded myself, and then run away here [to Balliol] to clear up. Still, it is not defeat, and that is all gain."[52]

Toynbee's response to the intense new pressures of the family catastrophe was twofold. On the one hand, he buried himself in schoolwork, more intensely than ever. As he wrote to his confidant: "I love this work, both learning and teaching, because it has been my best friend, and helped me through bad times since my Father's illness; though it is often a tussle to keep one's mind clamped onto it, when one is staring stupidly at some worry like a hypnotised toad."[53] Fanatic and, on occasion, frenetic dedication to work eventually became second nature to him, largely because it offered a refuge from personal difficulties. His experience in school had shown what dedication to books could do, for mastery of book learning eventually rescued him from the desperate loneliness of his first school days, making him a universally admired prize winner. As we saw, his Balliol career followed exactly the same pattern; but his father's collapse in 1909–1910, and the irremediable strain that set in between Arnold and his mother as a result, gave new and special impetus to the young Toynbee's work habits, establishing a pattern that lasted all his life, with only minor fluctuations in its intensity. His diligence and *Sitzfleisch* had always been exceptional; after 1909 they became phenomenal.

Yet, for reasons we will see more clearly in the next chapter, Toynbee's second response to the collapse of his parents home actually ran contrary to the simple pattern of retreating into the world of books. For exactly in the years 1910–1911, he strove, more actively than ever before, to cultivate some of the graces that befitted an Oxford undergraduate. He took an increasing role in college activities, including such whimsy as "ragging" benefactors in a speech at a public banquet—"the most amazing and original thing I have ever perpetrated. It contained jokes, too, and they were laughed at."[54] He played tennis and took riding lessons, though he never acquired a firm seat;[55] above all, he went in for long-distance running. After a two hour run, he wrote: "It is good to be weary and panting and thirsty and muddy: it keeps one hard." And again: "Running is magnificent. I do it several times weekly here [at Balliol], you know, in the Winter."[56]

Travel, first to Sweden and then to Ireland, offered another escape. "I have done hundreds of miles of bicycling . . . and I have been making hay with immense energy. Irish are glorious people to make hay with."[57] He soon gave up running, but travel remained a passion throughout his life. An extraordinary appetite for seeing new places, talking with local inhabitants, and using his leg muscles to climb up hills and explore the countryside stayed with him well into his old age.

Toynbee's combination of intensified book work and intensified effort to conform to the social and athletic behavior of his fellow undergraduates sufficed to overcome the disaster to his family life, at least outwardly. As he had done previously at Winchester, he concluded his undergraduate career in a glaze of glory. In March 1911 Balliol went head of the river, outdistancing all the rival colleges at rowing, and when exam time came round Balliol did just as handsomely. As Toynbee explained: "Meanwhile, I have got a first in Greats. . . . The College got seven, and are hugging themselves."[58] He also won the coveted Jenkins prize (commonly referred to as the Jenk or Jenks), which allowed him to travel in Italy and Greece for the better part of a year, 1911–1912, without having to worry about how to pay for it.

His tutors were proud of him. Cyril Bailey, who had helped bring him to Balliol, wrote: "My hearty congratulations on the Jenk: I think your performance all round was the best I ever remember.''[59] A.D. Lindsay chimed in: ''I must offer my warm congratulations, not only on your getting the Jenk but on your doing so well in the exam . . . the standard all round this year was very high and you were quite easily first. Someone put you first in every paper except the philosophy.'' Lindsay went on to express the hope that Toynbee might become a don at Balliol. ''My hope is that we shall be able to have you. I thought you ought to know this in case any other prospects come before you.''[60]

Even before the examination results were published, Lindsay's hope was fulfilled. In May 1911 the college invited Toynbee to stay on as tutor, with the understanding that he would take up his duties in the autumn of 1912 after using the Jenk to finance a trip to Italy and Greece. Toynbee was well pleased, for he had nurtured the hope of becoming an Oxford don for some time. ''If, oh if, I become a don,'' he wrote in November 1910;[61] by May of the next year he could burst forth: ''I have got great news—I am going to be a fellow of Balliol, take on Ancient History. . . . It is work after my own heart, the very job I would have picked out, but had never dreamed of getting.''

His new status altered Toynbee's perspective on the community around him rather abruptly, for the letter goes on: ''These fellows of Balliol have a most marvellous and beautiful devotion to the College: I suddenly saw it in the way they talked—I had only dimly realised it before. Meanwhile, do you really feel in the mood for wandering? For next Winter I am going Eastward—Athens I am pretty sure, with a bit of Italy, I hope, on the way. I want to live several months at the British School, and tramp over the length and breadth of Greece. If you have any possible ground, occasion, reason or excuse, come across and do some of it with me.'' The letter concludes: ''The thing I want to see is whether I shall possess any power or influence over these men at Balliol: if one achieved that, one would be like the gods—and that is why I want to get to Balliol as quick as may be, and leave All Souls [for which he had been invited to compete] out of the reckoning.''[62]

To be godlike, shaping young men who in turn expected and were expected to shape the public life of Great Britain, of the British Empire, and of the world, was not to be Toynbee's destiny. World War I changed all that, and his affection for Balliol and for the life of an Oxford don evaporated almost as quickly as it had kindled in his breast. Disappointment at having to teach students who fell far short of his expectations played a part in this shift of Toynbee's mood and ambition. But changes in his private life and the general upheaval of World War I had far more to do with his abandonment of an Oxford career that seemed so firmly his in 1911.

III

Grand Tour, Donhood, and Marriage

1911–1914

Between 22 September 1911 and 6 August 1912 Arnold Toynbee traveled in Italy and Greece, looking at ancient sites, studying the landscape, imagining how the ancients had lived in the geographical settings spread before his eyes, and letting his historical imagination range freely across all the centuries of Greek and Italian history, from 2000 B.C. to the present. He traveled on foot, with a water bottle for thirst, a raincoat for protection, a spare pair of socks, and enough money to buy the food he required from villagers along the way. He slept out under the stars, or on a cafenion floor, and walked a total of almost 3000 miles, following goat paths for the most part, but sometimes cutting cross-country, either heading toward a lofty vantage point from which to survey the lay of the land or seeking a more direct way to some ancient site.

At first he traveled in the company of other British classicists, visiting the most famous ancient places. Then, as winter set in, he and Robert S. Darbishire headed off to wilder and more remote regions on their own. After March 1912, when Darbishire went home, Toynbee was usually alone. Early on, he tested himself by going from Athens to Sounion and back in a single day—forty miles of walking on relatively good roadways. Two weeks later he left the well-beaten paths: "I did five days tramping of the most delightful sort," he reported to his mother, "for one had no luggage, and never knew where one would sleep the night."[1] On this occasion when his exhausted companions headed for home, he took a few extra hours to explore an alternative route across Mt. Kitharon between Attica and Boeotia. This was completely characteristic. Whether alone or in company Toynbee always excelled in the range of his curiosity, the vigor of his roving, and the fullness of knowledge and imagination he brought to bear on the scene before him.

His travels were liberating for him in at least two senses. First and most obvious, he removed himself from the intense family worries and intractable difficulties that had haunted him ever since his father's collapse in 1909–1910.

Correspondence kept him sporadically in touch with family problems, but for weeks at a time, when walking through remote landscapes, he was out of reach. Other members of the family therefore made key decisions without him. Second, his walking freed him from books. "I am learning an extraordinary amount of history by legs and eyes," [2] he wrote from Italy at the beginning of his travels: two months later from Delos he declared: "It is marvellous expanding one's comprehension like this—it gives one's 'historical imagination' real strong meat to crunch. Pompeii may be more perfectly preserved—but that is a shoddy watering place of the first century A.D.—while the place I am in now was the chief mart of the world for a century: here one is at the centre of things (even if there are boats to Mykonos only twice a week)." Contrasts between surviving fragments from "antique religious Delos" and traces of the Hellenistic slave mart it subsequently became stimulated his imagination to see "how cosmopolitan Hellenism swirled back and engulfed the simple, unpretentious local little Hellenism which produced it." [3]

Such simultaneous liberation—emotional and intellectual—is rare indeed, and Toynbee savored it to the full. As he wrote to his erstwhile companion soon after returning to the complications of ordinary existence: "Anyhow, the greater the freedom, the greater the happiness, and I have never been so free as walking about in Greece." [4] He treasured the memory throughout his life, and in the book *Experiences,* which appeared in his eightieth year, devoted more than twenty pages to describing what he called his "second" Greek education, his exposure to the landscape and people of the country instead of merely to ancient texts. [5]

The venture was important to him in still another way. Until he crossed to France, Toynbee had never had occasion even to try to speak a foreign language, or to arrange the practical details of life—where to stay, what to eat—for himself. School and family had lapped him round; fully autonomous adult roles had been left to others. He was aware of this fact and promised his mother in the first letter he wrote back from France, "Remember . . . I will come back more useful and compleat and altogether manly than I start out." [6] On first arriving in Greece, he was still inclined to rely on others to make practical arrangements, especially since noisy haggling over prices was repugnant to him; by the time he left, he had become expert in getting his money's worth and had survived encounters with wild dogs, shepherd robbers, and arrest by suspicious gendarmes. "I have also learned the rich man's point of view, for here I am very rich," he wrote to his mother. "Really, the richer people, who hunt and shoot and travel from their youth up as a matter of course, have a great pull over one: they learn so much more of the world, just by the way, while I have had to do it all at one go. Three months ago I had never mouthed a foreign language or used foreign money: now I can find words in French, Greek and Italian fairly naturally without feeling the effort. . . . Really a year of this is worth ten of liberal education." [7] Altogether, the everyday world opened before him as never before. In many senses of the word he did become a man during the year of his grand tour, just as he had hoped.

But of course family troubles did not disappear, and those who stayed behind had to make decisions about Harry Toynbee's care and how he should be treated. Since the sick man was showing no signs of recovery after nearly two years of derangement, in October 1911 the family arranged his transfer to a new and more expensive hospital at Hillingdon. Both Edith and her son hoped for what the young man called "a true improvement,"[8] but by the end of the year their hopes had been disappointed. Toynbee sought to reassure his mother by writing: "Don't set yourself down as a failure till you see how we children turn out." And in the same letter, "Leave final views about your life and father's to God."[9] The harsh fact was that Harry's higher hospital bills at Hillingdon could be met only by special gifts from his wealthier brothers; this seemed to Edith, and to Arnold as well, to involve too much dependency on them. "I am very glad you agreed with what I said" he wrote to his mother in March, "I am improving. I did not lose my temper at the uncles this time. My nerves have steadied down wonderfully since I have been here. I suppose it is open air and exercise and leisure. . . . Or perhaps it is only that I have grown up into a man."[10]

In May 1912 Harry was again transferred, this time to St. Andrew's, Northampton, where care was cheaper. He remained there until he died in 1941. This arrangement achieved a penurious financial independence for Edith, thanks to payments from the Toynbee Testimonial Fund, together with some contributions from Arnold, and her own earnings (eventually she was employed to arrange Florence Nightingale's papers). But that small victory concealed a larger defeat, since, when her husband went to St. Andrew's, she surrendered all effective hope of his recovery. Her son urged her on: "Much better to depend on the Testimonial and ourselves entirely. That is what I think about it; and I believe it is what you feel too. Mother, I know how bitter it was for you to admit that Hillingdon, even, was not going to cure his mind. . . . I had put a lot of hope in it too."[11]

Ironically, at almost the same time that this final determination was reached with respect to Harry Toynbee's care, his son contracted a fateful but very different disease by drinking from a contaminated stream in the Peloponnese on 26 April 1912. In later years Toynbee liked to think that he survived World War I because of the dysentery that resulted from his incautiously drinking bacteria-laden water.[12] This was partly true, as we shall see in the next chapter; but in the short run, the principal effect of his illness was to cut short his carefree exploration of the Greek countryside and compel him to come home a little earlier than he had planned because the local cure (drops of arsenic taken with sugar lumps) proved inadequate to overcome his ailment. As a result, when he got back to England on 6 August 1912 his mother put him in hospital where, after weeks on a liquid diet, [13] he was released as cured at the end of September.

As one would expect, Toynbee's eager, youthful mind seized on his experiences in Italy and Greece to generate new thoughts. Their subsequent importance for his life compare with the enhanced personal independence and lifesaving infection that he also brought back from his travels. Interestingly, what he remembered in old age and what he wrote about to his mother at the time do not correspond and often contradict one another, though both, presumably, are true in the sense that both sets of ideas passed through his consciousness, some seeming important at the time, others attaining importance later on. Moreover, there is a common denominator. Both his contemporary reactions and his subsequent reflections indicate that Toynbee was fumbling his way toward recognition (which became full and explicit in 1920–1921) of a civilizational difference between the England he knew and the Greek–Italian peasantries he encountered on his Grand Tour.

But to begin with, the discrepancies between upper class English manners and the behavior of Italians and Greeks offended him profoundly. His distaste found expression in the term "dago," applied indiscriminately to everyone he met. As he explained to his mother: "There is no moral difference between the educated and the uneducated, as there is in Europe. Here education is confined to shoddy externals. The 'black coats' are really abominable—on how I sympathise with Anglo-Indians who want to kick babus. . . . I can now understand race prejudice. It is a thing to be fought down like any other form of nausea—but as for giving Indians self-government!" [14]

Yet Toynbee recognized that what made Greeks into "dagos" was not race but social circumstance. As he wrote within a week of his arrival in Athens: "Were ancient Greeks like the modern? I think not: probably worse off materially . . . but it is the moral difference that counts. Then they were the centre of the world, and everyone came to Naxos [where he then was] for trade and poetry and sculpture: now they are hangers-on of Europe, and come to us for their models of everything—and their best is always a second rate imitation of our second best. We shall be 'dagos' too when civilization centres in China— so let us meanwhile arm to the teeth and fight to the death in order to remain top dogs and the centre of the universe, and govern India and exploit West Africa and colonise Canada, as the Greeks exploited Thrace and colonised Sicily and governed Asia—what a bad moral to end with." [15]

But the intellectual insight that cultural configurations altered across time and dictated contrasting forms of behavior did not prevent him from expressing angry, emotional reactions to conduct that offended his sense of propriety and correctness. Thus, for example, after being arrested on suspicion of spying for the Turks, he tried to use his status as an Englishman to avenge himself upon the officials who had interfered with his wandering near the Greek–Turkish border. As he wrote back home:

I went to the [British] Legation yesterday and talked to the Minister, and they [the Greek officials] will all get heavy punishment, especially the chief of police. I fortunately managed to keep my temper all through. Contempt for the dago has a wonderfully calming effect. . . . A fixed law without respect for persons is as strange to him as a fixed price. It is the struggle for existence uncurbed—the man with power uses his power to crush the man without it, and every affair of buying and selling is a battle. . . .

I think the greatest result of this year has been to make me appreciate the value of England (or rather of the small group of countries . . . which can be called civilised.) We are very small and very precious; and O the folly of it if we and

Germany destroy each other. Also the soundness of race prejudice, and meaninglessness of the "Rights of Man". . . . And now I see by experiment that there is
no "genus" Homo Sapiens, with certain inalienable endowments, but only an
infinite number of phases between chimpanzee and Supermen. . . . I would like
to study the dago deeply, but one would have to see South America and Sicily
first. Unlike the barbarian, he is a parasite—he can only grow under the shadow
of a vigorous civilization—his nature is unsuccessful imitation. For instance, I
don't suppose these people were dagos till they came into contact with Europe at
the revolution [1821–1830]. Even now the remoter villages, and all the shepherds
in the mountains, are not dagos yet, but white-skinned savages. . . . It is not a
matter of race . . . it is largely a matter of climate, and I think malaria is essential
to dagodom. But no Mohammedans are dagos: there is something in their religion
which saves them from imitativeness. . . .

However the chief thing is that these little men at Lamia [where he had been
arrested] will be dropped on heavily enough to prevent them playing the fool with
the next archaeologist who comes along. . . . Well, I shall religiously preach mis-
hellenism to any philhellene I come across. . . . But still I have not solved the
great question: "Were the ancient Greeks like them?" [16]

It is instructive to contrast this angry account with his recollection of the
same event as recorded in *Experiences:* "I was conducted up the whole ladder
of military bureaux till I reached the regional command, and at each higher
rung I became more indignant and more rude. However, they all took this
good-humouredly, and the regional command quickly gave me a clean bill of
health and let me go." [17] He cited the incident to illustrate how he resisted
lessons "living Greek men and women, who were highly intelligent, alert and
vocal" had tried to teach him about international affairs, which were then moving toward the outbreak of the Balkan wars (1912–1913). In effect, Toynbee
in old age thus entirely reversed the moral position he had taken at the time,
concluding the story with the words: "At Athens I made a fuss at the British
Embassy *[sic]*. They tactfully showed sympathy, but prudently took no action."

Near the core of Toynbee's negative reaction to Italian and Greek manners
were his own fixations about money. He disliked being cheated, and disliked
bargaining to establish a just price almost as much. This put him profoundly at
odds with Greek and Italian peasant manners, for they viewed a vigorous, vivacious interchange over price as part of any proper exchange, humanizing and
giving savor to an otherwise heartless market relationship. From childhood, his
parents had instilled into him a fear of penury, the need for economy, and the
importance of keeping accounts. Toynbee accordingly recorded every expenditure he made on his tour, and totted everything up on his return down to the
last penny. Every time he spent money, it felt like a loss since there was less
remaining for his grand adventure; and that sense of loss was magnified by the
nagging suspicion that he was paying more than he needed to or should because
of his awkwardness in bargaining.

Yet, within a few months of returning home, Toynbee's dislike of sumptuous
living as a don at Balliol provoked him to begin the sort of reassessment that

came to full flower in his old age. As he wrote to Darbishire, "Now that I am not exasperated hourly by lies, procrastinations and failures to do what they undertake, I am perceiving the virtues of the dagos: frugality, absence of snobbishness (the real, solid virtue, I think, of the black coat). If you come to think of it, Jesus could have gone about doing good in modern Greece, but he would find no foothold in Oxford." [18]

But at the time, and under the circumstances of his first encounter with a culturally different society, Toynbee remained locked into a narrowly limited range of social sympathies. Thus, for example, he objected to having Margaret Hardie, a student from Cambridge, accompany male students on trips to famous ancient sites, because she found it difficult to walk at his pace. "Women are horribly handicapped, and it is no good their butting their heads against us men" he wrote, most undiplomatically, to his mother, who as a young woman had in fact butted her head against the world of men with some success. "Their quarrel is really with God: of course, the kingdom they are trying to take by storm these days is really a men's kingdom, and when they get a footing in this male world, they find they don't really fit into it—which I expect is a bigger tragedy than many of them will own up to." [19]

Similarly, he was repelled by the Roman Catholic clerics he encountered in Rome, "singing greasily and mechanically" [20] at St. Peter's. He thought no better of the Orthodox monks of Mt. Athos because, he said, their piety and religion, lacking "the moral and philanthropic clothes of Europe," had become "purely parasitic." He concluded: "The sooner we see Austria at Salonika and Germany running Anatolia and the Bagdad railway the better. . . . May whatever Power does come down here put an end to this historical museum, Athos. It is a wonderful place, but everything ancient does not deserve preservation, and I am sure this community deserves to die." [21]

Such prejudices were incidental and trivial, though mildly surprising in view of his later ecumenical views. What really mattered for him at the time was the geographical and topographical setting of Italian and Greek history. Everywhere he went he surveyed the landscape with an eye to how human beings had put it to use in times past—ancient, medieval, and modern. Military defenses and routes of march particularly interested him. "I could almost turn specialist in that line," [22] he confessed toward the end of his travels, and the bulk of his letters home consist, in fact, of descriptions of sites he had visited, frequently illustrated by freehand maps and neatly drawn sketches of ancient strongholds.

On three occasions, if his later recollections are to be taken at face value, as I think they must, he experienced remarkable mystic encounters with the past. On 10 January 1912, when contemplating the site of the ancient battle of Cynoscephalae, he "saw" in his mind's eye how the Romans had defeated the Macedonians there in 197 B.C., and the reenactment was so vivid that Toynbee asked himself more than forty years afterward: "Can the dreamer really have sunk, for that instant, those twenty-one centuries deep below the current of Time's waters, on which he now finds himself riding, once again, in his normal waking life?''[23] Three months later, he had a similar experience in Crete when he caught sight of an abandoned Venetian villa—mute testament to the Turkish victory of 1669; and for the third time, visiting the Moreote citadel of Monemvasia, he "fell again into the deep trough of Time" on seeing abandoned bronze cannon littering the site.[24]

These three unusual experiences were not the first time that Arnold Toynbee's vivid imagination carried him beyond the ordinary state of consciousness. When preparing for his examinations at Oxford in 1911, a few words from the Epitome of Livy describing the death of one of the leaders of the so-called Social War against Rome sufficed to provoke a similar vision—a "transport" that was "infinitesimally brief" yet intensely poignant, and remembered for the rest of his life.[25] On two subsequent occasions, other texts and other sites provoked a similar entry "into a momentary communion with the actors in a particular historic event." This happened to him six separate times by his own count.[26]

What can more prosaic minds make of such experiences? Toynbee himself clearly felt that his capacity for visionary communion with the past was real, in some sense or other, else he would surely not have recorded these encounters so meticulously in the final volume of his great work. Yet it is worth noting that these visions came to him at times when he was under special personal stress. Obviously, worries about his father and his problematic relations with his mother haunted him in 1911 and 1912, even though his walking tours partially insulated him from family concerns. His later visions, too, occurred at moments of special anxiety, as we shall see. Perhaps, therefore, these mystic experiences should be understood as dramatic, extreme expressions of the habitual way in which he took refuge from personal difficulties by burying himself in the past, concentrating his faculties, often by an act of will, on his work as a historian.

Buy Toynbee's reactions to the landscapes of Italy and Greece were seldom mystical. Everywhere he went, he asked himself straightforward questions about how and why men had built dwellings and forts where they did. His curiosity was by no means restricted to antiquity, but ranged across the whole landscape, down to and including the plains villages built after the Greek War of Independence (1821–1830) on land once owned by Turks. On 23 May 1912, looking down on the valley of the Eurotas from the Byzantine stronghold of Mistra, he reflected on how political control of that valley had shifted back and forth between centers in the open plain, where unwalled ancient Sparta had been located, and the forts commanding passes into and out of the fertile lands lying beneath them. From his elevated vantage point he could see traces of ancient, medieval, and modern human occupancy of the Eurotas valley—and the ruins' locations were visual proof, for him, of a perennial tension between hillsmen and plainsdwellers, shepherds and cultivators.

The landscape also led him to recognize a repetitive pattern in the local past. For he could see both an ancient cycle (bodied forth in the remains of a Mycenaean hill fort, visible across the valley, and the site of classical Sparta in the open plain) and a second cycle, like the first (evidenced by the remains of

the medieval hill fortress, which surrounded him, and the modern village beneath, erected after 1821, where ancient Sparta had been).[27] Subsequently, he
generalized this insight by applying it to the historical geography of Greece as
a whole in a brilliant series of lectures in 1922.[28] And, of course, the concept
of a civilizational cycle became central to his mature concept of human affairs,
though only rudiments of his later ideas were present in his mind in 1912.

None of these experiences appear in his contemporaneous letters, whereas in
autobiographical passages in his later works he emphasized these isolated, largely
visual, images. Presumably he subsequently abstracted them from the hurly-
burly of his Grand Tour memories because they continued to resonate in his
mind and acquired new and enhanced significance as his view of history took
mature form. Many topographical details, meticulously recorded in a diary,[29]
were of no use to him later. But in retrospect, as at the time itself, his *Wanderjahr* remained a time of wonder and delight, when new thoughts and new
experiences swarmed in tumultuously upon a more nearly independent adult
self. No ten month period, before or after, was so fertile, though it took several
years before Toynbee reduced all the things he had experienced in Italy and
Greece to a satisfactory intellectual order. Nevertheless, the young man who
returned to England in August 1912 and began tutoring at Balliol in October
was different in important ways from the student who had departed less than a
year before. He had seen the world—or an important part of it; and he had
juxtaposed his book learning with classical Italian and Greek landscapes, viewed,
however, as perdurable throughout recorded history, with medieval and modern
as well as merely ancient components. No wonder it took him a while to digest
it all!

As we saw, on returning home Toynbee spent six weeks in hospital recovering from his dysentery, and was released just in time to start the autumn term
at Balliol in October 1912. He was both eager and reluctant to take up his new
life. "I am impatient to get started at Balliol," he wrote to Darbishire from his
hospital bed, "and find out what the job really feels like. You were disillusioned about the dull men—cannot they be poked up? I will try while I am still
enthusiastic, and not be cynical if I fail. . . . I suspect, though, that discharg-
ing large pouches of one's mind for duty, which one has been so far rather
painfully bottling up, is demoralising."[30] A month into his first term his ambivalence had not disappeared: "I am rapidly becoming at home in the other
camp, of those who live forever and ever and whose life changeth not, instead
of those who come and pass. Sometimes it comes into my head that I may be
doing this identical job when I am 59. The don crew are very friendly, and
there are fewer fossils embedded among them than I had expected. But why
need they fare so sumptuously? . . . I shall get myself made junior bursar,
and feed them on bread and water. I want to smash it and melt the College
plate withal. The whole business is piggish. . . . That is why I run."[31]

Yet there were compensations. Toynbee vigorously set out to communicate
to his pupils everything he had learned. "My job in teaching history is to make
people know a different life and civilisation from ours, from the bottom and
with different openings for good. . . . If I can get my men inside the Greeks,

mentally (though that is Sandie's [A.D. Lindsay's] job), physically and morally—with the imagination of limestone and pines and blueness thrown in—I
shall have done a good work.'' But only three of his fifteen pupils seemed to
have had any real enthusiasm for their studies. And, as he explained in the
same letter: ''I expound too much. I pour out stuff, trying to kindle their minds
and put a living picture into them. One ought to take up what they have actually written and worry about that. Overheard in the quad: 'What do you think
of Toynbee?' 'Oh, I think he is good.' 'There is one thing. He talks to you so
much himself, that you don't have to do any talking.' So I must change my
tactics. But it is good work teaching. . . . I don't think I should become dried
up and withered here. Meanwhile I shall get time enough to go on assimilating
history, which is the driving force inside me now, as it has been for some
years—nor do I see any signs of its abating. I want knowledge because I want
it.'' By December he could say, to the same confidant: ''I have enjoyed myself
this term thoroughly, even down to keeping the minutes of the College.''[32]

Toynbee's enthusiasm was not wasted, and at least some of his colleagues
approved. A.D. Lindsay wrote: ''I don't yet know what is going to happen to
all our papers in Greats, but you ought to know that their history shows an
immense improvement this year, both Greek and Roman, and considering what
a middling lot they are, you are to be congratulated.''[33]

Teaching and college duties were not all that engaged him. During his first
term at Balliol as a don he ''concocted a great paper on population, military
organization and mobilization at Sparta, which I hope to publish in the J.H.S.
[Journal of Hellenic Studies].''[34] This essay became Toynbee's second published work when it appeared later in the same year. It is a lengthy, learned
article seeking to thread a path through contradictory ancient evidence about
the size of the Spartan army and the legal status of its members, whether citizens, perioeci, or helots. Toynbee's distinctive contribution was to argue ''what
must have been the case'' with respect to relative numbers on the basis of the
detailed topographical survey he had made of the regions of the Peloponnese
once dominated by Sparta.[35] As such, it was a very creditable harvest from his
year in the field, and a fine start for a conventional career as an ancient historian.

But Toynbee was not content to become a conventional ancient historian.
His mind was running still on a ''philosophy of history'' that would deal with
all the links and uniformities of ancient, medieval, and modern times that had
come to his attention as he traveled in Greece, and as a step in the direction of
achieving a synoptic overview, he eagerly accepted an invitation from Gilbert
Murray to write a history of Greece for the Home University Library. In 1913–
1914 this project ballooned into a survey running from prehistoric times to the
Byzantine era; and Toynbee circulated summaries of the project to a number of
distinguished experts, asking their advice. Gilbert Murray responded: ''I think
your sketch of Greek history quite wonderful,'' though he followed that encom-
ium with a long list of points on which he thought Toynbee should alter his
tone, by adopting a more enthusiastic view of Greek achievements.[36] J.B. Bury
of Cambridge responded: ''You know too much. I have read all with the great-

est interest." and concluded with "a few criticisms."[37] Walter Leaf, representing an older generation of scholars, replied: "I have read your MS with even more interest—even excitement—than I had anticipated." But, he continued, "Your Homer I must leave to your conscience with a sigh."[38] This project was interrupted by the outbreak of World War I in August 1914 and resumed many years afterward. Its stark contrast to the minutiae of the article on Sparta demonstrated Toynbee's characteristic thirst for big views and general perspectives, superimposed upon impressive mastery of details. And details continued to fascinate him. "I am writing an absolutely immense paper on the Homeric Catalogue [of ships] to be read to the Philological Society at the end of this month. . . . It is growing vast protuberances in every direction, and I hope to print the whole lot, when it is finished, as a small book."[39] This essay, however, never achieved publication and left no trace in his surviving papers.

On top of his scholarly projects, the young historian became intensely interested in the campaigns and diplomacy of the Balkan wars. These wars were indeed dramatic. Greece, Serbia, and Bulgaria first combined to defeat the Turks (1912), and then quarreled over the spoils, starting a second Balkan war (1913) in which Greece, Serbia, Rumania, and the Turks combined to defeat Bulgaria. In 1912 he wrote: "I bore everyone I meet by talking of nothing but the war. I rejoice that they [the Greeks] are winning. And they are coming out so finely, dagos and all. There is deep patriotism and self sacrifice, and it is a righteous cause, if any ever was." The Great Powers, on the other hand, struck him as "incompetent cowards" in their efforts to end the wars by negotiation. As for Britain: "I hope we shall have the grace to cede Cyprus to Greece before the propaganda there gets feverish."[40]

In 1913, when a Greek delegation came to London for negotiations with the ambassadors of the Great Powers, Toynbee was able to meet them and reported: "I had the joy of airing the only foreign languages I can speak to . . . Venizelos' secretary, also to Venizelos [Prime Minister of Greece] himself. I had wanted to see what manner of men they were. I accepted the 'Albanophone Hellenes' argument with becoming enthusiasm."[41]

Current events were soon to impose themselves even more forcefully on Toynbee's attention, altering the whole pattern of his life and the lives of everyone else in Oxford as well as in much of the rest of the world. But in 1913 this was still unimaginable. What did occupy the forefront of his mind, threatening to distract him from his teaching and historical research, was the old problem of relations with his mother, transformed and no doubt intensified by the fact that a new object of his affections, Rosalind Murray, had appeared on the scene.

Toynbee first saw Rosalind Murray in 1910 when she was not yet twenty years of age. Though he encountered her only briefly and in the company of others, she impressed him mightily, seeming more like a faery princess than a merely human creature. Witty, talented, sophisticated, aristocratic, and altogether adorable, she was also unattainable—at least as long as her secret admirer remained a mere undergraduate, unable to support a wife. Yet his mother may have sensed a rival. At any rate, Toynbee went out of the way to assure her, when first starting on his travels, that he had "no intention of marrying till an indefinite time hence, or never."[42] And a few months later, when his mother reprobated the moral tone of Rosalind's newly published novel, *Moonseed*, Toynbee came to Rosalind's defense, saying: "She does not at all strike one as the disillusioned sort when one talks to her."[43]

A sharper collision occurred in August 1912 when Toynbee had to refuse an invitation for a month's stay with the Murrays because his mother insisted on sending him to hospital to cure his dysentery instead. As he explained to Darbishire, "I had not seen her [Rosalind] since the day after Greats, at a picnic which I ran. I was going to stay with them this summer, but lay in bed all the time, till just before term began."[44] Irritation at not being able to see Rosalind as he had hoped may have well have made real reconciliation with his mother impossible, despite the fact that in taking care of him so solicitously she was resuming the role she had been unable to fill when the crisis of her husband's collapse was fresh.

Yet Toynbee must have wavered. "I think acquiescence in invalidism is a natural and laudable state," he wrote from his sickbed.[45] Acquiescence in invalidism, under the circumstances, meant acquiescence in his mother's management of his daily life. But even when ill, he found full submission to her hard to accept. During Toynbee's absence, Edith had moved from London, so the familiar surroundings of his childhood were no longer there to sustain and anchor accustomed behavior, on his part or on hers. Instead, Toynbee felt the indignity of dependence. "If my Mother catches me still writing, I shall be in disgrace, for I am an invalid, or at any rate a convalescent, and in her hand" he wrote to Darbishire before entering the hospital.[46] When his health was restored and he started teaching at Balliol, he eagerly resumed the personal independence he had savored in Italy and Greece.

Decisive alteration in Toynbee's emotional alignment came after he helped to settle his mother in a new house near Winchester during the 1912 Christmas vacation. He found her unreasonable, "hovering near a breakdown: in the state when one can't bear to do either one thing or another. However, we are going into the house—there being an excellent girls' school in the place, and we having spent much money in putting straight the drains, etc. Nerves and morbidness are an evil thing to fight against, because they are infectious and destroy judgement when one wants to be wisest."[47] After this bruising encounter, return to Balliol was a relief, and the charms of Rosalind and of the Murrays seemed all the greater because of the sadly raddled state of his own family relations.

Toynbee's adoration of Rosalind fed on that fact. A few weeks earlier he had confided to Darbishire: "I am in love with Rosalind Murray, Gilbert Murray's daughter. I have been for more than two years, but the thing increases by great jerks. . . . I saw her for an hour or two at supper the other day, in the useless way one meets people in a crowd. She is living in London on her own, writing, and working in the East End [at Toynbee Hall, the settlement house named for his uncle]. . . . It is a weird thing not to know really what she is like: and yet to be in love like this. It is only guessing, yet one is quiet certain she is like that. And it is filthy impertinence and onesidedness! I am certainly just one of the casual crowd of people she comes across, to her. But this is all foolishness when I write it down. I am going to bed. You will just have to submit to having this vast egotistical screed shoved out upon you. You are somehow the natural victim because I write to you without shame, like talking.''[48]

By March 1913, with Easter vacation ahead, Toynbee's attraction toward Rosalind and irritation with his mother had come near the bursting point. His own words explain the situation clearly.

Dear Rob:

The spirit very much moves me to talk to you, and having finished my collections [student papers] and totted up all the marks, and notified that I will give them back tomorrow morning, I am free. I did collections solidly yesterday . . . and then I dined at the Murrays, and there was Rosalind Murray and since that the inner parts of me have been whirling round and round somewhere at the back of the parts that have been correcting these foolish collections, and I am moved to write to you because I want to talk and certainly can't bend myself to Aeneas Poliorceticus [a classical text about siegecraft]. I object to being run away with by myself, and not to be able to do the job I want to at the moment. I have only seen Rosalind Murray three times since the June before last: namely, $2\frac{1}{2}$ hours last night, $2\frac{1}{2}$ hours one Sunday night last term, and about $2\frac{1}{2}$ seconds in the distance at a meeting the other day. . . . Last June I worked a canoe with her for a day up the Cherwell. What is the use of five hours in a year and a half? All the same, I think I know her fairly well—pick up where I left off: but I want more. I am always being crossed. Whenever they have asked me to stay with them, I have always been in bed, or abroad, or obviously needed by my Mother. I was almost tempted to falling the other day, but when my Mother has just moved into this new house at Southwold, it is up to me to go and keep her company. But I might have gone three weeks to Italy with them. It is simply damnable to do one's proper job, and then feel sulky about it inside. But these stingy hours are no use: you want to live with people for a time in order to get further with them. Better luck in the summer perhaps. It is exasperating not to know what is in the other person's mind, or rather to know their mind more or less, except just so much of it as concerns you: it is 5 to 1 they think just nothing of you at all: but if only you knew. But you can't do that without seeing something of them. How monstrous it is, to stalk about this quad being gnawed at inside, while the person you want to be stalking about with is just 15 minutes off: and then not to be able just to go and say: "Come and stalk with me": moreover she is only in Oxford a few days at a time—otherwise in London, or with a brother up at Barrow.

Meanwhile back to the books, and write a great history, and get made professor, and gain moneys. . . . and meanwhile see her when you can: that probably won't be for three months now: meanwhile work, and don't gnaw yourself like a fool! I am going to adjourn, because I seem to be writing to myself instead of you, Rob. . . .

Well, goodnight, and write again soon, even though I have been in arrears lately.

Affectionately,
Arnold J. Toynbee[49]

The next letter to Darbishire reported more cheerfully: "Altogether it was a very successful vac"[50] thanks to family visits with friends in Sussex followed by a visit to Oxford by his mother and sisters. Then after mentioning some of his professional concerns, Toynbee confessed:

> But all this is not what I have really been doing all this time, for all the time I was in love (though not in the condition fortunately, in which I wrote you a sort of prolonged screech after seeing her at the end of last term: for then I was sick and weary): meanwhile I have seen her two and a half hours of a Sunday evening, and am still more so than before, if possible. Seeing people an hour or two like that, with half a dozen other people there, cuts you to pieces: they might as well be in a glass case: which is why I have delayed writing, for fear I should begin to screech again. . . . But being entirely in love and also entirely ignorant whether there is a ghost of a chance of her being in love with you, can't be borne indefinitely. I mean, given one chance in a hundred of her loving me (there is not the remotest probability that she does), the sooner I try the chance the better. But I shall have to talk to Murray first: for my Father's illness and her having been tuberculous, are two very big rocks, and I have got to work out first, whether I have any business to try my chance at all. If I have not, then the one joyful thing is that I won't have affected her, for, as I say, I don't suppose she has ever thought twice of me: and if I have, then it is plain sailing, for I can go speedily and find out, and it is only I that will stand to lose. Meanwhile, the more guts I put into these other [professional] jobs, the better: which, after all, are my jobs, and will be, whether this private business prospers or not. As for my corporeal guts, they are really mending their ways: I got sick of the tail end of dysentery and went to a very fine doctor about six weeks ago, who forbade bread, vegetables, fruit and all things up-dished: I flourish thereon (I mean on the things I do not eat), and if I am really recovered by September may rush to Bosnia with Wace and Cheesman [young classical scholars]. But I just must know about this other thing . . . as soon as term is over: it is just damnable to be quite in the air.
>
> Affectionately,
> Arnold J. Toynbee

Toynbee did indeed adhere to that schedule, but before bringing things to a head he told his mother of his intentions. This provoked the following letter from her:

> 30 April 1913
>
> My dear, dear Arnold:
>
> I am so glad you told me and I do thank you for it more than for anything you have done for me. Although I knew, I can't tell how. Except that I think we have always been so near to each other that I divined your heart by instinct. . . .
>
> First, I must say God bless you and this which has come to you. I always thought you would love early. You have Father's loving nature which must satisfy itself and find its natural outlet. And I think your love will be like his, deep-steady.
>
> I want very much to help you caught in this crisis of your life. I wish it were all easy and straightforward for you, but you know it isn't and its very seldom it is.
>
> First, you don't seem to know if she cares for you. Not at all? I know one

mustn't think that all youthful books are autobiographical, but they generally are rather so. So you mustn't be hurt if I say that I think she has had her heart touched before. . . . That's no cause for despair. People make essays at loving, as in other things, but when the real thing comes it is so different from the trial trips as to be unmistakeable—which is autobiographical. . . .

It is the tragedy of Father's illness that he who would have died to save you a heartache should now be the chief obstacle to your going straight ahead to try to win her.

Arnold dear, you mustn't try to win her till you have had it out with her father and mother. . . . Dr. Craig told you that to counterbalance the sensitive nerves which Father and I are handing down, you would be wise to choose a very even nerved, placid, strong mother for your children. I know so little, nothing of Rosalind herself. But I fear, I fear that her parents are very nervous highly strung [sic] and that there is much delicacy in the family. About her tuberculous tendency, I feel confident that her parents would not let her marry if she should not.

About minor matters I do not think there could be difficulties. They are democratic and would not shy at a "middle class son in law". Nor would . . . means be a great difficulty. You can earn as much as most. You need have no burdens. I can manage anything extra for Father, and the girls are not dependent on you. . . .

It's no good saying I wish you could have kept your heart free for three years more, since you haven't. . . . Whether this is going to be the greatest joy or a great sorrow to you, we cannot tell. . . . You'll think me a cold blooded calculator. But remember I have scarcely seen her. . . . But no fear of my not loving anyone who made you happy, and I remain

Your
Mother[51]

What happened next is best recorded in Toynbee's own words:

13 June 1913

Dearest Rob:

No, I have not written to you for a month, or rather for ten years: meanwhile, she loves me, and is going to marry me. I saw her several times towards the beginning of this term, more often than I had ever seen her at all since I came back: and I just could not stick it any longer. One Wednesday, it must have been four Wednesdays ago [14 May 1913], I knew suddenly that at 5:15 next day I was going to her Father and Mother and clear it up. I thought she was tuberculous, she is not: I also thought my Father's breakdown might affect me: it does not: I went to see this doctor about that next week, on the Thursday [22 May]: Rob that is the grimmest thing I have ever had to do, and I hope I shall never have anything like it again: you see, they trusted me absolutely to find out the truth from him (they only went to see him themselves *after* she had found she wanted to marry me) so it was not merely that one had to hold oneself from pleading one's own cause, but I had to be on the lookout all the time to be sure he was not hedging, or in the least uncertain. I was sure, though (he is a very jolly, straightforward man, with a smile, and that helped): only he said I must never overwork, or even do extra work, as my father's case did show I had a tendency to nerve-exhaustion, though it is not a mechanically hereditary thing: it depends entirely on myself—on whether I play the fool or not. I thought that dished me: I had meant to make extra money by extra work: you see, my Mother and sisters are a first charge on me, and I am in my Father's place towards them. This just stopped that: it was in the morning: I had been going on to talk to her that afternoon. (I had told her the Sunday before, that I should be in London on business that day, and would come to tea with her if the business allowed: the business was the doctor, of course: it was awfully hard to get that out in a casual sort of voice: I think I almost funked it): then I wandered about the Parks of that town trying to make myself go back first to see her people and tell them the doctor's opinion did dish me after all: though it was on the money and not the health question: I was hideously tempted to speculate on my energy and power of solid horse power work, knowing I was now free from that horrible hereditary shadow. It was quite childish. But about lunchtime I arrived at the Admiralty, and made a man there called Gleadowe [a friend of Rosalind's] come out and have lunch with me, and I talked it out with him (or at him) and that screwed me to the point of telegraphing to put her off and coming back here by the next train. Then they told me that the extra work did not matter, because she had moneys, and that I might go ahead: I had not thought much about moneys before, because the health question loomed before everything. So the next morning [23 May] I went up again to see her (she lives in a flat near Baker Street): I went straight on as if I were wound up by clockwork, and said just what I wanted to say. She was absolutely taken by surprise, and could hardly speak a word: I had not expected that: I had thought she would be calm and very kind, and that I should go to pieces, unless she saved me from that by dealing very kindly with me. But it was like to knocking her down. I have told you before that I have kept this absolutely to myself during these three years, I had not somehow reckoned on moving her. I just went along, thinking that would help her most: it lasted three minutes. I came straight back to Oxford: that was three weeks ago today. Then I waited. . . . On Sunday afternoon [25 May] came a letter from her Mother, saying there was no hope: on Monday morning a letter from herself: but I had got over the worst by then: first of all one can do nothing but look back, and that is terrible pain: it is folly, but one can't help it. I walked round New College cloister that evening: the shadows of the open work are wonderful on the grass, when the sun begins to get low and it is a real Summer evening. I got to sleep all through because I was very tired, but waking up early was terrible, on that Monday morning especially: one is not quite master of oneself at those weak hours before dawn, and pain has its way with one. The rest of Monday I was dead: I could not look forward or backward: and just piled on as many pupils as I could, and waited for the time to pass: I thought it might last months: but when I got out of bed on Tuesday morning, I suddenly looked ahead and saw I could not live in college for ever. . . . So I determined I would become a thorough master of this job, get made a junior dean, become a trained teacher and get a grip of history, so as to have a real trade, and then do work like my Father's, or something where human life and sorrow comes in. I wrote to [Alfred] Zimmern, and told him he must help me get a footing in some such work within the next half-dozen years. It was curious: I myself was altogether looking backwards, but I saw that as long as you were actually alive you could always build up work ahead of you, which would keep your forces in working order, even if your life faced the other way. I had to begin from the very foundations: you see, on Monday my work went absolutely: I saw it was not the centre of my life, nor ever could be (I mean, History): that discovery alone was perhaps worth that week. The rest of the week I spent in that canoe [paddling on the Cherwell], and took care never to be alone. Nobody in

Balliol suspected there was anything amiss with me. I asked my Mother to come next week, and she arranged to: I knew by then that I could hold out till she came, and that after that I should be on my feet again: I was going to do things instead of learning things. Then on Saturday (the Saturday before last) [31 May] there was a note from her Mother saying she wanted to see me again. My mind was so turned the other way (for I knew that if my mind allowed itself to hope for a change, I should do nothing else all my life but go on hoping) that I did not realise what that meant. I came to their house that evening: then she just came into the room and said she had changed her mind. She had gone back to her flat in London that week, and on Friday her mind had come clear: she had had as bad a week as I. At first we neither of us knew where we were: we knew it was real, but we could not take it in. Now we have, and it is real and solid and absolutely good. I am not the same man who wrote you last. I lost her first and then found her; and those two changes have made me a different person: this is beginning again, only I am looking forward now as I never have done before.

The College have been delightful: I expected a grim fight with them, like fighting dogs in Greece, but they suddenly seemed full of affection and good-will, and they have made no obstacles at all: not a trace of grumpiness, though I am making an awful nuisance to them. And everyone else is the same. This added love and kindness on top of the wonderful thing itself is almost overwhelming.

She has been spending the middle parts of the weeks alone in her flat again, in order to recover her breath: she is coming down again this evening. My dear Rob, there is nothing more to write for you can't put feelings into words, not unless one is a poet, like you: it is horrid not being able to draw outlines of Greek mountains when you see them and suddenly want to: but there it is: so you must just do the imagining of it yourself.

You shall now hear from me at normal intervals, nor, Rob, will you hear less or be more distant from me. And as soon as possible you must know her: we must come to Kentucky, or you hither.

> Your affectionate, as always,
> Arnold J. Toynbee[52]

She was 22 last October, I 24 in April, so we are the right sort of age.

***

Who, then, was Rosalind? And what led her to change her mind and accept Arnold's suit for her hand on 31 May 1913?

Adequate answers to these questions are not forthcoming from existing documents, for in her old age Rosalind systematically destroyed whatever letters and other private papers had survived earlier house-clearing efforts, preferring, as she said, "to burn them—requiescat in pace—and that is that."[53] She kept a diary, which remained intact at the time of her death, but it was destroyed in a drunken haze by her lover, Richard Strafford, and her son Philip. As a result of these efforts to preserve her privacy, confident reconstruction of Rosalind's personality and point of view is impossible. Yet her importance in Toynbee's life requires a biographer to make an attempt, using whatever escaped destruction.

Outward matters are simple enough. Rosalind Murray was born on 16 Oc-

Castle Howard
Photo by Lawrence Toynbee

tober 1890, the eldest child of Gilbert Murray and Lady Mary, his wife. She was named after her maternal grandmother, Rosalind, Countess of Carlisle, and spent much time as an infant and small child under her grandmother's eye at Castle Howard in Yorkshire. The grandeur and aristocratic magnificence of the Howards, and of her grandmother's older and therefore even more aristocratic family, the Stanleys,[54] deeply impressed the young girl. Being her grandmother's namesake, and the eldest daughter of an eldest daughter, she readily imagined herself as rightful heir and successor to her grandmother's estate. This will-o'-the-wisp, as we shall see, remained very much alive in Rosalind Murray's thinking until 1921, when her grandmother died. Even after the dream had burst, she retained an acute awareness of her aristocratic descent, and sometimes struck others as snobbish.

Rosalind, Countess of Carlisle, was, indeed, impressive, and the hereditary position that allowed her to dominate everyone, or almost everyone, around her must have made her granddaughter's dream of matrilineal succession almost plausible. For the Countess controlled Castle Howard and the broad estates that supported its magnificence with a whim of iron. She supervised the lives of her tenants, negotiated rents, arranged elections to Parliament for suitable Liberal candidates, and in general tolerated no opposition. She was a fluent, forceful public speaker, and took leading roles in temperance agitation, women's suffrage, and Liberal party affairs, especially in the north of England.

Her husband was a mild-mannered man of artistic bent, whose principal interest was landscape painting. In 1885 he parted company with his wife over her support for Irish home rule. Thereafter, he lived in London and left Castle Howard and the management of his estates in Cumberland and York entirely to her. When disagreements between the Countess and Charles, their eldest son, became troublesome, the Earl of Carlisle portioned off his heir by assigning him a stately residence at Naworth. This did not prevent further quarrels between mother and son over disposition of furniture which she removed from Naworth in order to furnish a nearby house, known as Boothby, for herself.

The Countess of Carlisle's domineering personality made relations with her other children almost as difficult. All but one of her six sons predeceased her, two killed in war and the others dying of natural causes. One of her five daughters died in infancy; the others all survived their formidable mother, but not without quarreling with her, sometimes over trifles. On the other hand, the Countess was capable of reversing herself abruptly, accepting what she could not undo by pretending nothing had happened. A sudden reconciliation with a daughter who had married without her approval was the most spectacular example of her weathercock moods; but there were many lesser examples of the old lady's always emphatic—but uncertainly fluctuating—likes and dislikes.

When the Earl died in 1911, the bulk of the estate, including Castle Howard, passed to his widow, to do with as she pleased. She probably did toy with the idea of giving Castle Howard to her namesake, the young Rosalind; but as her temper became more and more erratic, the favorites of one moment became scapegoats the next. As a result, when she died she left discrepant instructions about how to dispose of the property, but no legally binding will. These the family decided to disregard, and instead divided the estate more or less evenly among the surviving children. Castle Howard itself went to Geoffry, the only surviving son.[55]

Lady Mary, Rosalind's mother, was a lesser version of the redoubtable Countess. Like her mother, she was charitable, Liberal, and imperious, but unlike her mother she deliberately eschewed taking any part in the intellectual and political affairs that absorbed her husband's energies. A teetotaler and vegetarian, she dressed frumpily and lacked all sense of humor. True to her Whig ancestry, Lady Mary was inclined to radical opinions on public questions, while remaining unbendingly conservative on matters of private behavior and propriety.

Toynbee offered Gilbert Murray's biographer the following retrospective appraisal: "Gilbert Murray married Lady Mary, not only as herself, but also as a member of Lady Carlisle's 'establishment'. This was hard on Lady Mary. It was the penalty of being a member of a family with a very strong clannish solidarity, dominated by one powerful personality." After referring to "Lady Carlisle's demonic character and tragic history," Toynbee continued: "Lady Mary made invalids of members of her own family because this gave her power over them. (She was quite unconscious of this motive; she was utterly incapable of introspection.) Over health she was queen, while in Gilbert Murray's and Rosalind's intellectual life she was an outsider."[56]

Such a mother was difficult to grow up with, if only because she expected rigid adherence to her idiosyncratic code of manners by those around her. But involvement with her other children (Rosalind had one sister and three brothers, the youngest of whom was seventeen years her junior) deflected most of Lady Mary's attention from her eldest daughter, who seems to have spent more time with her grandmother than with her mother and father.

Yet, even though the family pattern meant that they were often separated, by far the most important person in Rosalind's childhood was Gilbert Murray. When she was only one and half years old, Murray set out "to teach her to talk by making her give me orders—'Run', 'Stop', etc."[57] "It is a great help she is so intelligent," he informed her grandmother in 1893; and a year later he remarked "Rosalind has become perfectly bewitching since her illness."[58] He cultivated her (undoubtedly precocious) literary skills, asking her to write him poems when she went away to school and offering delicate, lighthearted criticism of the childish verses she sent back.

Soon after she turned seven, young Rosalind was diagnosed as suffering from "rheumatism affecting the heart."[59] Real or imaginary, her delicate health persisted, with diverse diagnoses—anemia and tuberculosis among others—for the next decade and more. Treatment required wintering in a mild climate, so she went off to Italy for long periods of time. "We are simply pouring away money on Rosalind, and I think we are getting the money's worth. She does seem in a slow and doubtful way, to be really conquering her disease, whatever it is," Gilbert Murray wrote from Italy in 1906.[60]

Yet at times her health seemed perfectly robust. "I collected Rosalind at Liverpool Street. She looked big and brown and mischievous and rather pretty, and bursting with health. The wild life (and wild it was! Mary's hair would stand erect at the tales Rosalind told me yesterday) has suited her to perfection. . . . She seems to have associated freely with everybody on Walberswick beach, and some of them were certainly cautions! Bathing before anyone was up: jumping broader and broader ditches until she fell in, up to her armpits in mud; meeting a strange donkey and proceeding to ride him, are typical of the kind of things they did. It has been, in fact, a regular bursting loose after too much limitation."[61] Rosalind had stayed with a certain Mrs. Newberry, whom Gilbert Murray described as "a wise sort of person in her way," and it may not have been accidental that her health became normal when her venturesome spirit was given a freer rein than was permitted at home by her mother.

Venturesome and imperious she certainly was, when given half a chance. In a letter to her father George Bernard Shaw described Rosalind's childhood, with characteristic exaggeration, as follows: "Rosalind has always filled me with awe. When I first saw her, I divined a tragi-comic horror: a completely mature spirit and character imprisoned in the figure and person of a child, and thereby condemned to live in tutelage to two much younger children, Gilbert and Lady Mary, to wit. What Rosalind must have suffered from being ordered about and sent to bed and soaped and combed and short-frocked generally in an institution in which her natural place was that of matron, God only knows. . . . At all events, she has made good her right to publish a book, which I am persuaded she could have written quite well at the age of four."[62]

The book to which Shaw referred was Rosalind's first novel, *The Leading Note*, published in 1910 before her twentieth birthday. It was dedicated "To My Father," and deservedly, since Gilbert Murray had a great deal to do with its publication. In hope of drumming up literary support, he sent the manuscript to the leading novelist of the day, John Galsworthy, who found it "most promising and interesting. It has true spirituality, and technically it's extremely good for so young a girl."[63] A month later, after rejection by Methuen, a lesser London firm, Sidgwick and Jackson, agreed to publish the novel in an edition of one thousand, provided that the Murrays pay half the printing cost of £28!

The novel has a certain grace and charm. It tells how two visitors to Italy, an English girl and a Russian revolutionary, fall in love; but in the end, nobly and coldly, he refuses to let her accompany him back to Russia, lest she fall into the grasp of the Tsar's police. Along the way, the young people agonize over life and the inescapable choice between art and action. The tone of quiet desperation on which the story ends was, according to Rosalind herself, modeled on Turgenev.

The novel must have been quasi-autobiographical. Gilbert Murray asked his daughter, when Sidgwick and Jackson first agreed to accept the work, whether it was wise to publish after all, since she might later "regret having written a love story at your tender age" and risked having people who knew her say "that all the delicacy of your health was due to an unrequited passion for Prince Kropotkin."[64] Murray also started to use the mocking salutation "Dearest little Miss Petkoff" in letters to his daughter—a minor improvement, perhaps, over the "Pussidora" he had favored from about 1905.

But whatever attachment she may have had for Russian revolutionaries soon passed, yielding to another, and, from her parents' point of view, more dangerous attraction to a "Cambridge crowd" of aesthetes, whose leading spirit was the poet Rupert Brooke.[65] Rosalind's second novel, *Moonseed* (1911), may be taken as the literary expression of this attraction. It is a silly story of decadence and disappointment, telling how an English girl, in love with an English morphine addict, married a Frenchman instead, only to discover that he had been guilty of helping to drown a friend. Reviews of her first book had been highly laudatory, and only partly because of her youth; but *Moonseed* was a failure, having been written at speed by a mere poseur. All too clearly it lacked the basis in personal experience that had sustained her first novel.

Literary failure and her parents' disapproval no doubt diminished the appeal of self-conscious aesthetic decadence for Rosalind. As a sort of rebound, she next decided to flirt with an ultrarespectable, quintessentially British outdoor way of life by chasing foxes. But this, too, aroused quite extraordinary dismay in her father's breast, for he utterly abhorred blood sports.

On her twenty-first birthday Gilbert Murray wrote to "Miss Petkoff" as follows: "It is very strange to think of you being a grown up woman, though in a way you have been like a grown up companion to me since you were about eleven."[66] That, indeed, was why establishing real independence of her father was so hard. Gilbert Murray had shared literary interests with Rosalind which he could not share with his wife; and when she began to show precocious skill he undoubtedly transferred some of his own disappointed literary aspirations to his daughter. (Murray was related both to Rudyard Kipling and to W.S. Gilbert; and he once confessed that as a young man he had been jealous of Kipling's literary success.)[67]

Lady Mary, characteristically, had no doubts. On one occasion she traveled to Switzerland to rescue Rosalind from the clutches of the dreadful "Cambridge crowd." Gilbert Murray, on the contrary, both wished to see his daughter become independent and hated to have her do so. "We used to be rather specially close and intimate," he wrote to her in 1912, "and agree in our interests and ideas. Probably I caught you young and over influenced you. And now you are thinking for yourself in all sorts of ways and differing from me, not so much in views as in general feelings. That is all right. Only the process of breaking asunder is a necessarily painful one: it has been so for me and I imagine that it probably has for you. And I know I have sometimes been sarcastic and unkind. . . . And, dear, I am very sorry and won't do it any more."[68]

But a few months later, when Rosalind informed him that she intended to take up fox hunting, he wrote an angry letter, denouncing blood sports in the most emphatic way, and concluding: "It looks to me as if hunting was a flag to wave: a sort of indirect way of telling us that your moral standard in life was different from ours, and that you meant to act on it.

It may be that . . . I am no longer any good to you. . . . I think about you constantly, I admire your work, and I love you. I think, in trying to readjust our relations from parent and child to friend and friend, we have both made a lot of blunders. It is a difficult job and we were sure to do so. But it would be rather humiliating if you and I had not enough brains and sensitiveness to avoid the most ordinary pitfalls of life.

"About coming here, I also had my doubts, and was thinking of going away when you came." But "I had much sooner that you should come and try to get a little nearer to some understanding. I had a lot of stories to tell you." [69]

She did come, and perhaps listened to some of her father's promised stories. Then in January 1913 she got away from it all by going on a winter cruise to Jamaica with her eldest brother, Denis, who was having troubles of his own. By mid-month, having heard nothing from her, Gilbert Murray sent off a marvelously witty set of letters, each a polite reply to news of Rosalind's imaginary engagement with variously inappropriate persons: a Negro, a Spaniard, an American. [70]

Obviously, then, in the following May, when Arnold Toynbee proposed to Rosalind, both parents, and Gilbert Murray in particular, were eager to see their daughter settle down to marriage. Moreover, the name and prospects of the suitor must have been particularly welcome. [71] Her initial rejection may well have provoked intense family discussion; and it was Lady Mary who summoned Toynbee back with this characteristically imperious note:

31 May 1913

Dear Toynbee

Rosalind wants to see you again. Last Sunday she did not tell you the truth. She was not sure, but she thought it better to leave no doubt in your mind and I was thankful she thought that.

Now she has wrestled all week and can't leave you with less than the truth as her answer.

I tried to get you earlier, for she is very frightened and you had better come to her—but not at dinner time—you could come at 6 or at 8:30—and you shall see her alone.

She is afraid your mother will think ill of her.

Let her tell you everything—there is so little to tell that you don't know.

Mary Murray [72]

A week after Rosalind had accepted Toynbee's suit, Gilbert Murray wrote to a friend:

I have been waiting day by day to tell you our news, and now I may, though it will not be public for a few days yet. Rosalind is engaged to a young Fellow of Balliol, Arnold Toynbee (a nephew of the Toynbee Hall man). There was a week of emotion and indecision; then a sudden clearing of the skies, and now she is serene and happy with a light step and shining eyes.

He is one of the men we have liked quite the best here and has had a permanent invitation for Sunday evening supper for the last two years or so. But we never thought of him in connection with R. They scarcely knew each other and seemed to have hardly any common interests. He is an ancient historian, very able indeed. He bids fair to be one of the most distinguished historians of his generation—in academic circles I mean. And he is a delightful companion; there is no one with whom one had better talk when he turned up on Sunday evenings. Health rather middling; always tending to overwork. Anyone would think we had sought for a man exactly according to our own prescription, whereas in reality we were simply astounded when he told us he loved her. It looks like extraordinary good fortune, though one cannot help but feeling, as it were, tearful and troubled and half frightened. Only the silly subconscious part of one of course. I could not have chosen anyone I should like better.[73]

Murray may have been "tearful and troubled and half frightened" at the "good fortune" of the match, but his mixed emotions were surely far weaker than Rosalind's. She had changed her mind, presumably at least partly under her parents' pressure. Her prospective husband obviously represented all the virtues her parents most admired, and marriage to a young don with brilliant prospects at Oxford committed her to exactly the same life career her mother had accepted in marrying her father. But that meant shelving her romantic attachment to aristocracy on the one hand, and abandoning rebellion against her parents' moral code on the other. Yet she continued to feel rebellious, at least with part of her mind.

This is amply attested by her third novel, *Unstable Ways* (1914), drafted in haste during her engagement and put into final shape after her marriage. Although dedicated to A.J.T., *Unstable Ways* presents an utterly devastating portrait of Toynbee (or of an obvious lookalike, an academic suitor named Freddy Furze). The heroine, Giacosa St. Claire, having to choose among three different men, accepts Freddy, hesitantly; then, after a visit to his family, decides it was all a mistake and drowns herself rather than go through with the marriage! A few quotes may illustrate the tone: "They passed the spot where he asked asked her to marry him and for the first time, a strange desperation came upon her." Two pages further: "If only he would take her by force, if only he would storm her citadel of criticism, if only his love were a less timid, bashful thing. He was too considerate of her, too gentle. She was ashamed of her own grossness, for she acknowledged it as such, but the hunger for a less ethereal lover was upon her." And again: " 'You know,' she said suddenly, 'the world, the flesh and the devil have got possession of me; you are free of them. Will you be able to free me?' Freddy laughed: 'I don't think that would be a very hard job,' he said." And finally, a few pages before drowning herself, Giacosa says: "I have never cared about anybody or anything except myself."[74]

How literally to apply these sentences to Rosalind at the time of her engagement and marriage is impossible to decide. Obviously, she exaggerated. In real life, she did not drown herself, but instead went through with the civil marriage her parents arranged (after quizzing Toynbee on his religion)[75] on 11 September 1913. Still, there were difficulties along the way—and for all concerned. Edith Toynbee wanted a long engagement, but was overruled. Rosalind shocked her parents by proposing to dispense with the rituals of legal marriage. Lady Mary was indignant, confiding to her husband: "She really does feel that everything we think good is mere instinct or only Mumbo Jumbo. . . . Her friends Rupert [Brooke] and Dick [Gleadowe] have had a pretty bad effect on her."[76] A civil marriage, without benefit of clergy, was therefore a compromise between Rosalind and her parents.

Rosalind Murray ca. 1910
Chalk drawing by Gwen Raverat. Collection of Lawrence Toynbee

Toynbee seems to have obeyed Rosalind's wishes, as he was to do in all practical matters throughout their married life. Yet that did not guarantee her good pleasure. A few days after the engagement, Rosalind wrote from London to her fiancé: "I haven't been nice to you these days, not as I want to be—I couldn't help it somehow. . . . I am afraid because if things are very good, better than anything one has known, it is hard to believe them." And the letter concludes:

> I miss you so, and that surprises me, which is absurd, but to suddenly miss some-one one has lived quite happily without all one's life is strange you know—but then it is all so strange.
>
> Goodnight, my dear
Rosalind[77]

On Toynbee's side, his engagement provoked acute strains also. The master of Balliol wrote: "I don't pretend to like it, but if so it must be, there's an end."[78] His aunt Charlotte remarked: "You must be fonder of those big places than I am. I hate them: but then I believe I am more democratic really than you are!"[79] But of course the real tussle was with his mother. Her distress at losing her son's allegiance was barely disguised by conventional politeness, and when she heard that the Murrays planned a civil ceremony, she declared that she could not bring her daughters to a secular wedding. "It is very troubling," Rosalind wrote to Toynbee, "and I don't see what can be done; you see, from her letter it looks as if she would have preferred the Castle Howard wedding after all—but from her first letter it looked as though she would have minded us changing back, for no particular reason. Anyhow, that is how it is now."[80] His mother's behavior was indeed contradictory, for in an earlier letter she had declared that "she would much rather we kept to the Registry if we feel it is more truthful."[81]

Then just as events were moving to the climax, Toynbee fell ill and the wedding had to be postponed for five days. Rosalind reassured him: "Don't worry or think I mind. I am sorry you should be ill—very sorry—but otherwise what does it matter? . . . I am not even thinking it is a bad omen. . . . In some ways I think it actually better to be married a day or two later. I have been getting so silly and tired this last week that I felt I should be horrid to you—not actively horrid, but just through tiredness."[82]

But eventually all the alarms and excursions were lived down, and on 11 September 1913 a certificate of marriage was duly drawn by the Registry Office, District of Erlington, County of Norfolk, where the Murrays were vacationing. Gilbert Murray, Edith Toynbee, Mary Murray, David Davies (Toynbee's friend from Winchester), and Charles Roberts (Lady Mary's brother-in-law) signed as witnesses.

The newlyweds went to Castle Howard for their honeymoon. There, if Gilbert Murray's replies to letters from Rosalind are any guide, they were blissfully happy. "You seem like a princess in a fairy palace just at present," he told her two weeks after the ceremony; and a few days later: "It gives me more pleasure than I can say to think of your having reached such real love and happiness.''[83]

Yet there were some flaws: Rosalind's novel accompanied them to Castle Howard, and in view of passages like those quoted earlier, it must have made strange company on the honeymoon. "Giacosa doesn't get on," she explained ahead of time, "which is annoying as Dad has made a clear list of alterations and I ought just to have been able to get through. I must bring her to Castle Howard after all and really conquer her there. The sight of these two bundles of paper is becoming increasingly hateful to me."[84] Moreover, the formidable Countess of Carlisle took some getting used to on the part of her new grandson-in-law. As Gilbert Murray wrote: "Don't let Arnold be frightened. He must of course keep his temper and above all keep his manners. But I do not believe—as some authorities do—that there is any advantage in behaving like a worm. Let him just be himself. . . . Grandmother will like and respect him. As for you, be wise as a serpent and harmless as a dove, as you usually are."[85]

In October the new term began and they returned to Oxford. As an ordinary tutor "on the establishment," Toynbee was still required to live in college. Not until the autumn of 1914 was he permitted to set up a house of his own in town. In the meanwhile, Rosalind lived with her parents during term time, and they went together to Castle Howard for each vacation. Rosalind became pregnant early in the new year and promptly subsided into invalidism under Lady Mary's watchful eye. When the long summer vacation arrived, Toynbee reported back to Lady Mary: "She stood the journey much better than at the end of last term and had a long stretch in bed, supper and breakfast included, and now has been sitting or lying outdoors all day. . . . It is a wonderful joy to feel I shall never go back into College again—one can hardly realise it yet."[86] A few days later: "She is much better. No more tears." And again: "She sleeps well, and has an appetite. . . . She is really much better than during the term."[87]

Plans for a modest little house they had rented at 5 Park Crescent, Oxford (requiring, nevertheless, the services of a "cook, parlourmaid, gardener (Mondays), plus Mrs. Harris for mending and extra cleaning and Mrs. Massey, laundress"),[88] were rudely interrupted by the outbreak of war on 3 August 1914. They could not return to Oxford until 6 August, when space became available on the trains. Then, because in Lady Mary's estimation a suitable doctor could not be found in Oxford, Rosalind removed to London, where she stayed in Lady Carlisle's town house, attended by the most approved physicians, until the baby was born on 2 September 1914.

IV

The Great War
and the Peace Conference

1914–1919

The outbreak of war in August 1914 did not interrupt the rhythms of Oxford University life all at once. Consequently, as soon as Rosalind had time to recover from the birth of her eldest son, Antony Harry Robert Toynbee (Tony for short), in London on 2 September 1914, the family returned to Oxford. When the new term began in October, Toynbee started to teach Greek and Roman history once again, almost as though nothing had happened. He remained there until May 1915.

Yet the war, and the drama of its first weeks—the German army almost succeeded in crushing the French—profoundly changed Toynbee's life, along with the lives of almost everyone else in England. First of all, a new and very poignant question confronted him: should he, being in the prime of his manhood at age twenty-five, volunteer as a soldier? The frenzy of patriotism that prevailed in Britain in 1914 made it difficult not to join up. Yet in Toynbee's case two influences prevented him from doing what was expected of him.

On the one hand, he disliked and perhaps even dreaded what he knew of army routines. As a schoolboy he had looked on military drill as equivalent to the team games he so hated. Voluntarily to subject himself to military training, and to the muscular sort of comradeship that went with it, was profoundly repulsive to him. Yet duty called. Or did it?

On this score he was torn between conflicting pressures. His mother's stalwart and unquestioning patriotism simply assumed that her son would become a brave and efficient platoon commander, as befitted his age and station. "Couldn't you have some training soon?"[1] she asked when the war had just begun. Fourteen months later she exhorted him to heroism, as follows: "If you go, you've got to be a good, not an indifferent platoon commander: and you've got to hope very much not to be killed but to help get the better of the Germans. Everything is against you; your poor physique, your consequent lack of high spirits and animal courage: everything which arises out of the physical vigour of life which we've always unduly despised and which now we see to be so important when one gets back to pristine struggle. But you'll do it, if you have to go. You can always throw yourself into a thing thoroughly: and Greece showed you had your grit and courage when danger came. Don't dwell on your unfitness. I'm not afraid that you would show the white feather or fail."[2]

Toynbee enthusiastically endorsed the heroic ideal. Indeed he carried it to extraordinary heights, seeking always to excel, always to conduct himself in such a way at to command admiration and outdo what was expected of him by others. But from early youth his principal heroic achievements consisted of wielding words, not swords. Nonetheless, he admired military heroism: witness his fascination with the wars and battles of antiquity. Indeed, the whole tenor of his classical education endorsed the idea that a man's worth was tested and expressed primarily in battle.

This notion of what mattered most in human life had been first set forth by Homer and later elaborated, but never questioned, by later Greek and Roman writers. The behavior of the British, German, and French educated classes in 1914 would have been inconceivable without the prolonged exposure to classical ideas that dominated their schooling. Toynbee shared that training, and the moral code it imparted, to the full. His behavior in Greece in 1911–1912, replete with extraordinary physical feats of endurance and some real risk-taking, including an encounter with robbers, showed that the heroic ideal was not a mere academic exercise for him. In fact, it was central to his self-image, and the importance of his Greek experience lay very largely in the fact that there for the first time he was able to supplement his accustomed academic excellence with a tangible expression of physical heroism as well—outwalking everyone while suffering hardships and taking risks others shrank from.

But in 1914 and throughout the war, Rosalind and her Murray and Howard relations all pulled in an opposite direction. Rosalind's view was simple and direct: she wanted her husband close at hand for her own satisfaction. Her sexuality was very strong. Marriage allowed its legitimate expression, and she was unwilling to contemplate breaking off what had become a matter of great importance to her—and to him, for Toynbee reciprocated her ardors.

A few letters archly indicate something of the sexual warmth that enveloped the early years of their marriage—a warmth at odds with the reservations Rosalind continued to feel about the real and entire acceptability of her husband. Even in her most personal letters to Toynbee those reservations peek through in a contorted use of the third person to refer to herself and to him. Thus, for example:

25 May 1916

Dear Pussmaster:

It did want its man last night when it was in bed and felt so poor and cried. Post is just going, so Goodnight and all her love.

Pussit

Or again:

World War I: Toynbee and Tony, his son

2 August 1917

Darling

She wants him badly. Last night was in a grief about his being conscripted.
. . . Tell one what happens about it, and if you are likely to go to Stockholm
after all.

Puss[3]

Rosalind's wish to keep her husband close at hand was reinforced, from the
beginning of the war, by the political conviction that it should never have
started, or at least that Great Britain should have stayed out. This was the view
of Castle Howard, where the Countess of Carlisle's fiat prevailed. The Toynbees were staying there when the war broke out, and the Countess's political
view of the war therefore nicely coincided with Rosalind's personal preference.
This coincidence remained in place as the war years dragged by, reinforced
from time to time by the deaths of relatives and friends (like Rupert Brooke)
who had enlisted. Toward the end of 1916 Rosalind wrote to her cousin Bertrand Russell: "The longer the war goes on, the surer I feel that it is wrong. I
think wrong to have started in the beginning, and anyhow wrong to go on with
still."[4] Her political opinions thus reinforced her personal desires, and vice
versa. Moreover, such views made it (more or less) moral for Toynbee not to
enlist, and even allowed him to look for ways of avoiding the draft after parliament decreed compulsory recruitment in January 1916.

Yet Toynbee agonized over the problem. His inclinations attracted him to
Rosalind's side, but yielding to them meant betrayal of his heroic self-image
in a doubly dastardly way—avoiding self-sacrifice in order to indulge sexual
appetites that were already at least partly guilty. He had long been accustomed
to suppressing ordinary inclinations in pursuit of heroic achievement. But his
heroism was of the lamp. That became totally inadequate in the autumn of 1914
because, as always, he needed the respect and admiration of those around him—
which in 1914 required young men to volunteer. Toynbee resolved this dilemma by volunteering, but arranging that when he offered himself to the army,
he did so with a doctor's certificate in hand stating that his dysentery would
probably recur if he were exposed to harsh conditions. In October 1914 he
wrote to his mother: "Just a line to let you know that they have rejected me
because of my dysentery: it is very bitter, though undoubtedly right, I have
come to believe, and other people [Rosalind, obviously, chief among them!]
will be relieved. I am going to get some drill all the same."[5]

Toynbee was not telling his mother the truth in claiming that his medical
rejection was "very bitter." The bitterness was there, all the same, due to the
self-betrayal implicit in his behavior, for if he had not called attention to his
dysentery, he would surely have been passed for active service. He was of
prime military age, after all, and physically vigorous. His infection was over
and gone by 1914 and never recurred in later life. What persuaded the doctor
to write what he did remains hidden, but it is hard to believe that his medical
opinion was not influenced by family pressure.

A year later the issue arose again. By then, manpower for the British army was running short and Toynbee, reluctantly, offered himself for military service once again. He explained to his mother that the question was: "How far did my special capacity for doing work of the present [kind] outweigh my considerable incapacity for commanding a platoon? Didn't the urgent need for more men volunteering to go and be killed . . . really outweigh everything else?" But his decision to volunteer was "a harrowing issue to face, and it has given us both [Toynbee and Rosalind] a bad time. It really means very deliberately putting off one's own life, and one's common life, which is so much more precious and wonderful than any individual existence of my own."[6] The next day he confided to Gilbert Murray, "I rather feel myself that it is only a question of time—sooner or later. Rosalind holds to the idea that I shouldn't go (which is furiously out of character with her instinct to disarm trouble by going out to meet it)."[7]

Nevertheless, in the event, the medical papers he brought with him resulted in his rejection. This second medical excuse did not come easily. When a doctor named Gibson decided that his history of having suffered from dysentery in 1912 was not a disqualification for military service in 1915, Toynbee contrived to get a different opinion from an anti-war Quaker, Dr. H.T. Gillett, who wrote: "I do not need to see you for I suppose I would get no fresh information; but I feel sure that your diarrhoea (plague) would begin again as soon as you were exposed to damp and cold or undue fatigue.

"I have no hesitation in my own mind about it so it is easy to reply. I think it would be difficult for anyone to judge your case unless he had seen you when you were ailing, and I feel sure Dr. Gibson would agree with me if he had the opportunities I have had of seeing you."[8]

Dr. Gillett's opinion sufficed to excuse Toynbee a second time. "I was very much surprised because I gathered they were accepting everybody," he wrote to his mother. "I can't say that I am not glad, though, for I certainly believe that I should have been invalided out, sooner or later, and sent to a depot to do far less good work of a clerkly kind than my present."[9] Later, in 1916 and again in 1917, when volunteering gave way to a nationwide draft, Toynbee was twice called up for military service, but each time he secured exemption by asking his superiors to certify that the work he was doing for the government was vital to the war effort.

In later life, Toynbee often dwelt on how the dysentery he contracted by drinking from a contaminated stream in Greece in 1912 saved his life in World War I. He also justified the furious pace at which he worked as the only way he could pay off a debt incurred through his own survival to those who had died in the trenches. Yet if he had been willing to join the army, he could certainly have done so. Indeed, if he had not sought out a sympathetic doctor in 1915 and substituted his advice for the opinion of Dr. Gibson, who had found him fit for service, Toynbee would presumably have been accepted as a volunteer; and if he had not mobilized his superiors to get an exemption in 1916 and again in 1917, he would have been drafted.

Thus his public explanation of what happened in World War I was a half-truth at best, and he knew it. As a result, a sense of having failed to live up to the heroic ideal he took so seriously haunted him throughout the war and, in muted and half-suppressed forms, throughout his later life as well. From October 1914 onward, he had a secret to hide. Rosalind's imperious and self-indulgent will prevailed. Their life together, however precious, was also guilty, since it kept him safe at home when others were away at war.

Toynbee's agonized struggle over whether to enlist did not remain entirely within the bosom of his family. In November 1915 an anonymous reviewer of Toynbee's first book, *Nationality and the War,* declared in the Oxford Magazine that the author would be much better employed at the front than in writing such stuff. This provoked Gilbert Murray to spring to Toynbee's defense, pointing out that he had indeed offered himself for service and been rejected. Others, too, among them his admired mentor Alexander Lindsay, tried to assure him of the rightness of his ways. "I know it is very hard to be engaged in a thoroughly safe occupation," Lindsay wrote him, "but you must swallow your spiritual pride and stick to what you are doing. . . . If you don't feel happy about this, the thing to do is to try and get past the test. But I hope you won't do this. I think it would be silly of you if you do." [10] Toynbee was grateful for such support, no doubt, yet because his mind was so divided, he continued to harbor "the morbid idea that the people who tell you you are more useful at home are the devil, tempting you to the easier course." [11]

Anxiety about what was right and honorable to do in wartime therefore cast a nasty shadow over Toynbee's personal life and family relationships from August 1914 onwards. He resolved the problem in his usual way, by intensifying the pace at which he worked. Instead of pursuing ancient history and writing the book he had promised Gilbert Murray for the Home University Library, Toynbee decided to take up the question of nationality as manifest in recent European and Near Eastern affairs, hoping to explain to the British public the complications that lay behind the assassination at Sarajevo, and prepare the way for a just and durable peace by informing public opinion about all the disputed frontiers that would have to be redrawn at the peace conference.

Toynbee began systematic work on this book when Oxford's term started in October, and completed its 511 pages in four and a half months. [12] He was teaching all the while, and continued to perform other college duties of a sort that kept ordinary men fully occupied. Yet somehow he was able to acquire and organize a vast body of information about eastern Europe, most of it quite new to him, and then to write it all down so as to produce a coherent, readable volume in time to see it published on 1 April 1915.

Toynbee's extraordinary capacity for rapid composition and for gathering and digesting vast amounts of disparate information about contemporary politics was here demonstrated for the first time. This capacity sustained his subsequent career as author of annual surveys of international affairs during the interwar period. But the fact that he later repeated it, habitually, does not make the effort less astounding, considering the distractions surrounding him at the time. A newborn son, a wife with servant problems at home and real or presumed bad health, and a mother who was still struggling to regain her composure after losing her husband to the madhouse and her son to Rosalind made a

distracting background for his college duties. Yet by paying little or no attention to such personal problems, working prodigiously long hours, and concentrating on his chosen task, Toynbee got the book written at top speed.

His feat was costly nonetheless. In particular, his relationship with Rosalind
lost its initial, all-encompassing effervescence. In a way, this suited them both.
Toynbee minimized his domestic role, expecting and assuming that Rosalind
would manage the home and provide him with all the services—meals, house
maintenance, and love—that he had been accustomed to receiving from his
mother in childhood. This therefore allowed him to resume something like the
behavior pattern of his childhood and youth, with the difference that his sexual
impulses now found ample outlet. From Rosalind's side, her imperious temper
and passionate nature lent themselves to this division of labor on two conditions: that suitable servants could be found to do the menial work of the household, and that she had enough spending money to sustain the style of life to
which she was accustomed.

Unfortunately for Rosalind's peace of mind, neither of these conditions prevailed after the war began to bite into English society in 1915 and succeeding
years. Toynbee could do nothing about the servant problem; he could and did
try to remedy deficiencies of income by undertaking all sorts of journalistic
writing to supplement his salary. This meant working even harder, reducing
Toynbee's free time with his family to the vanishing point, and in general
hardening and confirming the sharp division of labor between himself and Rosalind—she in complete charge of the household, he reduced to marginality at
home, though still fulfilling for her the important roles of sexual partner and
provider of extra money. This pattern of family life became standard for them
thereafter. It meant that Toynbee spent very little time with his sons when they
were young, while acute and often exaggerated financial worries clouded his
relationship with Rosalind. The raptures of their early marriage thus swiftly
faded into a mutually acceptable and, for Toynbee, very strenuous, yet oddly
isolated routine.

Nationality and the War was important for Toynbee, therefore, inasmuch as
the intense effort he put into its preparation established a mode of family life
which was, in essence, a reversion to a pattern he had first established in youth.
The great expectations that had surrounded him in his early years required great
accomplishments now that manhood had come; and since he deliberately, though
somewhat shamefacedly, shied away from what had become the standard, military test of a young Englishman's prowess, Toynbee found himself compelled,
more than ever before, to excel at desk work. And the only desk work that
could fit his need had to be somehow connected with the war.

Nationality and the War filled the bill, as its title showed. The book leaves
an odd impression now, after more than seventy years, during which two world
wars have come and gone and the sovereignty of Western Europe over most of
the earth has disappeared. Its spirit is that of Liberal, upper class Edwardian
England, combining a concern for principle with a sublime confidence that
enlightened English opinion, and the benevolent interests of the British Empire,
would (or at least ought to) prevail. Some of Toynbee's predictions turned out

to be wildly wrong. He was, for instance, very much alive to the Yellow Peril, declaring: "The fundamental factor of world politics during the next century will be the competition between China and the new [British] Commonwealth. All the threatened nations—Canada, the U.S.A., the South American republics, New Zealand, Australia—will draw together in a league of nations to preserve the Pacific from Chinese domination. Japan will probably join their ranks. . . . Russia will actually be the chief promoter of this defensive entente." [13]

Within Europe and the Near East, Toynbee was convinced that the principle of nationality required radically redrawn political boundaries. Peace could be assured only if local peoples were allowed to set up a government of their choice. But this general rule had its exceptions wherever a need for economic outlets to the seas and other strategic considerations contravened local preferences.

Two features of his recommendations are worth noting. First and most striking: he was very kind to Germany, arguing that "the only way to convince Germany [that war is not in her interest] is first to beat her badly and then to treat her well." [14] Treating Germany well meant partitioning the Hapsburg monarchy and allowing Austria and Bohemia to unite with Germany, while also shearing off portions of Alsace and Lorraine in the west and some Polish lands in the east, all in accordance with local opinion as indicated by plebiscites. Such a peace settlement would make Germany supreme on the continent of Europe, but that did not bother Toynbee since a generous settlement in accord with the principle of nationality might be expected to convert the Germans and other Europeans from "national competition" to "national cooperation," particularly in view of the threat from China that he anticipated.

In the second place, Toynbee remained distinctly tepid about the future of any sort of international organization for peace. Such a body might, on the model of the Berlin Conference that divided Africa in 1885, "adjust equilibrium among nations" by apportioning "international areas"; but "it cannot aspire to the regulation of War, and it is a waste of ingenuity to propound any international machinery for this purpose." [15]

In view of his subsequent involvement in Foreign Office deliberations about the partitioning of the Ottoman Empire, it is interesting to notice that he felt in 1915 that Smyrna was "marked out to be the capital of a diminished Turkey," and that the best (but impractical) solution for the future regime of the Black Sea Straits would be to entrust their administration to the United States. But "of course they will not, and of course Russia will" assume control. [16]

In later life Toynbee dismissed this book as mere "juvenilia," and it deserves this epithet in the sense that it reflects a prewar outlook, as yet unmodified by the long tribulation of 1914–1918. Yet concern for international politics, and an (often excessive) readiness to foresee a future radically at odds with present circumstances remained characteristic of his mind throughout his life. In that sense, *Nationality and the War* was quite in line with his later work and foreshadows the annual surveys of international affairs that he produced between 1925 and 1939.

The book was politely reviewed but attracted little attention, mainly because the war did not come swiftly to an end, as almost everyone had assumed would be the case. By the time it came out, public attention fastened on how to fight the war, not how to end it.

A trivial, yet poignant, by-product of its publication was the scolding he got from his Uncle Paget for using the name Arnold Toynbee on the title page. That name, Paget declared, belonged to Paget's dead brother. His nephew and namesake, Arnold J. Toynbee, had no right to it. The widow of the wronged Arnold Toynbee wrote: "Your uncle Paget is a most thundering tactless person. Why he should bother you and the aunts with a trifle like that is impossible to see. . . . I have sometimes thought you might be rather bored or annoyed by frequent mention of your uncle Arnold." Still, she, too, agreed that "you ought to sign yourself A.J. or Arnold J. simply and solely because that *is* your name." [17]

From Toynbee's point of view, completion of *Nationality and the War* meant that his duty of teaching ancient history to Balliol undergraduates was no longer leavened by work that could plausibly be counted as contributing to the successful prosecution of the war. Given the intensity of his personal dilemma about whether really to try to enlist in the army, such a life was not worth living. He therefore looked around for something that would help with the war and use his talents to better effect than the army was likely to do.

Teaching had become intolerable. "I became immensely depressed soon after I left you," he wrote to his mother, "in reaction, I suppose, from the immense effort of the book. . . . I am still depressed about the war (not kilometres in Flanders, but the future of Europe) . . . [and] I was depressed about myself. So far my philosophy has been 'be true to your work' (my real work, history, I mean) and now I find that it is only one strand in many, and that one's life isn't so nakedly simple. I felt I was . . . altogether going to seed by being engulfed in the pursuit of knowledge. Then I got this job, which I am going to do partly as a contribution towards the war, partly as medicine, partly as experiment." [18]

The job in question was with a government propaganda outfit, newly established in London, aimed primarily at affecting opinion in the United States. Toynbee started his new work on 1 May 1915, midway through the spring term at Oxford. By then, wholesale enlistment of students and of young dons had already reduced the university to a shadow of its former self, and the college authorities at Balliol entirely approved of his departure. They even allowed him to retain his fellowship and its stipend, despite the fact that he gave up his college duties when he moved to London. However, the lease on his house in Oxford remained in force so he had to pay rent there, and at first Rosalind and her son simply moved into the London mansion maintained by the Countess of Carlisle at 1 Palace Green, Kensington. The new job, "for the duration of the war," paid £330 a year—actually a small increase over his university salary of £310; his fellowship income of £180 from Balliol continued unchanged. [19]

In private, Toynbee soon dubbed his office the "Mendacity Bureau," but he found the work interesting. "I am very glad to have this job," he wrote to Gilbert Murray. "Pillaging the American press is fascinating (I produce a re-

sume of it for the Cabinet every week) and when I look at twenty enormous
files of cunningly classified cuttings, I feel that I am doing something bulky at
any rate.''[20] Toynbee also solicited and then placed suitable articles about the
war in American and other foreign publications, and answered individuals who
sent letters to the British government asking questions about its policy in various parts of the earth. He learned to dictate to a typist, and boasted that he
could answer twenty letters an hour.[21]

In October 1915 his propaganda work took a new turn. "They have turned
over to me Bryce's evidence about the Armenians, to make up into a report. It
is quite beyond one's range—the horrors of it I mean. . . . There can't have
been anything like it since Assyria."[22] Turkish massacres of Armenians took
a large share of Toynbee's attention in the following months. Its shock value
for American and British audiences was similar to the shock he himself felt in
gathering information about all the bloody and brutal actions of Turkish soldiers and irregulars against more or less helpless Armenian populations. But it
was gruesome, dismal work all the same. "I don't half like the job," he confided to Gilbert Murray, "but whether that is because I think it evil, or because
it is at present without form and void, and I can't see what shape to put it into
or what meaning to give to that great slough of vile facts, I don't in the least
know."[23]

Attention to the fate of the Armenians became a war issue in Britain after 6
October 1915 when Lord Bryce[24] made a speech in the House of Lords deploring the forced removals and attendant massacres then beginning in the central and eastern part of Anatolia. His information came mainly from American
missionaries who had established a number of schools and medical centers in
Turkey during the nineteenth century. Though the initial missionary goal had
been to convert Moslems, in fact Armenians and other Christians were their
principal clients. When the Turkish government decided, as a war measure, to
uproot millions of Armenians, who were suspected (with considerable reason)
of disloyalty to the Ottoman regime, the missionaries were therefore firsthand
witnesses. They were completely appalled by the unofficial massacres that preceded and accompanied compulsory removal of the Armenian population to
distant and inhospitable borderlands. But until 1917, when the United States
became a belligerent, Turkish authorities dared not interfere with resident
American missionaries, who therefore continued to favor Lord Bryce, and anyone else who was in the least sympathetic, with long and detailed reports of
what was happening.

Some months after Lord Bryce's speech, British propagandists decided that
efforts to publicize Armenian sufferings would help to counteract German news
reports from the eastern front describing Russian atrocities against the Jews in
Poland. The United States was the main target, for the British calculated that
they needed to offset the widespread sympathy for Germany among American
Jews, who were well aware of Russian pogroms and other manifestations of
anti-Semitism. By blaming the Germans for tolerating worse atrocities in Armenia than anything happening in Poland, the Allied cause might be served, at
whatever cost to the truth about German responsibility for Turkish actions.

Accordingly, Lord Bryce agreed to compile an authoritative record of Turkish atrocities, but asked for an assistant to do the work. Toynbee was picked for that role, and continued to keep track of Armenian sufferings until the end of the war. This soon made him into an expert on Ottoman affairs in general, and on the Armenians in particular.

Neither Lord Bryce nor Toynbee was privy to the calculation behind the decision to publicize Armenian sufferings. They were both deeply committed to liberal values, including truthfulness, and made systematic and sustained efforts to be sure that everything they reported about events in Anatolia was indeed correct. The pipeline to American missionary observers on the spot made this comparatively easy; the systematic genocide launched by the Turks in fact provided them with overwhelming evidence of atrocities on a mass scale—so much so, indeed, that the bulky volume Toynbee compiled and published, *The Treatment of Armenians in the Ottoman Empire, 1915–1916,*[25] exhausts the reader with repetitious instances of brutality and bloodshed. Over 700 pages in length, it constitutes an unusual example of war propaganda, since Toynbee was at pains to make sure that each case he recorded was in fact true. He sometimes withheld particulars about his informants, especially when they were Armenians; but he systematically tried to evaluate the reliability of any given piece of information. Consequently, the bulky tome is, in fact, a scholarly compilation, however ugly its subject matter, and the accuracy of Toynbee's account of the destruction of the Armenians has never been questioned.

What he left out was why the Turks distrusted and disliked Armenian and other Christian minorities so much. Later, Toynbee came to feel that this lopsidedness was a betrayal of historical truth. His sympathies, in fact, reversed themselves, partly, at least, because he felt he had been unjust to the Turks, and needed to make atonement. At the time, however, though the human depravity he described was deeply repugnant to him, his conscience was clear. Emphatic denunciation of Turkish barbarity seemed fully justified, based as it was on carefully evaluated evidence.

Vituperative denunciation of the Turks, and of their presumed German masters, formed the substance of a series of shorter propaganda pamphlets that Toynbee prepared along the way. *Armenian Atrocities: The Murder of a Nation* appeared in 1915 in two successive editions, and carried Lord Bryce's name. Then in 1917 Toynbee published *The Murderous Tyranny of the Turks,* this time with only a preface by Lord Bryce. Both pamphlets were prepared in haste, and printed in enormous editions, intended to strengthen public conviction of the righteousness of the Allied cause. Their tendentiousness, especially in claiming that the Germans were responsible for Turkish atrocities, makes these pamphlets far better propaganda than the big book, and rather more deserving of the regret Toynbee later felt for his participation in the wartime distortion of truth.

Thanks to the energy and literary prowess of his youthful assistant, Lord Bryce did so well in reporting Armenian atrocities that he was invited to inquire into German atrocities on other fronts as well. The result was a series of pamphlets, all of them written by Toynbee, whose titles reflect their tone: *The*

*Destruction of Poland: A Study in German Efficiency, The Belgian Deporta-
tions, The German Terror in Belgium,* and *The German Terror in France.*
These appeared in rapid succession in 1916 and 1917.

By the time he was through, Toynbee was fed up with this sort of journalism. "Thank heaven I have done with atrocities," he wrote to Darbishire.
"There is a 'Terror in France' out to complete that damned 'Terror in Belgium', but that is the last." [26] Four months earlier, on 7 May 1971, he had
been assigned to a newly established Political Intelligence Department. An ad-
junct to the Foreign Office, it was staffed by a remarkable group of academics,
including such persons as Lewis Namier, Alfred Zimmern, Edwyn Bevan, and
a pair of Australian-born brothers, Rex and Allan Leeper, who, however, made
their careers in the Foreign Office instead of in academic life as the others were
to do. The most senior member of the group was James Headlam-Morley, a
classicist and onetime official of the Board of Education, who, like the
Leepers, remained in the Foreign Office after the war.

Nicknamed Ministry of All the Talents, the Political Intelligence Department
was youthful, assertive, and above all well informed. Because they knew so
much, argued so well, and had few doubts about their own abilities to foresee
the future, the Political Intelligence Department exerted considerable influence
on British foreign policy during the last phases of the war, and some, including
Toynbee, continued to serve as expert advisers at the Peace Conference. [27] Assignment to such a group was a notable promotion for Toynbee, offering him
the prospect of exerting real personal influence on the course of public events.
It constituted a rich reward for the vast effort he had made to master the flow
of information from Turkey in connection with his war propaganda.

Not surprisingly, he became responsible for political intelligence pertaining
to the Ottoman Empire; but, with the collapse of Russia, his expertise was soon
applied to the Moslems of Central Asia as well, and from there he went on to
explore the risks of confrontation between a newly self-conscious Islamic world
and a weakening British Empire—a clash which would affect India, Arabia,
Persia, Afghanistan, Central Asia, and Africa as well as the lands directly subject to Ottoman administration.

Toynbee organized all the tumultuous detail that crossed his desk in the next
two years around this vast theme. The notion of a confrontation between insur-
gent Islam and a decadent British Empire, doomed to reenact in the twentieth
century the nineteenth century role of the Hapsburgs with respect to east European nationalisms, was his own invention, and, as it turned out, he exaggerated the rapidity of public mobilization in Moslem lands. Still, Toynbee's key
insight, that the right of self-determination could not be confined to European
and Christian nations but would also have to be applied to Moslem peoples,
proved accurate in the long run, even if it turned out to be only half true in
1918–1923.

Concentrating, as he did, on events in Turkey and Russia, where the Bolsheviks came to power in November 1917, Toynbee soon developed an apocalyptic vision of what was at stake. "What do you think of the Bolsheviks?" he
asked Darbishire on Christmas Day, 1917. "I think they hold the world in their

hands—perhaps for six months only, but for six decisive ones. But the censor will blot this out, as the Bolsheviks aren't in favour here among official circles.''[28]

Two and a half weeks later he analyzed the cataclysmic changes that he judged to be already in course in a lengthy 'Memorandum on the formula of 'the Self-determination of Peoples' and the Moslem World,'' dated 10 January 1918.[29] After pointing out that both Germany and Britain had, like Aladdin, inadvertently unbottled the genie of political self-determination for Moslems through their war propaganda—the Germans on behalf of the Turks, the British by backing the Arabs against the Turks—he went on:

> The Bolsheviks act in the name of the European labouring class, which has everywhere been ruled from above, until it became "self-determining" and even dominant in Russia, under the Bolshevik regime. The Bolshevik policy is to bring about the same revolution in other countries too, and naturally the workers in these countries—in Germany and elsewhere—feel a certain sympathy with Bolshevik designs. This vast, instinctive, hardly formulated internal support is the Bolsheviks' strength. But the Islamic consciousness behind the All-Russian Moslem movement is a force of precisely the same kind. The Moslems of Russia, like the Russian labouring class, are anxious to "make Russia safe" for the rights they have won, and to win the same rights for their brothers in the rest of the world. And what is more important, they believe themselves to be face to face with the same enemy—namely "Capitalism," or in other words the European Middle Class, which they regard as the exploiter of the labouring class in Europe and of the Moslems in the East. . . . The entente between these two forces in Russia is likely to be a lasting phenomenon which will have an effect outside the Russian frontiers.

Some pages later, Toynbee viewed Britain's position in the revolutionary turmoil with alarm.

> We are in danger of drawing upon our own heads the Tsardom's whole heritage of hatred in the Moslem world, and at the same time seeing Russia and Islam both go over into Germany's camp, though the course of the war has shown how much we need both of them on our side against her. The situation in Russia and Turkey, which between them occupy the land bridge between Europe and Asia, is in flux, and for the moment we can only count upon a friendly island here and there— Armenians, Georgians, Cossacks and a doubtful Ukraine. By using these *pieds-à-terre*, we may save something from the general debacle; but it is submitted that Revolutionary Russia (which is Bolshevik and likely to remain so) and politically conscious Islam are the two main factors in the situation, and that if our policy is not based on an entente with these, the whole land bridge is bound to pass out of our influence and come ultimately under some form of German control.
>
> An entente with both is all the more necessary if, as was suggested in a previous section of this Memorandum, Bolshevism and the Moslem Movement are gravitating towards one another. But the question of policy towards the Bolsheviks is quite outside the scope of this Memorandum, which is only concerned with our relations with Islam.

After arguing that the only practicable policy for Great Britain in the Near East was to persuade the United States to become trustee for the Straits and other Moslem lands, Toynbee concludes: "It is a question now of winning back both Russia and Islam to our side, or seeing them both go over together to the German camp."[30]

Wilsonian and Leninist rhetoric was hard enough for old hands in the Foreign Office to adjust to, but to have an apologist for such revolutionary doctrines preaching at them from an upstart Political Intelligence Department added insult to injury. Yet the men at the top continued to tolerate, and sometimes even attended to, radical views emanating from the young don from Balliol who knew so much history and geography and seemed never at a loss for words. All the same, when Toynbee began to attend meetings of the Labour Party's Advisory Committee on Foreign Politics, he came under special scrutiny. Senior officials of the Foreign Office eventually decided that it was all right for him to do so on exceptional occasions in order to explain "some topic of importance," but he could not serve as a regular member of the Committee.[31]

From Toynbee's point of view, to find himself so close to the seat of power was tantalizing. On the one hand, he delighted in associating with the great and powerful. The graceful exercise of power was what a Winchester and Balliol education was meant for, preparing its graduates to rule, and to influence rulers. But since Toynbee's advice was usually rejected, and often not even listened to, he could never suppress his sense of being an outsider—a role he had cherished throughout his school and university career. This kept him at a distance from those he advised, especially from those at the very top. "Oh, by the way," he wrote to his mother, "Rex Leeper and I were summoned the other day to a Conference—first at 10 Downing Street with [Prime Minister Lloyd] George, and then at the War Cabinet with Carson and Cecil and Milner. There were we and Buchan and half a dozen red hats [generals] and the Consul General from Moscow. Nothing exciting was settled. . . . It was interesting, though, to see them close, and talk to them across a table. Not great men, any of them, I think, and only one of them good."[32]

Immediately after the Armistice that ended the war on 11 November 1918, Toynbee reported: "The work is becoming terrific, but I am getting more and more into the thick of things—as far as concerns my part of the world—and am going to have the time of my life. . . . It is a race between food and Bolshevism—in other words Hoover and Lenin: a curiously matched pair of combatants."[33]

Having the time of his life at work meant that he spent less and less time with Rosalind. He worked late, and scarcely saw his offspring, except asleep. Once, when Rosalind was suffering from an attack of measles, Toynbee went off to Switzerland on a month-long secret mission. Ostensibly he was a journalist, writing about the activities of the International Red Cross; but in fact he went to try to persuade British Secret Service agents, code named 'C', to share information about Ottoman affairs with him. He saw some of 'C's' political and economic reports but was refused regular access. On getting back home, he asked Gilbert Murray to intervene with "the grandees at the top" in hope of persuading them "to take some action which would have a permanent effect." "The Foreign Office has definite claims on 'C,' and there seems no reason why these claims should not be insisted on."[34]

Clashes with military intelligence officers were, in fact, recurrent, and when army spokesmen challenged his judgment and information, Toynbee was capable of vigorous riposte. In July 1918, for example, a General McDonough of the Military Intelligence Department accused Toynbee of having exceeded his authority by recommending a policy of nonintervention in Central Asia which was "tantamount to saying we should hand over the whole country to Germany." Toynbee briskly defended himself, replying: "I see no indication, on the evidence, that the Central Asian Bolsheviks are prepared to hand over the country to Germany," but since "these Bolsheviks have come into violent collision with the Moslems" (contrary to his expectation of the previous January), military intervention on either side involved unacceptable costs for longrange British interests. He concluded: "To sum up, I admit that my political information about Central Asia is scanty, and it is regrettable that, if the DMI [Director of Military Intelligence] had fuller information it was never communicated to us." His Foreign Office superiors backed him up, a bit tepidly, agreeing with McDonough that Toynbee should have avoided specific policy recommendations, but endorsing his "careful reading" of the situation in Central Asia.[35]

His "careful reading" was assuredly colored by his own growing radicalism and dismay at a social system that could provoke and sustain such a war. Alienation from the middle and upper class way of life to which he had been apprenticed attracted him to the Labour Party, and he became a member in 1918. Rosalind joined too, carrying the Liberal radicalism of her grandmother, the Countess of Carlisle, to what no doubt seemed its next logical stage. "I find myself inclining steadily towards the social revolution," Toynbee told his mother. "The middle class have had their fling for a century and produced this [war]; now let the working class have their try. I am for nationality at one end and internationalism at the other, as essential parts of reconstruction, and if existing states and their traditions cannot square with them, let them go to the devil, the United Kingdom and the Dual Monarchy and all of them."[36] Such language must have shocked his mother, whose patriotism was even firmer than her Anglican piety. It might well have shocked some of his colleagues and superiors in the Foreign Office, too, if they had been privy to Toynbee's inward thoughts.

One can understand what drove him to embrace radical views. Having failed to "do his part" in the war by enlisting in the army, he justified his personal behavior by condemning the criminal folly of war more violently than he might otherwise have done. Virginia Woolf recorded her impression of his dinner table conversation in January 1918 as follows: "Arnold outdid me in anti-nationalism, anti-patriotism, and anti-militarism. . . . . I like her [Rosalind] better than Arnold, who improves though, and is evidently harmless, and much in his element when discussing Oxford. He hasn't much good to say of it and will never go back. . . . He knew the aristocratic heroes who are now all killed and celebrated, and loathed them; for one reason they must have thought him a pale blooded little animal. But he described their rows and their violence and their quick snapping brains, always winning scholarships and bullying and . . . admitting no one to their set.''[37] As it turned out, his social radicalism was transient, and Toynbee soon stopped denouncing the upper classes. But detestation of war remained central to his thought and feeling for the rest of his life.

Moreover, even at its most extreme stage, Toynbee's alienation from the establishment was tempered by the fact that he also nursed quite positive hopes for himself and for a reform of international affairs. Everything turned on the question of whether his labors in the Foreign Office, supplemented by a stream of journalistic writing and speechmaking designed to inform the public, would contribute to the establishment of a just and enduring peace. If so, then the ambiguities of his wartime behavior would achieve a satisfactory resolution. This made it supremely important for him to have a tangible, personal role in defining peace terms. Toynbee's own self-esteem, in effect, required that his voice should prevail when it came to advising the British government on policy toward the Ottoman Empire. Only so could his secret shame at the way he had escaped military service be dispelled by the knowledge that he had helped to create a good and lasting peace.

In 1914–1915, as we saw, Toynbee had argued that Germany must be beaten on the field of battle before a suitably generous peace settlement could be devised. After the failure of the Somme offensive in 1916, with its unprecedentedly heavy loss of British troops, Toynbee decided that the war should be ended by compromise. He was willing to give Germany Mitteleuropa, breaking up the Hapsburg monarchy in the name of self-determination. After President Woodrow Wilson's reelection in November 1916, Toynbee wrote: ''I hope and sometimes dare believe, Wilson will be mediating between us this time next year.''[38] But the collapse of Tsardom and of autocracy in Russia in February 1917 gave the war a new significance for him. Thereafter he saw German militarism pitted against the Allies' commitment to liberty at home and at least a tepid endorsement of the sort of national self-determination abroad that Toynbee had advocated in his book *Nationality and the War*. Continuation of the fighting therefore came to have a new meaning, and Toynbee renewed his commitment to victory over Germany. That, presumably, was why he called the Russian February revolution ''the most wonderful thing since the Marne—or rather it dwarfs the war, if one judges by effect on the future of the world.''[39]

No doubt Toynbee's reaffirmation of the necessity of defeating Germany was sustained by his transfer to the Political Intelligence Department of the Foreign Office in May 1917. After all, the Allies had to win the war before he could play the peace-building role he aspired to. In the meantime, he had the satisfaction of being in a position to influence British policy by giving information and advice to his superiors in the Foreign Office with respect to the affairs of the Ottoman Empire and the realm of Islam as a whole. In theory, Toynbee and the other members of the Foreign Intelligence Department were only supposed to provide correct and up-to-date information upon which senior officials of the Foreign Office could rely when it came to recommending policy to the government. But in practice, information and policy were inseparable. Indeed, the position papers that Toynbee and his fellow analysts of the PID wrote used a format that required an explicit formulation of policy alternatives, so that the busy men at the top could make swift and informed choices. Sometimes, as we have seen, Toynbee departed from his assigned role by suggesting the proper choice; and always, of course, the way alternatives were formulated and the language with which the actors and factors in foreign affairs were described implied and often prejudged British policy with respect to them.

As the struggle finally tipped against Germany in the late summer of 1918, his job therefore gave Toynbee ground for hoping and even believing that he might personally affect the peace with Turkey, and with luck, might even forestall the threatened clash between politically awakened Islamic peoples and the British Empire—but only if he were in fact listened to by the men at the top.

In practice, this turned out to mean one man alone: Prime Minister David Lloyd George. Lloyd George was thrown into the role of Britain's supreme diplomat and peacemaker by the fact that President Wilson decided to attend the Peace Conference in person, instead of delegating responsibility to professional diplomats and foreign ministers, as had been customary at European peace conferences in the past. Wilson's presence dictated the presence of other heads of government, since a mere foreign minister could not expect to deal with an American president on an equal basis.

Lloyd George readily lent himself to the role, for he was supremely confident of his own abilities, disdainful of experts, and entirely unwilling to listen respectfully to advice from subordinates. Toynbee's ambition collided with this fact. In wheeling and dealing to achieve some sort of compromise among the Allied governments—almost any compromise would do—Lloyd George paid scant attention to advice coming from the Foreign Office. Toynbee's wisdom and knowledge were therefore wasted. Accordingly, when he left the Peace Conference in 1919 he carried with him a fierce, intensely personal grudge against Lloyd George, and, as we shall see in the next chapter, lent a willing hand to postwar newspaper campaigns against the Prime Minister's Near Eastern policies.

Beforehand, of course, this disappointing upshot was not clear. Instead, Toynbee approached the Peace Conference with excited enthusiasm. Rosalind arranged to go to Paris as assistant to the *Manchester Guardian* correspondent, and for three and a half months, from November 1918 to March 1919, she and Toynbee lived together in a Paris hotel, without children and servants to distract them. It was "a lovely, wonderful time" [40] for them both, even though Toynbee's long hours and frenetic pace of work did not relax, but reached a kind of crescendo when it became necessary to try to find out what the United States was prepared to do in the Near East with respect both to immediate questions and to longer range commitment to an Armenian mandate, for example, or to participation in a future regime for the Black Sea straits. Toynbee was also thrown into contact with spokesmen for various interested parties— Armenians, Greeks, Arabs, and Jews, all of whom had claims upon Ottoman lands. Every day he read all available current information on what was happening in Moslem lands, wrote minutes on official telegrams that passed across his desk, and from time to time produced a more general appreciation of the situation, setting forth alternative British policies in view of the rapidly changing circumstances. It was a busy, dizzying climax for a man who turned thirty on 14 April 1919, exactly at the time when key decisions about the treaty with Turkey were being made.

Letters to his mother reveal something of the strain and excitement he felt: "I am working a lot with Laurence [sic]—the man who cut the Hejaz railway and marched round through the desert to Damascus . . . I have also been working with Smuts," he reported early in the conference.[41] And later: "The work is very heavy and I don't see the end of it, but it is extraordinarily amusing. . . . The most amusing part of all, of course, is attending the sittings of the Council of Ten. I have been there several times, when they have been 'hearing the cases' of people from my part of the world—my business being to sit in a small gilded chair behind the big plush sofa containing the British plenipotentiaries, and hand them maps and documents at the right moments. Clemenceau sits with his back to the fireplace . . . the plenipotentiaries sit in a hollow square of sofas, facing Clemenceau on either side. People like me, and the people taking minutes, sit on small chairs behind the sofas, all round the edges of the room . . . a rather fine room of the 17th century style—good old Ancien Regime like the whole of France."[42]

By the end of March, however, Toynbee began to feel depressed by his inability to influence decisions with respect to the Near East and adjacent lands. "I am gloomy about the Conference," he wrote. "A sort of paralysis has come over it: the top gets more and more out of touch with the bottom (at any rate in our delegation and the American)—and the different Powers with one another. . . . The outlook for relations between Europe and the East is very bleak. . . . If the mandates turn out to be hypocrisy, we shall have a Moslem risorgimento (perhaps anarchic) with the British Empire for Austria-Hungary— a melancholy prospect."[43]

Early in April he officially confessed his impotence. "The Middle Eastern Sections of both the British and American Delegations hold the view that Smyrna and the surrounding district ought not to be separated from Turkey, but this view has, I believe, been overruled by the British and American plenipotentiaries."[44] But Toynbee, nothing if not stubborn, launched one last try to influence the course of events by calling in a better placed accomplice, Harold Nicolson, as collaborator.[45] The result was a memorandum, dated 15 April 1919, attacking treaty terms that had been tentatively approved by their superiors. These terms, incidentally, soon proved entirely unrealistic, assigning, for example, newfangled League of Nations mandates to the United States for an independent Armenia and also for Constantinople and its adjacent region.

After pointing out some of the defects of this scheme, the two young men took it upon themselves to try to undo the damage by offering a counterproposal. Its key passage reads as follows: "Having carefully considered the above objections in all their bearings, we question whether peace would not in the end be better served by some less elaborate, if more drastic idea, that is, by cleaving Europe from Asia, and by giving Greece Constantinople and the European shores of the Straits and Sea of Marmora, and by leaving Turkey in Anatolia and on the southern and eastern side of the water.''[46] Harold Nicolson later published a diary in which he summarized his part in the affair, concluding his account as follows: ''We put this down on paper; we sign it with our names; we send it in. It will not be considered.''[47] He was right, of course. Greek and Italian as well as French, British, and American ambitions were all clamoring for attention, and, from Lloyd George's point of view, any compromise, however unrealistic, once agreed upon, was too precious to undo simply because some clever young men argued against it.

But for Toynbee the defeat spelled failure of his whole effort to justify his personal role in the war. It hurt him deeply and contributed to a physical and moral collapse that afflicted him when he left the Peace Conference. He was bone-weary from overwork, to be sure; but that was the least of his troubles. Exhaustion provided a plausible public excuse for his collapse, but three other circumstances, coming on top of his failure at the Peace Conference, combined to provoke his breakdown. These were (1) dissatisfaction with his prospective position as a professor at the University of London; (2) a poignant reminder of his father's helpless condition; and (3) storms and stresses in connection with money, Rosalind, and the Countess of Carlisle. All three were interrelated, but it seems best to begin with the most important, his relationship with Rosalind.

Rosalind's wartime experience had been almost as strenuous and far less satisfying than Toynbee's habitual overwork. After a few weeks in her grandmother's house in London at 1 Palace Green, she decided she could not endure the place any longer. However splendid, it was staffed by servants who were not hers, and no doubt they made her feel like an intruder. But rented housing in wartime London was hard to find, and after leaving 1 Palace Green, the Toynbees moved repeatedly from one place to another, without ever being able to get a secure lease on a flat or house that Rosalind liked.

Practical problems of housekeeping multiplied; in particular, servants became scarce. Those who did not enlist or take jobs in war factories were often incompetent or careless. Rosalind simply could not maintain the array of household servants that had been taken for granted before the war, and she wound up doing some of the household work herself—preparing meals and washing dishes, for example. That she could do, but looking after an infant was too much. Yet satisfactory nannies soon became unavailable. ''It [sic] did not write yesterday,'' she confessed to Toynbee in 1916, ''because it was a rather collapsed Puss, owing partly to not resting when it should have, and partly to a general conclusion that Nurse is incompetent and incapable of managing.''[48]

Her difficulties were increased by ill health and repeated pregnancies. While she was still nursing her firstborn, Tony, she became pregnant again. ''It looks as if there were going to be a baby,'' Toynbee reported to Lady Mary. ''It is unexpected—we ought to have taken thorough measures—and I am of course very anxious . . . to know how she will stand the strain so soon, if it turns out to be true. It is also rather disconcerting.''[49] It *was* true, and their second

son, Theodore Philip, was duly delivered on 2 June 1916. Edith Toynbee came
to help with the housekeeping at the time of his birth, but a single house could
not contain two such strongminded women for very long, so Rosalind was soon
left to cope as best she could with two infants, aided by a series of nursemaids,
none of whom stayed in the house for more than a few months. She had two
subsequent miscarriages, one probably occurring in October 1918 and the other
a year later.[50]

Other ills afflicted her from time to time as well. Some were genuine, like
the measles; others were imaginary, or partly imaginary, like the bad heart her
mother attributed to her. "That unprosperous child!" Lady Mary exclaimed,
"She told me she couldn't get warm even in bed. . . . She can't be well, and
does not seem fit to cope with her children and house."[51] Matters reached
something of a climax in May 1918, when the boys both came down with
whooping cough and the current nursemaid went mad under the strain. "Rosalind and Arnold both look like ghosts," Lady Mary told the Countess.[52] Eight
months later, when Rosalind joined Toynbee in Paris during the winter of 1919,
her remarks about needing to be relieved from the cares of motherhood provoked Gilbert Murray to respond:

Beloved Miss Petkoff:
    Your last letter was very interesting. I think you ought to be freer, and propose
to put Tony and Philip up for sale in the Exchange and Mart, unless you telegraph
to the contrary. (Arnold too?)[53]

Yet Rosalind was, at times, a loyal and patient wife, as well as a warm and
loving mother to her two boys. She played the role of gracious hostess, wife,
and mother to perfection when Toynbee's sisters and his mother came to visit
at Christmastime, for example. "It was delightful of you to write from your
office today to say it was a lovely week," Edith wrote her son after one such
visit. "It was, it was. We were all so near together, and Tony is delicious
cement. . . . Dear Lad, you have a very adorable wife and child. Rosalind
with Tony is something to be thankful to have watched."[54] Her second son,
Philip, made his intense attachment to Rosalind very clear in a thinly disguised
autobiographical poem, *Pantaloon*,[55] and her letters show that the feeling was
sometimes fully reciprocated. On returning from Paris in 1919, for example,
she wrote: "I believe that much as I loved them tiny, I am going to find it
even better as they get more understanding. . . . I have always so grudged
their getting big so fast."[56] Physical intimacy was very much part of her moth-
ering, as she explained a few months later: "They are sleeping with me in
turns—very dear and warm they are and snuggle up and nearly push me out of
bed."[57]

Yet her mother love was capricious. She needed respite from the demands
her children made on her, and when there was no nursemaid to take them away
for most of the day, she found it very hard to bear. Her husband was almost
no help, being far too busy at his work. Rosalind's remonstrances had small

effect, even when she resorted to transcribing letters presumably dictated by Tony, of which the following is a sample:

> Dear Daddie, Darling Daddie
>   Daddie come back. We're writing a letter to make Daddie come back. . . .
> Draw a puffer [railroad engine] for Daddie to get in. Kisses and love from
> Little Tom.[58]

Holidays, usually spent at one or another of the Howard or Murray country houses, did bring Toynbee into touch with his sons and with Rosalind, and several letters record their joy at such reprieves from the usual round in London. "It really is a divine place," Toynbee wrote from Castle Howard in 1917, "perhaps for the babies most of all. . . . Altogether we are enjoying ourselves completely. . . . This place does seem to keep the war at arm's length. There is something very strong and alive and unassailable about it, like the eighteenth century living on triumphantly into 1917."[59]

The eighteenth century survived in Toynbee's life in another way as well, for early in the war Castle Howard had inspired him (or perhaps his wife on his behalf) to seek to replicate Edward Gibbon's career as gentleman of leisure and historian extraordinary. The first step was to resign from his fellowship at Balliol, which he did in December 1915, thereby cutting himself adrift from Oxford and the academic career he had begun with such high hopes just three years before.

Toynbee's reasons were complex. On the one hand, he had found it almost impossible to combine his duties as tutor with writing a serious book like *Nationality and the War*—and his central ambition remained always to write a great work of history. "I was being bad towards the College and bad towards Rosalind," he told his mother when he first left Balliol.[60] The obvious solution was to become a gentleman of leisure, like Gibbon or Lord Acton, and thus be able to write history and at the same time play a proper part in family and social life.[61] The only thing needed was a sufficient private income, and that, Rosalind believed, could be wheedled from her grandmother, the Countess of Carlisle, if she were properly approached.

Why, indeed, should Rosalind not hope to inherit Castle Howard itself, and all that went with it? She was her grandmother's namesake, eldest daughter of an eldest daughter; in the matriarchy that the Countess had created around herself that was no trivial heritage. The Countess had quarreled violently with all of her sons, and Rosalind was, or aspired to be, something of a favorite with her grandmother. Moreover, the Countess had complete legal right to dispose of her properties as she wished, and she may well have dropped hints from time to time about what Rosalind (and others) might expect from her will.

Waiting for the succession, however, was not enough: Rosalind and Toynbee needed more income if they were to leave the rent-free grandeur of 1 Palace Green and set up for themselves in London. Accordingly, at his wife's suggestion, Toynbee wrote a letter to the Countess explaining his financial situation and literary ambition, asking her to make him independent of his obligation to teach by granting them some sort of annual largesse. This request did not please the Countess. She could not understand why Rosalind was unwilling to live at 1 Palace Green. "We must not waste money now on quite unnecessary house rents," she told her granddaughter.[62] Her reply to Toynbee's letter was accordingly brisk and to the point, and only partially granted his request: "I think your income, even without the fellowship, will suffice, if, as seems pretty sure, you continue to make 300 pounds a year. We will assume that henceforeward my 250 pounds, promised to Rosalind before her marriage, will not be invested, but paid over annually to her and you for your free use. This will enable you to give up the fellowship when you like. I will pay the 250 pounds in advance on June 1, 1916."[63]

Income instead of capital was not quite what Rosalind had hoped for, but it did allow Toynbee to resign from Balliol with financial impunity. His explanation to Lindsay and other Balliol colleagues was that he could not put his whole heart into his teaching and other college duties, so should resign. Lindsay was angry: "You write as though you haven't any obligations at all and only personal feelings. . . . Don't you think anything or anybody has any claim upon you?"[64] On Christmas Day, Lindsay expressed the hope that "we may be able to continue mutual respect and affection while damning each other's principles,"[65] but in fact their friendship broke off and was never renewed.

His mother was outraged by what Toynbee had done—giving up the dignity and independence of a post at Balliol for dependence on Lady Carlisle's bounty. Apparently she lost her temper in the presence of the Murrays when the matter came up in conversation.[66] His aunt Charlotte also reproached him for not loving Balliol as her Arnold had done.[67] Thus Toynbee's abrupt (and unnecessary, since Balliol would have continued his fellowship indefinitely) resignation seems to have sealed a break with the Toynbee side of his family, tying him more tightly than ever to his wife's relatives.

In all probability, Toynbee's fundamental motivation was his need to insulate himself from the military version of the heroic ideal that pervaded Balliol, and which he shared without acting upon it. The thought of coming back to face all the veterans who had done their part in the war presumably appalled him. By not volunteering, giving up his fellowship and depending on Lady Carlisle's bounty instead, he took refuge with his Murray–Howard in-laws, who either commanded him to stay home, as Rosalind and the Countess both did, or recommended that he do so, as Gilbert Murray did. But there was a heavy cost, for he shared his mother's prickly financial punctilio, and found financial dependence on the Countess and her whims to be humiliating.

He was, in effect, caught between the formidable contending personalities of the two Rosalinds. On the one side his wife wanted more income so she could live comfortably in a house of her own and also see more of her husband. On the other hand Rosalind, Countess of Carlisle, wanted to keep full personal control of her wealth, and in return for her largesses expected an obsequiousness that neither Rosalind Toynbee nor her husband was prepared to render. They tried, indeed, repeatedly to butter her up, but refused to crawl at her feet.

It was an awkward position for anyone to be in, and particularly awkward for a man who had an almost pathological fear of running out of money.

Renewed efforts to negotiate an arrangement that would guarantee the Toynbees sufficient private income to allow him to give up paid work came to naught in January 1917, though Lady Carlisle did promise them an additional £200 a year.[68] But in August of that year, when Rosalind and her children were staying at Castle Howard, a fierce quarrel arose when Rosalind Toynbee took exception to the Countess's unmeasured denunciation of the Germans. This provoked an attack on Rosalind's character, and, she confessed to her husband: "I lost my temper and left the room—result is that I am in complete disgrace (which includes poor Tony) and am not spoken to, only glared at. It appears that I have never been forgiven for refusing to live at Palace Green— that the future drawn is of my refusing to consider your interest."[69] The two Rosalinds parted without reconciliation.

Next New Year's, not surprisingly, the money from the Countess failed to arrive punctually. Rosalind told her husband "if necessary, overdraw with Mr. Gillett, writing to him and asking to be allowed to do so for a month or two, as it is sure to come in before long. Try not to worry over it."[70] The Countess did come through eventually; and in December 1918 she transferred capital to Rosalind and Toynbee in the form of a deed of trust for £6000, yielding, it was estimated, £300 a year. This superseded her former annual payments and guaranteed the Toynbees against repetition of the crisis with which the year had begun.

But Rosalind's expenditures continued to press hard against income, and sometimes exceeded it. She tried, halfheartedly, to keep accounts but habitually spent what she felt she needed to spend without bothering about the bank balance of the moment. Her cavalier attitude toward money was something Toynbee could never accept. Instead, he worried constantly. His only resort was to try to earn more, and this he did by agreeing to any and every journalistic assignment that came his way, so long as a fee was attached. His principal connection was with a weekly journal, the *Nation,* for which he wrote articles on aspects of the European situation almost every month.[71] But he wrote for the *Times History of the War* as well, and for obscure publications like *Home-reading Magazine* and *Highway*—anything that paid a fee.[72]

But neither journalism nor largesse from the Countess of Carlisle seemed likely to provide enough income for Toynbee to attain the status to which he and Rosalind aspired, especially in view of the wartime inflation. Instead of becoming a gentleman of leisure, free for his literary pursuits, therefore, he had to entertain the prospect of working for pay. When this became clear, his first choice was the Foreign Service. In 1917, when Toynbee secured assignment to the Political Intelligence Department of the Foreign Office, such a future seemed within reach, even though his formal letter of appointment, when it came, expressly declared "that your services in the Political Intelligence Department will give you no claim to employment in a permanent capacity."[73] But plans were afoot to make the new department permanent, and some of its members did in fact stay on after the war.

Toynbee was not among them, for, as we have seen, he became disillusioned when his advice was disregarded at the Peace Conference. In addition, it seems probable that his radical views, especially arguments about the necessity of establishing an entente with the Bolsheviks, turned some of his superiors in the Foreign Office against him. Information to that effect may have helped him to decide against trying to stay on after the war. He therefore reverted, *faute de mieux,* to considering an academic appointment. He had already toyed with various possibilities for returning to Oxford, but without really pursuing any of them. The feelings that led him to give up his fellowship at Balliol remained too strong to allow a graceful return.

But in July 1918 the University of London announced a new chair in Byzantine and modern Greek history and literature, funded by Greeks of the diaspora living in England, with an additional annual contribution promised by the Greek government. Toynbee was reluctant to apply. "One might be in a false position towards the London Greeks," he wrote to Gilbert Murray, "if one isn't particularly Philhellene. . . . And it would mean giving up the Foreign Office." [74] But he did apply, while retaining his doubts, even, or perhaps especially, after being interviewed by the committee charged with making the appointment. On that occasion, Toynbee told the committee that he flatly refused to teach modern Greek literature, thinking this would probably disqualify him. "The one thing I am determined on, before taking it, is that the electors should know what I intend to do, and that I should know that they approve of my doing it. It was so miserable being in a false position at Balliol and I am shy of risking the same thing again." [75] But he needed a job, and Rosalind decided that the post at the University of London was the best thing available. [76] His refusal to teach modern Greek literature failed to disqualify him in the eyes of the committee, and on 29 May 1919 he received a letter of appointment as Koraes Professor at King's College, The University of London, with a salary of £600 per annum. The appointment was for five years, with the possibility of renewal up to the age of sixty-five. [77]

With his future thus assured, Toynbee might have been expected to rejoice. A full professorship at the age of thirty was unusual in academic life; and the University of London was not without prestige, even if it lacked the aristocratic patina of Oxford and Cambridge. But Toynbee was far from happy. "My particular mania is dislike of the professorship," he wrote to Darbishire. "I feel uprooted and bewildered. Everything is in such flux that I had to take the professorship. . . . I am in a loathsomely neurasthenic and self-centred condition." [78]

His condition was compounded further by the fact that he and Rosalind had just suffered a run-in with the Countess that closed off any remaining hopes they may have had of inheriting substantial wealth from her estate. The Countess's equanimity had been shattered in December 1918 when the so-called Khaki election resulted in a victory for Lloyd George and a government coalition dominated by Conservatives. But the real novelty of the election was the crushing defeat it brought to the Liberals, who were overtaken by the upstart Labour Party. The eclipse of her party was, for the Countess of Carlisle, an unexpected and intolerable turn of the screw. Those who had abandoned the Liberal cause and joined with Labour counted as nothing less than traitors in her eyes; and, of course, Toynbee and Rosalind were among the guilty. A month after the election, Rosalind told her husband: "The enclosed letter from Grandmother only reached me two days ago. . . . I read it as distress with the Labour party over election results, and that the breaking of fellowships applies to possible (and probable?) rupture with us. However, I have written thanking her very much and not referring to anything of that kind. . . . (I think by the way that it also means the answer to any after war possibilities—don't you?)"[79]

By the end of March 1919, Toynbee was overdrawn at the bank and the Koraes appointment was not yet settled. Rosalind pointed out "if we go to Castle Howard we can quite afford two months without pay, for we should have no expenses during that time except rent."[80] But this required an invitation, which was not forthcoming, even after repeated requests. On 2 May 1919 the Countess expressly declined to have them visit her. She railed at "the great catastrophe of the election" and made it clear that she now counted the Toynbees among her enemies. "And after all what have you to do with Castle Howard?" she asked rhetorically. "This beautiful place so lovable for those who accept it with a simple affection and clear conscience—but such a jarring false note, such a mockery for those who have joined the ranks of the people who have declared war on such as we, who dwell in great rooms filled with private galleries of books and pictures. Is it not anathema to your comrades? Then why come here?"[81]

Amazingly, this diatribe did not prevent Toynbee from responding: "As in your present letter you do not answer yes, you still do not say no, I am going to hazard being importunate and to ask again."[82] This time the thunder really rolled. The Countess responded:

My dear Arnold
   I think I have been too inexplicit for you to understand me, and I must be horribly matter of fact.
   I understand that you and Rosalind enrolled yourselves in the Labour Party last winter: that party is fighting hard the Liberals and will smash us if they can. . . . I have been an intense passionate lover of my Liberal creed and party all my life long. . . . If I were to have as my guests . . . those who belong to a party that seeks to compass our destruction, there could be no vivid, helpful, comforting talk for me. We should have to keep off political subjects and that would make intercourse very unreal and dry and very different from our old breezy, happy times.
                                              Yrs very Afftly
                                              Rosalind Carlisle[83]

But beggars could not be choosers, and after a two week interval, Rosalind returned to the assault:

Dearest Grandmother
   I feel ashamed to ask again what you have so clearly refused and for such definite reasons, but I do not really know what to do.

Dr. Wheeler has just been to see Arnold again this morning. He says quite definitely that he . . . must take a long and complete rest. . . . Castle Howard is the one place he longs for and feels he could rest in, and it is just this that has made me write to you again. . . .

He seems somehow to have lost all self-confidence and distrusts his own powers to do anything. The Byzantine Professorship to which, as you know, he was much looking forward, now fills him with depression and apprehension. He feels he doesn't know enough to teach the necessary history, and cannot face the prospect of learning anything new. This is so unlike his usual attitude that I feel sure it is only a sign of real exhaustion. . . . I believe that if he felt he had a rest at Castle Howard to look forward to, it would buoy him up. . . . I hope, oh so earnestly, that you will let us come.

> Your loving
> Rosalind[84]

This appeal succeeded in somewhat softening the Countess's heart, and she suggested that they might stay at her house at Boothby. The news improved Toynbee's spirits. "He has not been up yet," Rosalind reported to her grandmother. "He still gets very exhausted at the least effort, such as shaving or having a bath, but he is much more cheerful.[85] When they got to Boothby, he recovered abruptly. "Arnold already seems a different person," Rosalind reported. "I wish you could have seen the children when they ran out into the garden yesterday morning. . . . Arnold ran with them and seemed filled with new life."[86]

But Rosalind's cheerful letters and fulsome thanks to her grandmother could not heal the wound entirely. The Countess refused to forgive political treason; and the Toynbees had to face the fact that Rosalind's dream of living on a private income, in the style of eighteenth century grandees, would never be realized. His political views thus hurt Toynbee twice over: first by foreclosing a career in the Foreign Office, and now, a second time, by dissolving a dream of leisure and ease for writing the great history to which he still aspired.

A third factor had helped to unsettle his mind and provoke the physical and mental collapse of April and May 1919. This was a traumatic renewal of contact with his father, occasioned by the fact that Harry Toynbee caught the influenza that was then raging worldwide, and seemed about to die. When Toynbee and his mother visited Northampton in March 1919, it was only the second time that Arnold had seen his father since Harry's confinement ten years earlier. The effect was harrowing, and, for a moment, renewed the bond between mother and son. Toynbee expressly hoped that his father would die, arguing, in a letter to his mother, that "it would be so much better for him and for you. . . . We had a dear time together—it is always dear being with you—whether in common troubles or on a holiday like last Easter. I think I am writing you a sort of love letter: I remember your saying once, at some beginning of term, that I was 'almost a lover' to you as well as a son, and I have never stopped being that."[87]

She replied the next day:

My own dear boy
    How good of you it was to write on Saturday, and such a long and dear letter
too. Yes, we shall always be ''lovers'' as well as mother and son. If I could only
let you see what your coming to Northampton meant to me, you would even more
fully realise that. The night before you came all the terror and cruelty of the
universe had gripped me. When your telegram came I once more felt . . . at any
rate we hold each others' hands, and love each other to the end—and love of any
sort casts out fear. You brought back courage and dear normality to me—God
bless you for it.[88]

The longstanding tension between Toynbee's attachments to his mother and
to Rosalind was thus at an exceedingly exacerbated level in the spring of 1919,
when his financial difficulties and the quarrel with the Countess of Carlisle
coincided with his father's illness. In fact, Harry Toynbee recovered, and Edith
Toynbee returned to Oxford, where she had lived since 1917, occupying the
house Toynbee had rented when he was a tutor at Balliol. He continued to pay
the rent until his sisters had completed their education, and thus contributed to
his mother's support to a significant degree.
    But the surge of renewed intimacy between mother and son that burst forth
in March 1919 was not sustained. Toynbee preferred Rosalind and the Murray–
Howard connection, and all that was involved in that choice. His mother was
never comfortable with them, and he could therefore never be entirely comfortable with her after his marriage. The tug of war between wife and mother
was too fierce to be papered over easily, and the gap between Rosalind's aristocratic tastes and aspirations and Edith Toynbee's parsimony and pride made
ordinary social encounters awkward. The struggle between opposing ways of
life bothered Toynbee too. It exacerbated his financial worries and shadowed
his acceptance of the Koraes Chair, which would require him to teach and take
part in the busy work of university life, instead of being free to write the book
of his dreams.
    His collapse in the spring of 1919, with its multiple causes, also had a vari-
ety of manifestations. In addition to his physical prostration, he experienced an
extraordinary psychological encounter which he described, thirty-five years later,
as follows:

    In London, in the southern section of the Buckingham Palace Road, walking
    southward along the pavement skirting the west wall of Victoria Station, the writer,
    once, one afternoon not long after the end of the First World War—he had failed
    to record the exact date—had found himself in communion, not just with this or
    that episode in History, but with all that had been, and was, and was to come. In
    that instant he was directly aware of the passage of History gently flowing through
    him in a mighty current, and of his own life welling like a wave in the flow of
    this vast tide. . . . An instant later, the communion had ceased, and the dreamer
    was back again in the every-day cockney world.[89]

An odd experience, indeed, and one that presumably confirmed Toynbee in
his resolve to write a great work of history, somehow, by hook or by crook,
in spite of all the practical obstacles and distressing distractions created by the

necessity of earning a living. A sense of being set apart from ordinary mortals, destined to great things, had been instilled in him from childhood. The disappointments, trials, and tribulations that pressed so hard upon him in the spring of 1919 did not extinguish that conviction. His vision confirmed it. And within a few weeks, as the special crisis provoked by the visit to his father in March, his failure at the Peace Conference in April, and the clash with the Countess in May subsided, he was able to resume ordinary activity, and could even look forward, with some relish, to the possibilities of his new post at the University of London.

A new chapter of his life opened. The extraordinary efforts, psychological distresses, high expectations, and ultimate failures of his wartime years lay behind. Later, they were partly forgotten, partly transmuted, and partly suppressed, but the Great War and his experience of it loomed like a great cloud over all his subsequent life and thought. Nothing would ever be the same, for Toynbee, or for most of his contemporaries. Unimagined flaws in European civilization had emerged to threaten its extinction. Survivors ought to put things right—or at least try, paying special attention to the problem of preventing another war. That, accordingly, became the principal focus of Toynbee's professional activity in the interwar years. He had failed at Paris in 1919; he did not intend to fail again.

V

The Koraes Professorship
1919–1924

By October 1919, when he took up his duties as Koraes Professor of Modern Greek and Byzantine History, Language and Literature at King's College, The University of London, Toynbee had pretty well recovered from the depression and physical collapse that paralyzed him in the spring. The year before, in November 1918, Rosalind had located "a nice flat in Carlyle [*sic*] Mansions where Henry James lived. They have got one—subject to the sanitation being right."[1] The sanitation turned out to be acceptable, and on her return from Paris in March 1919, Rosalind superintended the moving of furniture and other possessions into the new flat. Her mother approved the result, calling the setup at 16 Carlisle Mansions "delightful." The Toynbees lived there for the next few years. Their housing problem had at last been solved. Things even picked up on the servant front when a French Protestant governess, engaged while the Toynbees were in Paris, finally arrived and seemed "just the right sort."[2] The subsequent summer at Boothby was restorative as well; and, more generally, the armistice and expected peace meant that the Toynbees could look forward to a less unsettled life than they had known in wartime.

Rosalind was able to resume novel-writing, and Toynbee was free at last to work out the grand philosophy of history of which he had dreamed since schooldays. He was determined to make his new professorship serve that purpose, and took pains, in his inaugural lecture on 7 October 1919, to emphasize the enormous scope of Greek history as he chose to interpret it. Toynbee, in effect, cast himself as successor to Herodotus. The Greeks, he argued, had been Europe's easternmost outpost throughout ancient, medieval, and modern times. Their history therefore registered with special poignancy the alternating ebbs and flows of eastern as against western styles of civilization. The geographical theater of this action and reaction extended from the Atlantic coast to the borders of India, and involved peoples of the northern steppe as well those of Arabia. Thus, from Toynbee's angle of vision, Greek history and world, or at least Eurasian history were scarcely distinguishable.[3]

The patriotic Greek donors and conventional English philhellenes in the audience may have worried lest the little Kingdom of Greece be unduly dwarfed in such a vision. Indeed, basic discrepancies of outlook between Toynbee and the Greeks of London, who had endowed the chair, were clear enough even at the time. The chairman at the Inaugural Lecture, Ioannes Gennadius, a former Greek Minister to Britain and a leading figure of the London diaspora, made this obvious. Instead of simply introducing Toynbee, he chose to deliver a lengthy speech, defending the historic greatness of Byzantium against the sneers of Edward Gibbon, which, he claimed, still distorted English understanding of Greek history, and ending with a diatribe against any use of demotic Greek by the occupant of the new chair. For him, at least, ancient, medieval, and modern Greece were one and the same—emanations of an unchanging Hellenism; and the role of the new professorship was to inform the British public about the enduring greatness of the Greek nation.[4]

At first, divergences between the donors' expectations and Toynbee's performance were not especially acute. He lectured on Byzantine history in 1919–1920; in the following year he took up the history of the Ottoman Empire and the emergence of the independent Kingdom of Greece in the 1820s. In the next academic year he capitalized on the topographical studies of his student *Wanderjahr* by giving a remarkable series of illustrated lectures on the historical geography of ancient, medieval, and modern Greece.[5] But in 1922–1923 Toynbee began to explore the wider reaches of what he conceived to be his subject with lectures on Near and Middle Eastern history, and he proposed to follow that with lectures on the history of the Eurasian steppes in 1923. By this time the donors were incensed by the pro-Turkish tone of Toynbee's writings about the war that had broken out between Greece and Turkey in Anatolia (1920–1922) and added to their indictment of political bias the charge that he was abusing his chair by lecturing on subjects far removed from Greek history.[6] Toynbee's worst fears before applying for the Koraes Chair were thus realized.

The fact of the matter was that Toynbee wished to pursue his own ideas and work out the details of his interpretation of history, whether or not his inquiries and conclusions fitted the bounds of Greek history, however defined. A principal key to his thought was a war-sharpened sense of cyclic repetition in human affairs. Drawing parallels between ancient and contemporary society had been commonplace among classically educated Europeans before the war, but they did so in diffuse and unfocused ways. Comparison with Hellenistic or Roman times came just as easily as comparisons with Athens and Sparta. For Toynbee, the war made the comparison crisper and far more compelling. His wartime sense of general human helplessness to ward off self-inflicted disaster echoed Athens' circumstances in the Peloponnesian War of 431–404 B.C. and Rome's desperate plight in the Second Punic War of 218–201 B.C. These parallels haunted him during the Great War with a power and poignancy that such comparisons had lacked in peacetime. "One is like a beetle under a steamroller," he wrote to Darbishire, "waiting till the machinery moves onto some other beetle's back. That sense of being destroyed by the machine you have made with your own hands is very terrible—the Greeks went through that. The worst of it is that civilisations take so long to perish after the mortal blow. . . . But history is as melancholy as the war itself.''[7]

Familiar passages, especially from Thucydides and Lucretius, coursed through Toynbee's mind in odd moments of leisure during the war years; and the idea of a cyclical pattern of history burned itself indelibly into his consciousness. As he put it more than thirty years later: "The general war of 1914 overtook me expounding Thucydides to Balliol undergraduates reading for *Literae Humaniores,* and then suddenly my understanding was illuminated. The experience we were having in our world now had been experienced by Thucydides in his world already. . . . Whatever chronology might say, Thucydides' world and my world had now proved to be philosophically contemporary. And if this were the true relation between the Graeco-Roman and the Western civilisations, might not the relation between all civilisations known to us turn out to be the same?''[8]

In this recollection, however, Toynbee elided phases of his thinking. Not until 1920 did he begin seriously to wonder about whether "all civilisations known to us" followed the same pattern of growth, breakdown, and dissolution. Before then, his reflections concentrated on the geographical segment of the earth with which he was already familiar—Europe and the Near and Middle East. And throughout the war years, the analogy with the Rome–Carthage struggle of the third century B.C. seemed as compelling to him as the analogy with fifth century Athens and Sparta. Only in 1920 did he have occasion to formulate anew his war-warped vision of Graeco-Roman antiquity and make the equation between the Great War and the Peloponnesian War canonical for his entire interpretation of history.

Before that time, plans for his great book remained inchoate. "I am going to write a short history of Greece and a much longer history of how Rome destroyed the world," he wrote to Darbishire in 1918. "I believe one can put one's experience of the war best in parables. At least I don't see myself writing a history of this war. I sometimes think of a history of European civilisation from the Dark Ages to 1914, treating it as a unity and not as a bundle of separate nation states."[9] To his mother he was more specific about downplaying nationalism, telling her: "I want to write a history of Europe treating European civilisation as a whole, with the modern national states as more or less transitory surface phenomena.''[10]

Later, when he had to face up to the failure of his hope of affecting the peace settlement in 1919, he wrote to Darbishire once more: "I want to read again and I want above all to write history. Tomorrow I am thirty and I haven't touched the Home University history of Greece which I began five summers ago. I am planning—have been at odd moments during the war—a big history of Rome, concentrating on the conquests and the social revolution—showing how the first brought on the second, and how between them they destroyed not only Rome but the whole of Greek civilisation."[11] In due course, Toynbee did indeed get round to his "big history of Rome," which appeared in two stout volumes as *Hannibal's Legacy* in 1965. But between the spring of 1919 and September 1921, when—while riding through Bulgaria on the Orient Express— he found himself "jotting down, on half a sheet of notepaper, a dozen headings which turned out to be the subjects of the principal divisions" of what eventually became *A Study of History,*[12] something obviously changed his mind about what he should write next, and how to go about it.

Available documents shed a somewhat fitful light on this crucial shift in Toynbee's historical vision. The earliest important source consists of notes for six lectures on ancient history delivered at the University of London in 1919– 1920. Toynbee began by saying: "I shall confine myself to *our* civilisation, and by that I mean the civilisation of Europe and the Middle East, which . . . cannot by taken piecemeal. In that field, the vital elements of civilisation are common to the whole."[13] He nonetheless distinguished three separate hearths of civilization: the Aegean islands, Mesopotamia, and Egypt. But, he claimed, Mesopotamia and Egypt merged into one after the Arab conquest of the seventh century A.D. so that "the Middle eastern civilisation of today is really still living on the immense store of energy accumulated in Egypt and Mesopotamia in prehistoric ages and released in the 4th millennium B.C.—the present life in the East being a feeble and mechanical repetition of the great original pulsation."

By contrast, the West twice collapsed: once in the fourteenth century B.C., with Minoan Crete, and again in the fifth century A.D., when Rome went under. "Each time the catastrophe, though terrible for the generations that experienced it, . . . endowed our civilisation with a new lease of life."[14] The East's changelessness reflected the fact that a "rigid and oppressive social order . . . strangled moral and intellectual progress, and soon, material progress as well." The Assyrians, in particular, administered a "mortal wound to Middle eastern civilisation" so that "the East had become too feeble and stiff-jointed to be cured even by the sovereign stimulus that Greece could give; and Greece had her own tragedies—the Peloponnesian war, and the almost Assyrian calamity of the conquest of the civilised world by Rome, which crippled her energies. The abortive western inoculation left the Middle East more exhausted still, and she reverted to the soporific—this time drugged with a religious ingredient applied by the universal empire and religious system of Islam. But the life was leaving the body, and the heart which had begun to beat so mightily 5000 years before died down in the darkness of the Turkish and Mongol conquests. . . . The Middle East still lies under the shadow of death, and we, just smitten by our Peloponnesian war, have shouldered the ancient burden of the Greeks."[15]

Thus, in the immediate aftermath of the Great War of 1914–1918, trying to give shape to his longtime preoccupation with the tangled histories of "east" and "west," Toynbee fell back on naive organic metaphor and spoke of civilizations as though they were living organisms. The West rose and fell and periodically reenacted its own past, according to this vision, whereas a changeless East merely slept after its inaugural rise to the level of civilization. It was not a very convincing version of the facts, as Toynbee already half realized, but his reflections on history were still limited by a grotesquely inadequate dichotomy of East versus West (otherwise translatable as Us versus Them) that he had inherited from Herodotus and innumerable nineteenth century predecessors.

His definition of the term civilization in these lectures was, however, relatively precise—rather more so than in his later years, when different layers of meaning tended to flow together in his mind and confuse his use of the term. In discussing the rise of civilizations, in the first lecture, he remarked: "I believe human development is a process in which human individuals are moulded less and less by their environment . . . and adapt their environment more and more to their own will. And one can discern, I think, a point at which, rather suddenly, the human will takes the place of the mechanical laws of the environment as the governing factor in the relationship." That point constituted the beginning of civilization and of history too, for "the impress made by human beings on their environment creates a record which enables later generations of human beings to reconstruct in their minds their predecessors' activities." [16] Civilization and human freedom are therefore closely akin. Both are opposed to "mechanical laws" of the environment—and Toynbee explicitly pointed out that what he called "environment" included mechanical laws of human society that allowed one class or group of human beings to oppress others. The Bergsonian heritage is obvious; continuity with nineteenth century English liberalism is even more transparent. There was, in fact, little that was new in this earliest sketch of Toynbee's effort at a philosophy of history.

The next important text attesting to the evolution of Toynbee's ideas about the past took the form of a lecture at Oxford, delivered in the spring of 1920. Subsequently published as "The Tragedy of Greece," [17] it attracted some attention. Gilbert Murray wrote: "I am altogether aglow and thrilling with The Tragedy of Greece," [18] and an anonymous reviewer called it "remarkable" despite its "traditional Oxford outlook" and failure to recognize the superiority of modern to ancient civilization. [19]

In a sense, Toynbee simply condensed the sketch of his planned history of Greece for the Home University Library into a single lecture; but the fact that the Great War had intervened between the time he first planned that book and the delivery of his lecture quite altered the thrust of what he had to say. Greek history had become unambiguously tragic in his eyes, and the tragic structure Toynbee now found in the classical past resulted from psychological destruction wrought on human consciousness by all-out wars.

Toynbee drew directly on Thucydides for this new view. Explaining the brutality of party massacres at Corcyra in 427 B.C. Thucydides had declared that war "proves a rough master, that brings most men's characters to a level with their fortunes." [20] It was this kind of psychological brutalization that Toynbee had in mind in attributing the breakdown of civilizations to unregulated warfare. As far as the ancient Mediterranean world was concerned, he concluded that the breakdown of Greek civilization occurred as a result of "the failure of interstate federation" during the Peloponnesian War (431–404 B.C.). Thereafter, the ancients experienced three distinct "rallies," each defeated in turn by renewed warfare, until a final dissolution occurred in the seventh century A.D.

Thus articulated, just four months after the League of Nations had started its work at Geneva, the lesson for contemporaries was easy to read. Moreover, at the beginning of his lecture, Toynbee made the parallel between ancient and modern times quite explicit, saying: "I suspect that the great tragedies of history—that is, the great civilisations that have been created by the spirit of man—may all reveal the same plot, if we analyse them rightly." And, again at the close: "When Ancient Greek civilisation may be said finally to have dissolved, our own civilisation was ready to 'shoot up and thrive' and repeat the tragedy of mankind."[21]

This lecture in effect provided an occasion for Toynbee to put the impress of the war upon his prewar mastery of Greek and Roman history and literature. He did so in such an elegant and convincing fashion that he never changed his mind again. To date the breakdown of classical civilization to 431–404 B.C., long before Greek philosophy and science had fully emerged, and to treat Rome as a mere caudal appendage to Greece, was surely extraordinary. To impose a simple dramatic structure on the entire sweep of classical history was no less surprising. But with these bold strokes, inspired and reinforced by the way remembered classical texts had resonated with current events during the Great War, Toynbee laid down all the fundamentals of what he later set forth at far greater length in A Study of History. It was indeed a remarkable lecture, giving enduring form to the classical component of his education and experience up to that time.

Whether Toynbee was aware of how much he owed to Thucydides in imposing such a tragic pattern on the history of the ancient world is an open question. In Toynbee's first undergraduate year, Francis M. Cornford, a don at Cambridge, had published a controversial book entitled Thucydides Mythistoricus,[22] in which he argued that Thucydides had, late in life, revised his view of public events by coming to regard Athens' eventual defeat after twenty-seven years of war as retribution for the city's overweening pride, or, in Greek, hubris. Thus, according to Cornford, after Thucydides had seen his hopes for the future of Athens dissolve in wartime extremism, he fell back upon one of the doctrines of traditional Greek religion to explain what had happened, though without expressly attributing the outcome of the war to the will of the gods. Nevertheless, Cornford argued, Thucydides came to view his native city as a tragic hero, which, like individual heroes, was destined to suffer punishment for acts of hubris, whether because of the nature of things, or because the gods really were jealous of human greatness.

Toynbee was well acquainted with this heretical view of Thucydides, and actually listed Cornford among those who taught him "methods of literary presentation" in the "Acknowledgements and Thanks" with which A Study of History comes to its close.[23] But whether he ever accepted Cornford's thesis about Thucydides is unclear; and nowhere does Toynbee acknowledge that in finding a tragic pattern for the entire sweep of Graeco-Roman history he was influenced by, or even considered, the analogy between his labors and those

that Cornford attributed to Thucydides. Any connection must therefore have
been unconscious, for Toynbee was meticulous in recording intellectual debts
of which he was aware.[24]

Nevertheless, Toynbee gradually shifted from his earlier Herodotean approach to history and came round instead to a view that can be labeled Thu-
cydidean. From childhood and throughout the war he had been professionally
concerned with what he viewed as the perennial interaction of ''east'' and ''west,''
within a geographical theater that only slightly enlarged upon what Herodotus
had treated from the same point of view centuries before. But after 1921 he set
out to organize the history of a much larger part of the world around a series
of parallel, separate civilizations, whose rise and fall conformed to the tragic
pattern he had anatomized for the first time in his 1920 lecture. Insofar as this
segmented view of history, treating each civilization as complete in itself, matched
the method Thucydides used in writing about the separate, sovereign polities
of the ancient Greek world, one may say that Toynbee abandoned the far-
ranging Herodotean approach with which he had begun his researches, accepting instead a tragic and compartmentalized view of the processes of history that
was characteristic of Thucydides.

A powerful factor in forwarding this alteration of Toynbee's outlook was his
encounter with the ideas of Oswald Spengler a few months after he had delivered the lecture at Oxford. Spengler wrote the first volume of *Der Untergang
des Abendlandes* during the war, and when his book appeared in 1918, on the
eve of the armistice, his prognosis of the decline of western civilization exactly
fitted the mood of the German public. It therefore attracted wide attention among
the educated classes of Germany and achieved a good deal of journalistic no-
toriety as well.

Hostility to all things German, together with Spengler's ponderous style and
the alien conceptual structure of his book, inhibited its immediate reception in
English-speaking lands. But Toynbee read German with ease, and when his
friend Lewis Namier lent him a copy in the summer of 1920, he read it at
once. Twenty-eight years afterward, Toynbee remembered the experience as
follows:

> As I read those pages teeming with firefly flashes of historical insight, I won-
> dered at first whether my whole inquiry had been disposed of by Spengler before
> even the questions, not to speak of the answers, had fully taken shape in my own
> mind. One of my cardinal points was that the smallest intelligible fields of histor-
> ical study were whole societies and not arbitrarily insulated fragments of them like
> the nation-states of the modern West or the city-states of the Graeco-Roman world.
> Another of my points was that the histories of all societies of the species called
> civilizations were in some sense parallel and contemporary; and both these points
> were also cardinal in Spengler's system. But when I looked in Spengler's book for
> an answer to my question about the geneses of civilizations, I saw that there was
> still work for me to do, for on this point Spengler was, it seemed to me, most
> unilluminatingly dogmatic and deterministic. According to him, civilizations arose,
> developed, declined and foundered in unvarying conformity with a fixed time-
> table, and no explanation was offered for any of this. It was just a law of nature

which Spengler had detected, and you must take it on trust from the master: *ipse dixit*. This arbitrary fiat seemed disappointingly unworthy of Spengler's brilliant genius; and here I became aware of a difference in national traditions. Where the German *a priori* method drew blank, let us see what could be done by English empiricism. Let us test alternative possible explanations in the light of the facts and see how they stood the ordeal.[25]

This account, like his recollection of how Thucydides affected him in 1914, elides a process of intellectual adjustment and regrouping that Toynbee went through between the summer of 1920 and the moment when he was able to scribble down key elements of the framework of *A Study of History* in September 1921.

Toynbee's first step was to try to clarify his philosophy of history by giving it precise literary form in the way his Oxford lecture "The Tragedy of Greece" had given firm and elegant form to the studies of his youth. Accordingly, he spent his summer holiday in 1920 sketching out his thoughts on "The Mystery of Man."[26] The manuscript survives in a state of great confusion, for the work did not flow smoothly. Toynbee made several different starts, often crossed out passages, and sometimes conserved paper by writing new versions on the back of a sheet already used for something else. Long afterward, he collected the loose sheets together and labeled them "Unsuccessful attempt at starting A Study of History, done at Yatscombe, in the cottage, summer, 1920."[27]

Ostensibly, at least, the manuscript takes the rather surprising form of a commentary on the second chorus of Sophocles' tragedy *Antigone*. As Toynbee explained, almost thirty years later: "This Greek poem does convey, superbly, the strangeness, the grandeur, and the pathos of human life, . . . but my medieval-minded approach to my subject by reference to a Greek masterpiece was too indirect to be practicable, as I soon discovered."[28] Yet the fact that he prefaced his text with a passage from Sophocles did not really affect what he had to say, and it was not his medieval-mindedness that made the effort abortive. Rather it was the thinness of his data for what he aspired to achieve. Toynbee set out to philosophize about the human past by generalizing about civilization and its characteristics; but to do so with any plausibility he needed to know far more than he did.

"An inquiry into civilisation" he wrote, "should therefore begin discursively with an examination of the several civilised societies which have regarded themselves as distinct from the rest of the human race. When we have noted a number of phenomena common to some or all of these, we shall have obtained material for an analysis of civilisation itself."[29] But Toynbee had as yet no familiarity with the history of most of humanity since China, India, Japan, and the Amerindian and African peoples remained almost entirely beyond his purview. Nevertheless, he rushed ahead, without fear and without research, listing ten characteristics of civilized societies, as follows: "(i) Civilisations are recent compared to the age of the human race; (ii) they have arisen at a few points only on the total habitable surface of the earth; (iii) they have arisen in areas not previously inhabited by uncivilised Man; (iv) they have been created by a few communities only out of all those which have existed since the genesis of mankind; (v) they have generally been created by several communities simultaneously; (vi) their rise has apparently been sudden; (vii) they spread from their original home to other regions and populations; (viii) their growth is subject to arrest, breakdown and extinction; (ix) these failures do not generally coincide with a scarcity or exhaustion of material resources; (x) communities and areas which have given birth to one civilisation do not bear a second harvest if the first fails." [30]

The rest of the thirty-eight page manuscript develops the first eight of these points, almost exclusively on the basis of Aegean, Egyptian, and Mesopotamian antiquity. Toynbee, in fact, added very little to what he had said in his lectures on ancient history during the immediately preceding term at the University of London, although he did offer a few surprisingly callow remarks about Amerindian "incapacity for civilisation," which he attributed to "some difference of quality in the communities which first settled respectively in the Old World and the New." [31]

The essay was simply broken off because Toynbee came to realize that he needed to know more than he did if he wished to generalize about human affairs in the way philosophy of history required. He explained long afterward: "I made a first shot at writing a comparative history of civilisations in the summer of 1920, but I could not find my way into my subject and did not yet command enough knowledge to make a start. I knew something about Graeco-Roman history and a little about Islamic, Byzantine and Western history as well. I now deliberately widened my horizons by making reconnaissances into the histories of India, China, Japan and the pre-Columbian civilisations in the Americas." [32]

Spengler pointed the way, for the German scholar had treated India and China as autonomous civilizations, analogous to the civilization of classical antiquity or of modern Europe. Another influence came from the works of F.J. Teggart (1870–1946). Long afterward, trying to remember how his ideas had developed, Toynbee told an inquirer: "You are quite right about my debt to Professor Teggart. I read three of his books at a time when, though I knew what I wanted to do, I still could not find the right starting point. Dr. Teggart's work, more than anyone else's, opened the door for me." [33]

Teggart was a rather isolated figure. He was of Irish birth and education but taught at the University of California, Berkeley, where he sought "to do for human history what biologists are engaged in doing for the history of the forms of life." [34] He argued that narrative was inadequate for this purpose. Instead "the analytical study of history must be founded upon a comparison of the particular histories of all human groups." [35] Teggart also explicitly required that India and China be included in the scope of comparison; discussion of the origins of civilization confined to the ancient Near East was, in his view, inadequate. [36] But, of course, this was exactly what Toynbee had done in his abortive essay. One can understand, therefore, why Teggart's programmatic statement commanded Toynbee's attention.

But to understand why Toynbee neglected other parts of Teggart's argument one must go back to Spengler, and to Toynbee's close encounter with brutalities incident to the Greek–Turkish war in Anatolia early in 1921. Teggart had also argued that "human advancement is the outcome of the commingling of ideas through the contact of different groups," since interaction with strangers may, in the right circumstances, provoke "the mental release of the members of a group or of a single individual from the authority of an established system of ideas."[37] Spengler, on the contrary, had asserted that separate civilizations were so profoundly different that effective communication from one to another was impossible. According to him, mixture of ideas or of anything else deriving from different civilizations was a sign not of human advancement but of degeneration, presaging cultural death and final dissolution.

On this question, Toynbee eventually opted for a modification of Spengler's view. He did so partly because the assumption that civilizations were intrinsically separate allowed him to use the parallels he had already discerned between Greek and Western civilizations as a model and guide for further comparisons extending across the parts of the earth he had previously neglected. But the clincher was his exposure to consequences of indiscriminate borrowing from the West among Greeks and Turks as manifested in the war between those two peoples in 1920–1922.

Until after his trip to Anatolia, Toynbee remained very much in doubt. His projected great book had somehow to address two themes: the resemblance between classical and modern European history, on the one hand, and the perennial encounter of "east" and "west" on the other. Having sketched a satisfactory answer to the first of these in his lecture "The Tragedy of Greece," Toynbee badly needed to find a similar structure for his second theme. But his effort in the summer of 1920 got nowhere, and both Spengler and Teggart had made him see the inadequacy of a philosophy of history based on the limited geographic range of his historical knowledge. After all, that range had been defined entirely by accident when, as a schoolboy recuperating from pneumonia in 1903, his uncle gave him a historical atlas showing both Alexander's empire and the Roman Empire in a single spread.

All the same, Toynbee's restless imagination could never give up the effort to understand the grand outlines of history, even before his reading provided him with a basis for generalizing about the whole world. Accordingly, intriguing ideas flitted across his mind from time to time. "I have spent a peaceful day," he confided to Gilbert Murray, "trying to . . . test a theory of mine that the Achaemenid empire is to the Caliphate as the Empire of Augustus is to that of Constantine, with a millennium of Greek penetration thrust in between and interrupting the natural course of Middle Eastern history. This interruption, and consequent unnatural prolongation of the 'universal empire' stage, which we got through so much more quickly in the West, may partly explain why the East has gone so wrong.

"I got at the idea by asking myself why one should take Islam as a dividing line any more than Xianity. One knows that B.C. and A.D. are a false antitheses *[sic]* and no division between cycles of civilisation; may it not be the same with ante and post Hegira? After all, the Arab culture of the caliphate, based on Syriac and Greek, belongs to the Ancient World, and the real break or dark age in the East comes with the nomadic invasions of the 11th–13th centuries A.D. Seljuks and Mongols. Do you think there is anything in this?''[38]

Germs of one of the least plausible of Toynbee's later interpretations—his explanation of the sudden appearance of the Caliphate, a "universal state" without any obvious broken-down civilization for it to rally—here emerge for the first time, though in a significantly different form from that which appears in *A Study of History*. The difference lay in this: in 1920, Toynbee still thought of the "east" as an undifferentiated opposite to and partner of the "west." By the time he wrote *A Study of History* he had discerned multiple civilizations within both "east" and "west." This allowed him to provide the Caliphate with a "Syriac" civilization, which, however, broke down and went "underground" for a thousand years before achieving its "universal state." What alone made such fanciful interpretation plausible was the assumption that civilizations pursued essentially independent, insulated careers, with meaningful interaction between them limited to specific occasions and forms—renaissances and what Toynbee was to call "affiliation."

This idea that civilizations were essentially separate and in normal circumstances were incapable of meaningful communication was, I suggest, the central idea that Toynbee eventually borrowed from Spengler. To be sure, he altered Spenglerian dogma by recognizing two specific forms of interaction between different civilizations—affiliation and renaissances. But with this modification, Toynbee's mature view, set forth in *A Study of History*, agreed with that of his German predecessor.

This, surely, was Toynbee's greatest mistake, since peoples and civilizations surely *do* interact, and, within limits set by communications, have interacted continuously ever since significant divergences first arose among human cultures. Teggart had pointed to this obvious fact, and proclaimed it the key to human progress. But Toynbee had found no way to conform to this part of Teggart's program by writing a worthy sequel to Herodotus' epic account of the ancient Greeks' encounters with the peoples of the east. He simply could not discern any satisfactory coherence amid the confusion of detail that he had mastered.

By falling back on the organic analogy and treating civilizations as essentially separate entities Toynbee simplified his self-appointed task of making human history understandable in its entirety. Nonetheless, the cost was very great, for by systematically disregarding the tangled reality of cultural interactions in times and places that did not fit as either renaissances or affiliations, Toynbee introduced a defect into his portrait of the past that his enormous erudition could disguise but not repair.

Toynbee was not converted to Spengler's easy recipe for making civilizations comparable all at once. What tipped the balance for him and made it possible for him to design *A Study of History* on the hypothesis of the separateness of civilizations was his experience in Anatolia in 1921, when he witnessed Greek–

Turkish cultural and military collision firsthand and encountered in person the sort of atrocities he had described at length five years earlier in *The Treatment of the Armenians in the Ottoman Empire*.

As far as Toynbee was concerned, these experiences clinched the case for the separateness of civilizations and the perniciousness of cultural borrowing because he attributed the behavior he witnessed to the breakdown of traditional civilized moral codes among Greeks and Turks. What made the struggle so destructive, according to Toynbee, was the desperate effort Greeks and Turks were both making to fashion nation states on the west European model. Aping the West in this way required the repudiation of an older and distinctively Near Eastern mode of political and social order which had sustained a complex local intermingling of different peoples and religions. Such indiscriminate and destructive borrowing, Toynbee concluded, signified the death rattle of the in- digenous, autonomous civilized traditions that had once regulated Greek and Turkish societies.

From his new point of view, the immoral violence he witnessed simply reg- istered the collapse of the moral restraints which had been at the heart of those dying civilizations, and was not, as he had assumed when writing about the Armenian atrocities, an indelible Turkish characteristic. The fact that Greeks perpetrated atrocities like those perpetrated by Turks proved that something common to both peoples was at work, erasing old constraints and permitting the naked manifestation of brute human nature.

His trip to Greece and Anatolia in 1921 thus provoked in Toynbee a consid- erable reorganization of his older ideas, and he approached what he saw with Spengler's doctrines about the impermeability of living *"Kulturen"* hovering in the back of his mind. What was in the forefront of his attention, however, when he visited the battle zones of Anatolia, was the memory of his failure to influence the peace settlement with Turkey, and Lloyd George's disregard for the expert advice that he and others had proffered. In particular, Toynbee bore the Prime Minister a grudge for disregarding his arguments against assigning any part of Anatolia to Greece. Instead, Lloyd George had found it convenient to persuade the Supreme Allied Council to invite a Greek expeditionary force to land at Smyrna. The Greeks were eager to comply and occupied the city and its immediate hinterland in May 1919, though not without provoking some riotous violence at the time of their landing.

A year later, in May 1920, the Allies occupied Constantinople and its im- mediate environs, setting up a Supreme Allied Council on which the British, French, Italian, American, and Greek governments all were represented. What occasioned this action was the rise of a Turkish nationalist movement, led by Mustapha Kemal. The Greek landing at Smyrna triggered Kemal's defiance of the Ottoman government in Constantinople, which did not dare to oppose the Allies. From the security of the interior, where the Allies' arms could not reach, Kemal and his associates set out to resist plans for partitioning Anatolia, and, in particular, opposed Armenian claims to sovereignty over parts of what is now eastern Turkey along with Greek claims to the area around Smyrna in the west. The Allied occupation of Constantinople, of course, reduced the Sultan's government to the status of a mere puppet regime, and it rapidly lost influence among Turks of every political persuasion. The Nationalists, on the contrary, set up a provisional government at Ankara in April 1920, and were gradually able to rebuild and re-equip a Turkish army.

Meanwhile, at the Peace Conference, definition of terms to be offered to the Ottoman government dragged along, owing to disputes among the Allies about what to do with Constantinople, and how to apportion the vast territories from the coasts of the Black Sea to the Persian Gulf, all of which were in the throes of rebellion and civil war. By the time the Allies agreed on terms, and Ottoman plenipotentiaries actually signed the resulting Treaty of Sèvres (August 1920), the Sultan's government and the thinly scattered British, French, and Italian garrisons that remained on Ottoman soil were no longer capable of enforcing its terms. Only the Greeks seemed able and willing to try. Lloyd George therefore called on them to launch an offensive against the Turkish Nationalists.

In the summer of 1920 the Greek army was able to extend its bridgehead into Anatolia almost at will, but the Nationalists were not much weakened by losing territory they had not really controlled before the Greek advance. Instead they concentrated on the east, where a deal with the Bolsheviks resulted in the collapse of independent Armenia before the end of the year. Simultaneously, the Bolsheviks were able to consolidate their control of southern Russia but failed to revolutionize Poland. Meanwhile, French and British diplomats clashed over the frontiers to be assigned to their newfangled League of Nations mandates in Palestine, Syria, and Iraq, and the Italians decided to withdraw from southwestern Anatolia because of troubles at home. As a result, by the end of 1920 the British government had lost all its ostensible allies in the Near East except the Greeks, and found itself dependent on Greek military resources in its effort to enforce the Treaty of Sèvres against the Turkish Nationalists.

Toynbee watched the unfolding of these events with intense interest. He wrote a stream of articles about the Near East, which appeared chiefly in a periodical entitled *New Europe* but also in *Contemporary Review, Review of Reviews,* and elsewhere. As Lloyd George and his government got into deeper and deeper trouble by trying to enforce the terms of the Treaty of Sèvres, Toynbee found it hard not to gloat at the price being paid for the neglect of his expert advice. More to the point, he believed that if he could help to bring a just and lasting peace to the Near East by dint of a journalistic campaign that would inform British opinion of the errors of Lloyd George's policy, then his dubious role during the Great War would achieve a sort of retroactive justification. That required leave from his professorial duties, which the Senate of the University of London generously granted without interruption of salary "for a period not exceeding two terms as from 1 January 1921 in order to . . . proceed to Greece."[39]

Despite the university's official language, Toynbee appeared on the scene of the Greek–Turkish war not as a professor, but as a correspondent of the *Manchester Guardian,* thanks to a deal he made with C.P. Scott, the *Guardian's* famous editor, which guaranteed him traveling expenses of £10 a week and a fee of 4 guineas per column for letters and telegrams. This arrangement allowed Toynbee to bring Rosalind along for part of the trip. When he got back, with characteristic precision (and satisfaction) he calculated that they had made a profit, since between 7 January 1921, when he left London, and 21 September, when he got back, he had spent a grand total of £259-1-10½, which was exactly £30-17-4½ less than the *Manchester Guardian* had paid him.[40]

Toynbee went first to Greece, where a new royalist government had come to power in December 1920. He interviewed the Prime Minister and other leading politicians, and, incidentally, tried to get the Greek government to pay a subvention to the University of London for the Koraes Chair—a subvention which had been promised by former Prime Minister Eleftherios Venizelos, and which his political enemies in the new government were reluctant to honor. He then went to Smyrna and made several trips into the hinterland, staying within Greek lines and traveling as an honored guest of the Greek military.

Toynbee's first impressions were favorable to the Greeks. "I have been bowled over by the Greek officers," he wrote to relatives at home. "I think they have been enormously influenced by the years they spent on the Macedonian front in contact with the British."[41] To his mother he wrote: "I must say I like the Greeks on renewing acquaintance. (I am, of course, charmed, as everybody is, by the Turks.) I somehow think they are different from what they were nine years ago—not nearly so Levantine. But maybe it is wrong to judge by the army."[42]

Greek civilians, too, attracted his sympathy. "Yesterday I had the greatest reception I shall ever have in my life," he wrote (mistakenly as it turned out). "I motored over to a town called Kula . . . and they had got it into their heads that I was a sort of emanation of H.M.G. [His Majesty's Government]. When we got within about a quarter of a mile, we saw crowds coming out with Union Jacks and Greek flags, and I was marched through the town with a procession of school children behind me, the priest on one side, the chief merchant on the other, and everyone shouting 'Zoe o Anglia' [Long Live England]. . . . The Turks came out of their houses and looked on. . . . In the afternoon a deputation of about six Turks waited on me and I managed to carry on a Turkish conversation. Poor Greeks of Kula! They are now just inside the Greek lines, and they seemed to think that I could keep them there."[43]

But Toynbee was aware that travel under the official auspices of an occupying army would not give him a reliable picture of how the Turkish population felt about the situation. After being treated by the Greeks "as a sort of cross between the Prince of Wales and H.M.G.," he reported from Smyrna, "I am making connections with the Turks. I simply must see their side before I go to Constantinople."[44] A Canadian named Alexander MacLachlan, president of International College, Smyrna, put him in touch with some leading Turkish professional men of the city, but they were very guarded in conversations with Toynbee, judging him to be a partisan of the Greek cause. When Rosalind joined him in Constantinople in mid-March, therefore, Toynbee had yet to establish any sort of easy or open relation with representatives of the Turkish

majority under Greek control, and remained favorably impressed by much of
what he had seen of the Greek occupying forces. His articles in the *Manchester
Guardian* faithfully reflected these impressions.

Early in April he visited the northern part of the front where a Greek offensive had just failed to achieve its objectives. He saw the battlefield, still littered
with corpses, and then walked some forty miles with the retreating Greeks,
whose courage and discipline he admired.[45] But Toynbee was still concerned
to see the Turkish side of the quarrel. He began to do so, with a vengeance,
when in May 1921 he and Rosalind accepted an invitation to go on a trip
organized by the Red Crescent, the Turkish equivalent of the Red Cross, to
pick up refugees from a little town on the easternmost shores of the Sea of
Marmara called Yalova.

As Toynbee later figured things out, after the limited offensive of March and
April had failed, Greek military plans for the summer called for evacuation of
some of the territory they had occupied around the Sea of Marmara in order to
concentrate their forces on a new, deeper thrust against Ankara. To protect
their flanks from harassment, Greek military authorities then encouraged irreg-
ular bands of armed men to attack and destroy Turkish populations of the region they proposed to abandon. By the time the Red Crescent vessel arrived at
Yalova from Constantinople in the last week of May, fourteen out of sixteen
villages in that town's immediate hinterland had been destroyed, and there were
only 1500 survivors from the 7000 Moslems who had been living in these
communities. The little vessel took 320 terrified refugees to Constantinople,
mainly women and children, but only after a two day struggle with the Greek
captain in charge of the town.

The Toynbees were angered and appalled at what they found at Yalova.
Rosalind wrote a lengthy and detailed account of their whole expedition, which
she sent back to her mother with instructions to have the letter reproduced and
circulated among influential Liberals and anyone else who might be useful in
mobilizing sentiment against Britain's acquiescence in such barbarity. As she
explained: "The official British policy is quite evidently to shield the Greeks
in every possible way . . . partly from the confused idea that they must be
better really than the Turks . . . and to avoid being left face to face with
Kemal."[46]

Rosalind had nothing but disdain for the British diplomat with whom the
Toynbees had dealt on their return from Yalova, and her portrait of the Greeks
was devastatating. "The Greek officer in charge was a Cretan captain, more
like a stage brigand than I could have believed possible—a face so full of
undisguised and uncontrolled malignity I have never seen, and hope I may
never again. He was absolutely furious at the whole proceeding and swore that
there were no refugees. . . . From time to time they [the captain and a lieu-
tenant under him] kept asserting, with faces distorted with rage, that they were
Allies, officers and civilised men." Her sympathy for the victims was unambiguous. "The Turkish women of that district still dress like the Virgin Mary,
long blue draperies and white veils round their faces, exactly like the typical

Italian Madonna, and there they sat, that second day, several hundred of them, white with terror, extraordinarily still and quiet . . . and all around them surged this crowd of diabolical 'Christians' threatening and jeering.''[47]

In the following two weeks, the Toynbees accompanied two other Red Crescent voyages, bringing other Turkish refugees from the eastern shores of the Sea of Marmara to the safety of Constantinople. But these subsequent encounters were far less traumatic, perhaps, Rosalind surmised, because the Greeks had been "frightened into behaving decently."[48] Toynbee had, indeed, telegraphed the Greek Commander-in-Chief, demanding that he order the Greek captain at Yalova to let the refugees go; and on getting back to Constantinople he had demanded of the British and other Allied authorities that they do something to protect the Turkish villagers from their assailants. He got no solid satisfaction from Allied or Greek authorities, but in ensuing weeks his actions on behalf of the Turks did open doors for him among the Turkish community in Constantinople and also in Smyrna, when he returned to that city for a few days in August.

Toynbee's reactions were a shade less imperious than those of his wife. "I have got an almost intolerable horror of the whole business," he explained to his mother. "Do you know I have personally saved 700 people?" But his ambitions for doing good did not stop there. "Here I am now about the only Englishman who is both in the confidence of the Turks and in touch with the FO [Foreign Office], and peace between London and Angora [Ankara] is the key to general peace out here, and to the end of this massacre, devastation and economic ruin. If I can do the slightest bit towards bringing this about, it is up to me to put everything else aside—as in fact I am doing."[49]

Toynbee's mode of striving to achieve this goal was twofold. On the one hand, his dispatches to the *Manchester Guardian* set forth full details of what he and Rosalind had witnessed, though in cooler and more controlled language than what she had used privately. The news shocked a great many people in Britain. C.P. Scott, the editor, had been brought up, like other Liberals, on Gladstone's denunciation of the Turks, dating back to the 1870s, and the well-documented Armenian horrors of 1915–1916 had reinforced the image of Turkish brutality. When Toynbee's first dispatches about Yalova reached Scott, he wrote back: "Your estimate of the situation puzzles me extremely. I have seen no evidence whatever that the Greek excesses approached in scale those of the Turks."[50] Nonetheless, he published Toynbee's articles in full, and a week later he wrote: "I was of course unaware of . . . the extent and character of Greek excesses."[51] He was, in effect, won over to Toynbee's view that the Treaty of Sèvres was unjust and unenforceable, and soon came to agree with his correspondent that a substitute for Lloyd George's disastrous foreign policy in the Near East was needed if Britain were to salvage both its honor and its interests in the region. The weight of the *Manchester Guardian* was therefore lent to Toynbee's personal campaign against the Prime Minister from June 1921 onwards.

Toynbee also had friends inside the Foreign Office, in particular Eric Forbes

Adams, with whom he corresponded in private, relaying peace terms proposed to him by Turkish spokesmen, and urging action. In July 1921 Adams was politely noncommittal with respect to Toynbee's suggestions of how to make peace with the Nationalists, but six months later he wrote: "I am sure the general lines are right and as I said they coincide more or less with the general lines on which we are working." [52] But whether or not he personally agreed with Toynbee's suggestions, Adams was quite unable to alter government policy. Lloyd George remained in office; the Greeks gathered their forces for a second lunge toward the interior of Anatolia, and the war continued. As before, Toynbee's advice about how to conduct British Near Eastern policy fell on barren ground.

The war reached its climax just as Toynbee was preparing to return to London. In July 1921 the Greek army started its advance on Ankara, but on 8 September the Turks counterattacked, winning a decisive victory on the banks of the Sakkaria River. A week later the Greek army began a general withdrawal, and the campaign ended twelve months later, when the defeated Greeks were compelled to evacuate Smyrna. Meanwhile, the Turks had adopted a policy of destroying Christian settlements and populations as they advanced, so a horde of desperate refugees accompanied and preceded the remnants of the Greek army when it abandoned Smyrna in flames in September 1922.

After this debacle, Lloyd George had no allies left who would or could defend Constantinople against the victorious Turkish Nationalists. British resources on the spot were inadequate and reinforcements unavailable. The Allied military administration of Constantinople (since 1920 really only British) therefore came to an ignominious end, and the city and its European hinterland came under Turkish sovereignty once again. The only concession to British interests that Lloyd George salvaged was a Turkish promise to demilitarize the Straits between the Aegean and Black seas. It was an enormous victory for Mustapha Kemal and the Turkish Nationalists and a defeat for British policy in the Near East. Everything, in fact, had turned out as Toynbee had predicted. Turkish rights to self-determination had been vindicated on the field of battle and Lloyd George had been proven wrong in neglecting Toynbee's advice.

All the same, Toynbee returned home to face an awkward situation as far as his professorship was concerned. His dispatches from Constantinople exposing Greek atrocities were not what Greeks in London expected from the Koraes Professor. Realizing this, Toynbee had offered to resign his chair when first he began to send back stories of Greek crimes. But initially his immediate academic superior, Ernest Barker, the principal of King's College, was reassuring. "What a horrible story is told in the document I return to you," he wrote. "I trust you to use your own judgement about anonymity. . . . I do not bother about the Greek government, or what it thinks of you. I know you are simply concerned to do right and to tell the truth." [53]

As for the Greeks of London, they were not organized to attack Toynbee effectively in 1921–1922. The committee representing the donors had not met since 1919, and some of the leading London Greeks were so bitterly opposed to the restored royalist government as almost to welcome anything that would discredit it. For the time being, therefore, Toynbee kept his chair despite the way he had publicly attacked the policies of the Greek government and the Greek military administration of Anatolia.

The human depravity he had seen in action at Yalova and elsewhere in Anatolia bit deeply into Toynbee's consciousness. He had spent years during the war chronicling massacres committed by Moslems against Christians. Now he had seen Christian massacres of Moslems, smaller in scale but no less lethal, no less deliberate, no less brutal. He had been aware of the negative side of cultural borrowing since 1911 when he first encountered westernized Greeks and learned to detest "all those strange hybrid characteristics produced by the jumbling of their civilisations with ours." Even before Yalova, as between Greeks and Turks, he judged there was "very little difference between them. . . . I feel pretty clear that the Near East Christian isn't superior to the Moslem, though again I do not think he is inferior. . . . Such progress as both have made lately is nearly all due to the influence of a finer civilisation— ours—and the Moslems have a greater barrier to get over before they can learn from us." But, though baffling, the Turks had "something in them positive and different from what is in us."[54] This tentative appreciation of Turkish "difference" was certainly reinforced by his subsequent experience of Christian barbarism.

Moreover, three other motives pushed him toward rapprochement with the Turks. One was a matter of realpolitik. Toynbee was convinced that only by coming to an understanding with the awakening Moslem world could the British Empire thrive in the postwar world. This had been the central thrust of his expert advice to the government during the last two years of the war, and subsequent events seemed only to prove the accuracy of his estimate of the future.

But Toynbee's inner feelings went far beyond cool headed and cold hearted calculation of national and imperial interests. He could not help feeling profoundly grateful to the Turkish Nationalists. Their victory belatedly vindicated his otherwise ambiguous role during World War I. He now could justify his failure to enlist, because, thanks to the Turkish Nationalists, he was going to be able to contribute to the establishment of a viable peace, despite Lloyd George's obduracy.

A third factor affecting Toynbee's outlook was his growing repentance for the lopsidedness of his presentation of the Armenian atrocities in 1915. Though all his facts were well attested, he now saw that he had painted only part of the picture in 1915—faithful to a superficial level of truth, perhaps, but disregarding deeper, greater truths. He had sins to atone for and wished, vehemently, to set about doing so. The result was a book, written at top speed on his return from Constantinople, entitled *The Western Question in Greece and Turkey: A Study in the Contact of Civilisations.*[55]

As the subtitle suggests, Toynbee used his old notion of a difference of civilizations to explain what was happening in the Near East—but he did so in a new way. First of all, Greeks were credited with a civilization of their own, different from that of the Turks. He called it Near Eastern (quondam Byzantine) and the Turks' he termed Middle Eastern (identical with Moslem). Both these civilizations, he declared, were in process of dissolution owing to the superior power and attraction of the West. Both were quite literally demoralized by that fact, eager to borrow, open to influence, and yet unable to master the alien civilization they aspired to. In a sense, all that happened was therefore due to the West's presence and preeminence. The Eastern Question, which had haunted European diplomacy since the 1770s, was, Toynbee argued, really a Western Question—the result of encounters between three different civilizations, and the disastrous breakdown of the two weaker partners.

From this angle of vision, the "true diagnosis of the atrocities might be that they were a prolonged epidemic to which the Near and Middle Eastern societies were subject from the time when they lost their indigenous civilisations until they became acclimatised to the intrusive influences of the West." But permeability to the West was itself a sign of decay. "So long as a civilisation is fulfilling its potentialities and developing in accordance with its genius, it is a universe in itself. Impressions from outside distract it without bringing it inspiration, and it therefore excludes them as far as possible from its consciousness. But no civilisation has yet found the secret of eternal youth, still less of immortality. Sooner or later, they are each overtaken by some irreparable catastrophe, which not only cuts short their growth but strangely transmutes their essence. The steel, formerly so hard and clear, becomes soft and rusty. It is a tragic transformation. Yet this rust, which in the craftsman's eye is a foul accretion, is revealed to the scientific vision as a subtle compound, in which unlike elements, miraculously blended, acquire properties foreign to each of them before their union." [56]

This eloquent metaphor shows how he combined the tragic view of civilizational history, which he had developed so persuasively in his Oxford lecture in the spring of 1920, with a Spenglerian view of the impermeability of civilizations to outside influences as long as they were experiencing growth and fitted the needs of their bearers. That assumption made what he had just witnessed in Anatolia understandable. Even more significantly, it made a new approach to the problem of anatomizing the whole record of human history feasible. Instead of butting his head against a tangle of detail with endless cross-currents among persons, localities, regions, and continents, Toynbee could now hope to make sense of it all by focusing on a limited number of separate and autonomous civilizations. He was in a position to test the hypothesis, advanced tentatively in his Oxford lecture, that "the great civilisations that have been created by the spirit of man may all reveal the same plot, if we analyze them rightly." [57]

Mulling things over on the train as he traveled back from Constantinople, he subsequently remembered being able to jot down "a dozen headings which turned out to be the subjects of the principal divisions of my future book." [58] No simple jotting of a dozen headings survives, but he preserved a twelve page outline, much worked over, which bears the inscription on the top of the first page "Drafted in Orient Express, September 1921." [59] This presumably is, or more likely derives from, the outline Toynbee made when returning from Constantinople. Yet its main headings do not resemble the outline of the final work. The entry developed at greatest length deals with civilizations and civilized institutions as machines. Toynbee's youthful flirtation with Bergson was still dominating his thought, to judge from such passages as these: "Without mechanisation, progress in control by life is impossible, but with each successive stage in mechanisation, control becomes more precarious, and the machine may run away with the pilot. . . . This is the tragedy of it." He continued: "Civilisation is an attempt by greater mechanisation to develop from man to superman. . . . Tragedy of civilisa[tion is] that no civilisa[tion] has succeeded in raising human nature to permanently higher level."[60] Only traces of these ideas got into A Study of History when it appeared between 1934 and 1954, and mechanization did not survive as a main heading of the mature work. This is scarcely surprising, for Toynbee had much to learn and some ideas to dismiss before he was ready to write.

Nevertheless, by September 1921 he had solved a problem that had baffled him previously. He had found a way to proceed toward writing his great book. Proof of this lies in a single small and undeveloped heading, "(iii) Comparison of Civilisations," that contains the germ of the twelve volumes of A Study of History. It lists, in very brief form, the following stages of civilization: birth, differentiation, expansion, breakdown, empire, universal religion, and interregnum. The terms are not quite the same as those he eventually used, but Toynbee's mature vision of the pattern of civilized history is here set forth in sketchy but clearly recognizable form, and the idea of making comparison across all known and knowable civilizations is implicit in this entry in the outline.

In the years that followed, this insouciant little item (less than a quarter of a page out of twelve pages in toto) expanded in practice to take over from nearly all the other themes that Toynbee had intended to explore in 1921. It did so because Toynbee devoted all his spare time during the next decade of his life to mastering the histories of China, Japan, India, and other parts of the earth, and in doing so "Comparison of Civilisations" naturally and inevitably expanded to fill his entire conscious horizon. His new task was certainly formidable, but less daunting than before since now he knew what to look for. Empires, barbarian invasions, and universal religions, characteristic of the final stages of a civilization's career, were the conspicuous signposts that would allow him to identify each of the separate civilizations that had ever existed on the face of the earth. As he followed this path "Comparison of Civilisations," originally only a small part of his plan, proliferated and expanded, incorporating some elements from the other headings of his initial outline and squeezing others out entirely, making A Study of History what it eventually became—less a philosophy and more a compendium of historical parallels and resemblances.

Thus, though the great book continued to alter and evolve in the following years, so that its thrust and proportions changed very markedly between 1921 and 1934—not to mention 1954—Toynbee's recollection of having set down its "principal headings" in the train coming back from Constantinople in September 1921, though wrong in fact, is right in principle. He had solved the organizational problem that had been haunting him for years, and now knew what he had to do to achieve his ambition. It was a notable breakthrough and he remembered it as such, while forgetting how much further his thinking evolved before the book actually achieved its final form.

The trip to Anatolia was significant for the Toynbees in another way. While they were in Smyrna in early August 1921, Rosalind's grandmother, the Countess of Carlisle, died. Rosalind hurried back and reported to her husband: "Grandmother was cremated at Golder's Green . . . and there have been great family councils over the will. Castle Howard and enough of the estate to support it were left to Mother and Dad, but they have passed it on to Geoffry [the Countess's only surviving son], as they feel it more suitable so, which I suppose it really is, though I can't help some lingering regret."[61] Her dream of someday inheriting Castle Howard and all its grandeurs had thus suffered a final blow, but she consoled herself with the thought that a house and farm nearby, known as Ganthorpe, would remain in the Murray's possession.

Indeed it seems clear that Lady Mary laid claim to Ganthorpe only to please her daughter, whose attachment to Castle Howard and all it stood for was no secret. Ganthorpe was nothing to compare with Castle Howard, but it *was* next door. Relatives living there could expect to have easy access to the splendors of the Howard inheritance without the responsibility of having to maintain it. Rosalind could therefore assure her husband that "Ganthorpe will be a delightful home. I do hope very much they will live in it and not let it, but that is not settled. I feel happy about it all. It does mean that our links with Castle Howard will not be broken a bit and Geoffry will certainly let any of us go anywhere in his part of the land—including the Long Gallery—and we will always be able to stay somewhere near."[62]

Working out details of how to divide the Countess's numerous estates took time, and it was January 1922 before Gilbert Murray could tell Rosalind "We succeeded in provisionally dividing up the property."[63] Yet two months later he asked: "Now do you really think that you specially want to live—and farm— at Ganthorpe? Is it not really Castle Howard that fascinates you? And would not that rather uninteresting house, and the two farms with it, be rather a bore, if not Geoffry but some strangers were living at Castle Howard? I ask because it would apparently be rather more convenient—not more than that—if Geoffry had all the land unbroken. We should, of course, have compensation elsewhere."[64] But Rosalind was not to be shaken, and the Murrays accordingly inherited Ganthorpe when the estate was finally distributed. They did, however, decide to rent it, so Ganthorpe was not at first available to Rosalind. Nonetheless, Castle Howard, where she had spent a great deal of her childhood, where she had honeymooned, and where she and her husband and sons had so often come for holidays, remained part of Rosalind's world, even if full possessory rights were denied her.

Not surprisingly, Rosalind's attitude toward her Howard heritage put a considerable strain on Geoffry's wife. In 1925, when Rosalind and her sons were living for the summer in one of Castle Howard's numerous apartments, Rosalind reported to her husband: "We had a long and to me rather surprising talk. She burst out rather suddenly about the difficulties of their position here, how it wasn't a bit the same as if they had inherited the place. They never felt it really their own, and asked me if I minded Mother giving it up to them and what I felt about it all. She was almost crying, and I was able to answer quite frankly that I had had a first feeling of regret, but that thinking the whole thing over, I realised it would never have done for Mother, and that being so, thought Geoffry was the obvious person. She said she felt dreadful sometimes about my being here in this 'back place' when it ought really to be us who were in Castle Howard and not them, and so on. It all arose through their having sold the Sir Joshua [Reynolds painting of] Lady Crowder (for a very large sum, to an American) and [Rosalind's aunt] Dorothy being angry when she heard."[65]

The Murray family experienced other upheavals that affected Rosalind less directly. Her sister Agnes died in 1922, after a flamboyant and rebellious career. She was living with a Balkan ex-diplomat in a remote rural region of France when an inflamed appendix burst, and her doctors were unable to curb the resulting general infection. Rosalind was pregnant at the time, and remained at home, but Toynbee was conscripted to accompany Lady Mary to the scene, in hope of being able to help the dying Agnes.

Rosalind took a violently censorius attitude toward her sister's breach of sexual propriety in living openly with a man whom she had not married, and Lady Mary almost broke with Rosalind over the issue.[66] Gilbert Murray played the role of peacemaker, writing to his elder and only surviving daughter: "She had, amid all her escapades and naughtinesses, a wonderfully loving and generous nature, and she lit up our lives. . . . My dear love to the Creature: the only Creature left."[67] Agnes' death was matched by recurrent bouts of irresponsible behavior on the part of Gilbert Murray's two older sons, largely due to alcoholism. The comparative normality of the Toynbee household was therefore something of a comfort to the Murrays.

This was especially true for Gilbert Murray, who kept himself furiously busy with League of Nations affairs as well as with teaching and writing, but within the family circle focused his affections and hopes more and more on Rosalind, his longtime favorite. He praised her new novel to the skies and could not understand why publishers refused it. "The book made on me the impression of a sort of genius," he told her. "I mean that there were things of extraordinary beauty and poignancy, and one could not see how they were done."[68] Two years later, he wrote: "I am rather glad that Arnold has written to propose a guarantee" to Rosalind's prospective publisher,[69] but when the book, *The Happy Tree*, eventually appeared, he was "a little disappointed in the reviews."[70]

Less prejudiced critics judged very differently, as an entry in Virginia Woolf's diary shows: "Rosalind and Arnold appeared with a kitten and the manuscript of her new novel. She is a wisp of a woman, with eyes of a kind sensitive thoughtful nature, which can't, I am afraid, produce much in the way of art. She can't by any possibility write a long book she said; and she only made ten pounds by her last; and altogether she seems rather skilless and defenceless, though she is Gilbert Murray's daughter."[71]

The Happy Tree is, like Rosalind's other novels, rather thinly disguised autobiography. The narrator, at age forty (Rosalind turned thirty-six in the year when the book appeared), tells how she grew up with cousins, one of whom she loved, in a country mansion dimly reminiscent of Castle Howard, and then met with a series of disappointments during the war. In the last lines of the book, the heroine summarizes her life as follows: "I was happy when I was a child, and I married the wrong person and someone I loved dearly was killed in the war—that is all." [72]

It is hard to doubt that Rosalind was thinking of Arnold and of Rupert Brooke when she wrote that sentence. Autobiographical touches are indeed unmistakable: the heroine's husband, a professor of ancient epigraphy, works too hard at the Admiralty, is prickly about money, and tried to enlist but was found unfit; her grandmother refuses to pay for a nursemaid; and the heroine gives birth to three children! Yet of all of her novels, this is by far the most accomplished, coming, as it surely did, very much from the heart. [73] It proves, if proof be needed, that Rosalind's restlessness in her marriage persisted, even when the conditions of her daily life returned more nearly to normal after the war.

Assigning three children to her fictional heroine was entirely true to life, for Rosalind, too, gave birth to a third son, Lawrence, in December 1922. Six years younger than Philip, Lawrence soon usurped most of Rosalind's attention and affection since Tony, the eldest, was already away at school and Philip soon followed. The effect on Philip was traumatic. Being hustled off to school at age six, almost immediately after Lawrence's birth, seemed like an outright rejection by his mother. [74] Philip reacted by showing off in every way he could think of. When precocity (and Philip *was* precocious) failed, the obvious way to get attention was to misbehave. He accordingly fluctuated between almost frantic efforts to win his parents' admiration, and bouts of provocatively outrageous behavior. The pattern lasted all of his life, and made him a truly extraordinary person. [75]

Tony reacted differently to the new family circumstances, bottling his feelings up within himself. As Rosalind's love and attention focused more and more on Lawrence, he withdrew into a world of his own. But when Tony and Philip came home for school vacations, their contrasting temperaments mingled successfully enough in play. Philip's erratic conduct and noisy assertiveness did not prevent him from admiring his elder brother's courage, pride, and icy self-control. [76]

Rosalind continued to play the lead role in the family. As Philip wrote in 1961: "How could anyone stand firm against the vast resources of that will, a mind which was always made up at the first scent of opposition, and never opened to the least doubt again until she'd had her way with the world and her family?" And again: "I never heard her admit to a mistake." [77] Clearly, Rosalind shared many traits with her redoubtable grandmother; and, as we shall see in the next chapter, the difficulties with her sons that had shadowed the Countess of Carlisle's mature years were replicated in the Toynbee family as soon as the two older boys' adolescence set in.

Toynbee remained a marginal figure in the home. He was always busy with desk work and did not know how to play with his sons any more than he had been able to play with his schoolmates. He gladly accepted Rosalind's management of daily affairs, though in financial matters he did try to exercise a restraining hand. But here, too, he was ineffective, since despite his professorial salary, significantly supplemented by income from journalism[78] and Rosalind's income from investments, family finances continued to teeter on the brink of insolvency. Lady Mary Murray, having inherited her share of the Countess of Carlisle's estate, was always there to help out. She paid all of Tony's school expenses, for example, and from time to time did something to relieve her daughter's overdrafts at the bank.

Toynbee found financial dependency hard to endure and fretted over it endlessly. In riposte, Rosalind accused him, with some justice, of suffering from a psychological "complex" on the subject of money. "The remedy that I have always been hoping for," he wrote ruefully to Lady Mary, "is to keep down expenses and increase income till one gets the budget to balance normally year by year. I have always turned down going to a doctor about it, first because I am sceptical of psycho-analysts and secondly because I think the whole spirit of the age is to say 'ill' instead of 'inefficient.' But I rather begin to see that my complex being there—whether it has a real basis or not—I have no right to let it be a nuisance to other people—first to Rosalind and now you." Yet he was nothing if not stubborn, and after promising to seek medical treatment for his "complex," returned to the case in hand by declaring: "As to the £100, I am really sure it is better that we should pay it back . . . by taking installments off what you are paying for Tony."[79] And repay it he did, little by little, across the next few years.

Rosalind's sovereign disregard for financial constraints stood in sharp contrast to the rigor with which Toynbee's mother and sisters managed their affairs. In 1919 Edith Toynbee vehemently refused Toynbee's proffer of help in paying for his sisters' education, preferring to use capital the girls had inherited.[80] And as soon as his sister Jocelyn secured a salaried position, she wrote to say: "The time has now come when I am going to insist on paying the rent of this house. . . . It would be an immense relief to all three of us to feel that you are no longer forking out £50 a year for us."[81] His mother's financial independence was finally secured in 1925 when another inheritance allowed her to purchase the house at 5 Park Crescent in Oxford. She and Margaret, her youngest child, lived there until her death in 1939; Margaret remained in the house until 1986.[82]

Ironically, Toynbee could not enjoy the financial independence he so craved and which his mother and sisters attained on far smaller incomes. Rosalind's habits and outlook stood in his way. Rosalind presumably felt that she had a legitimate claim on family income that happened now to be concentrated in her mother's hands. Spending more than was in the bank seemed harmless enough, since tradesmen's bills could always wait until family resources were mobilized to pay up. She had none of Toynbee's acute repugnance for indebtedness; he, on the contrary, could never reconcile himself to needlessly, or at least heedlessly, spending more than they had in hand.

Friction over money, fading sensual attraction, and Lawrence's mounting monopoly of her affections all combined to weaken Rosalind's ties to Toynbee. She remained outwardly dutiful, but the inward reservations she had felt toward her husband from the very beginning became stronger and came nearer the surface as time went on. Toynbee, on the other hand, remained enthralled by his wife, but though he adored and obeyed her, he never achieved real partnership with her in family matters.

They did achieve professional partnership occasionally, as happened so conspicuously at Yalova. But Rosalind aspired to an independent literary career, and, as we have seen, used her novels as vehicles for expressing her rebellious moods. The fact that her writing failed to win much attention, whereas Toynbee's book *The Western Question in Greece and Turkey* was very widely reviewed, and soon required a second edition, may have bothered her but did not deter her from persisting in her literary ambitions. This meant, in turn, that she had very little time for active collaboration with her husband in any of his writing. He undoubtedly would have welcomed her help, but he respected her ambition and admired her novels, almost as uncritically, and rather more blindly, than Gilbert Murray did. "I am very much excited about her new book," he wrote to Lady Mary in 1926. "There is such mastery in it—she can make you feel and see quite exactly what she wants. Whether it be popular is another thing, but I know it is good."[83]

Their busy, increasingly separate lives took a new turn in 1923–1924 because of the controversy that swirled around Toynbee's continued advocacy of peace with the Turks. The committee of London Greeks who had raised the money to endow the Koraes Chair had retained a shadowy existence if only because the original terms of the endowment provided that the Subscribers' Committee should have the right to be consulted in case the chair fell vacant and should regularly be informed of the professor's activities—courses taught, number of students, and the like. No one had bothered to inform Toynbee of these provisos, or that the Subscribers' Committee continued to exist, and, in fact, the committee seems not to have met between 1919 and 23 January 1923, when four indignant members got together to see what could be done about the way Toynbee was betraying Greek interests. The upshot was a letter demanding an account of his activities, in accord with the terms of the endowment. This came as a total surprise both to Toynbee and to Ernest Barker, principal of King's College. Nevertheless, they complied, giving full details of his lectures and other academic work since 1919. The committee responded with criticism of both the quality and quantity of Toynbee's academic service, cast doubt on his linguistic command of modern Greek, and denounced his "propagandistic" activities, which, they declared, were bound to "affect even his status as an historian." The committee ended by declaring that Toynbee was morally obliged to choose between the Koraes Chair and his advocacy of Turkish causes.[84]

Toynbee deeply resented these aspersions on his competency and character and felt that to resign after such an attack would be an admission of wrongdoing. Besides, as he had confessed to Barker as early as 1920 while offering to resign his chair at any time Barker might ask him to do so, "I sincerely hope that the possibility is remote, for I won't conceal that it would be a consider-able disaster for me."[85] Financial disaster was what he had in mind, for his professorial salary of £600 was his principal income, and, as we have seen, money was always short in the Toynbee household.

In fact, Toynbee was not a success in the classroom. His really remarkable lectures on the historical geography of Greece, for example, attracted only two students, and his graduate seminar that was supposed to analyze traveler's re-ports about Greece was ill-conceived and futile. He did find time to prepare two elegant little volumes of excerpted translations from classical texts, *Greek Civilisation and Character; Self-revelation of Ancient Greek Society* (London, 1924) and *Greek Historical Thought from Homer to the Age of Heraclius* (Lon-don, 1924). But in the eyes of his critics, these small tributes to Greek antiquity scarcely counted against his advocacy of the Turkish cause in the Anatolian war.

The Subscribers' Committee attack was timed to coincide with Toynbee's return from a second journalistic trip to Anatolia. In August 1922 the editor of the *Manchester Guardian* proposed that he should "go out again and get right through to Angora,"[86] but the trip was put off until April 1923, when Toynbee did indeed succeed in getting to the Turkish Nationalist capital, interviewed leading figures of the government, and even dined with Mustapha Kemal, whom he found "undoubtedly a great man. . . . You would swear that he was an Austrian or a German. He is sympathetic but not amiable . . . a little like a leopard preparing to spring." Toynbee's dispatches to the *Manchester Guard-ian* outlined Nationalist peace terms and sketched the unfortunate way in which mistaken government policy had pitted Great Britain against the awakening Moslem peoples of the Near and Middle East. "I have been seeing Afghans and Indians here as well as Turks, and there is no doubt that they are getting up momentum (momentum of will power to be our equals) and that the whole brunt is going to fall on the British Empire."[87]

Toynbee was, indeed, hammering away at the theme he had first seized upon during the war, arguing that Great Britain must come to terms with the Moslem peoples' thrust for equality with European nations. As unofficial mediator be-tween the victorious Turkish Nationalists and the British public and govern-ment, he hoped to bring peace at last to the devastated war zone in Anatolia; and since long-range pacification between Moslems and Europeans was far more important than the sufferings of the Greeks and other Christians on the spot (however real those were), Toynbee was prepared to forget and forgive all the crimes Turkish soldiers had committed—all the more since he had come to think of such crimes as a function of civilizational break-up, and not, in any simple way, the fault of the perpetrators.

But since he had made so much of Turkish atrocities in 1915, and of Greek atrocities in 1921, his callous disregard of the sufferings of Greek refugees in 1922 seemed inexplicable to many English philhellenes, as well as to almost all Greeks. As Turkish crimes against Greeks and other Christians of Anatolia rose to a crescendo with the Nationalist advance upon Smyrna in the autumn of 1922, Toynbee's silence began to seem blatantly partisan. This, more than anything else, turned his critics into white-hot enemies. Toynbee, on the other hand, was driven from within by his ardent wish to vindicate his role in the Great War by bringing about a good peace in the part of the world that had become his special responsibility. That meant that he must help to reconcile Britain with the victorious Turks, and he was unwilling to say or do anything that might postpone or hinder such an outcome. He therefore felt that silence about Turkish atrocities was both prudent and necessary, however regrettable.

Toynbee realized how much he offended Greek and philhellenic opinion, and when his continued tenure of the Koraes Chair became problematic he explored the possibilities of getting an appointment in the British Foreign Service in Turkey or some other Moslem country of the Near East. But his confidant in the Foreign Office, Eric Forbes Adams, told him there was no chance of that, since "Lord Curzon . . . would, I fear, hardly regard you as a persona grata— for no good reason doubtless except that politicians don't like successful prophets." Moreover, "Lord Curzon also thought that you too quickly attacked H.M.G. after your government service." [88]

Fulltime journalism was a conceivable alternative but seemed distressingly uncertain when compared to the dependability of a monthly salary check. Nevertheless, when the Subscribers' Committee returned to the attack, Toynbee decided that he would have to submit his resignation anew, though in self-vindication he also wished the university to reject the endowment and refuse to appoint a successor on the ground that the Subscribers' Committee sought to exercise improper control over the opinions of the occupant of the post. Certainly the accusation against him was unabashedly political. The committee charged in a letter of 24 October 1923: "Professor Toynbee, having accepted a Greek chair, founded by Greeks, with a Greek endowment, and graced by the illustrious name of Koraes, has used most of his energies, resources and spare time to conduct a virulent and sustained attack on the Greek nation in its hour of utmost peril. He has also gone to Angora and engaged in close and friendly understanding with our enemies, to the propagation of whose interests he had devoted his enthusiastic endeavours." [89]

Opinion within the University of London ran very high. Some championed Toynbee's right to freedom of expression and resented the donors' effort to control professorial activity. Others felt that Toynbee had indeed abused his post and was damaging the interests of the university, which had funded several new chairs in the School of Slavonic Studies on the strength of grants from foreign governments. [90] In the end, the University Senate accepted Toynbee's resignation, effective 30 June 1924 (when his initial term of appointment expired in any case), and instead of rejecting the endowment (as Toynbee had wished), arranged for modification of the terms of grant to assure the independence of future holders of the chair.

Toynbee was not pleased and sought to vindicate himself in a letter to *The Times,* describing the terms attached to the Koraes endowment and underlining the fact that he had not known of them until the controversy broke out, and would not have accepted the position had he been informed of the Subscribers' Committee rights to intervene. The implication was, of course, that the arrangement constituted an improper infringement of academic freedom. Toynbee's letter therefore provoked continued newspaper controversy and an angry private letter from R.W. Seton-Watson, who occupied a chair at King's College funded by the Czechoslovak government. "I am horrified," he wrote to Toynbee, "at your sudden plunge into public controversy, and am bound to warn you that I am not alone in regarding your action as an open declaration of war. . . . You have shown an utter disregard for the interests of the University and the College and your own Chair in the future, and have cruelly compromised Barker."[91]

Toynbee at first wanted to compel his colleagues to agree that he had done nothing wrong, and in February he drafted a motion to come before the University Senate, as follows: "Do the Senate recognize that, having regard to the conditions attached to the foundation of the Chair, by which certain rights were assured to the Subscribers, Professor Toynbee is fully justified in deploring the circumstances which led to his acceptance of the Chair in 1919 without knowledge of those conditions?"[92] Gilbert Murray had acted as confidant to Toynbee throughout the controversy and had done what he could to cool tempers on both sides. He now prevailed on Toynbee to withdraw this motion, in return for a conciliatory private letter from the Vice Chancellor of the university. With that the controversy came to an end, but, of course, Toynbee was out of a job.

The Murrays tried hard to cushion the financial blow. "I have spoken with Rosalind's father," Lady Mary wrote to Toynbee as the controversy mounted, "and we agree that it is a world of pities that you should unnecessarily feel . . . worried now in the prime of your lives and wait until we are dead to be cozy. We cannot well give you £1500 a year now, for many reasons, but suppose we hand over the Ganthorpe estate to you—that is, above £230 per annum gross. . . . If you lose the professorship, as you know, we hope to make up £500. I hope this will suit your views."[93]

The Toynbees, nonetheless, refused Ganthorpe, presumably because Lady Mary was then facing unusual expenses for rehabilitation of her sons. "If, later on, things are easier for you in other ways," Toynbee replied, "so that you can better afford to spare Ganthorpe, let it come eventually as a reward for having done without it successfully, which would be much better for us. It makes all the difference having the inside of a year's notice, as I shall now anyway have, and I don't think there ought to be any question of our coming down on you for that interim £500 a year, which you so generously offered when my resignation seemed likely to date from last June and to catch me without warning."[94]

A few weeks later, Lady Mary renewed the offer of Ganthorpe, provoking a second letter: "As to Ganthorpe, you do put too many temptations in our way! For we are both in love with it, and if someday we were to have it, and perhaps in the end to live there, we would get great happiness from it. (We certainly should not sell it!) But, again, we shouldn't like to take it now, as a way out of our difficulties. . . . Later on it might be different."[95]

Toynbee's resignation attracted sufficient attention in the academic world and beyond that offers of alternative employment were not slow to appear. Several

American universities were interested in hiring him to teach classics and ancient history, for example,[96] and the Turks offered him a post in Constantinople. Gilbert Murray had schemes for finding him a staff job in Paris with the Committee on Intellectual Cooperation of the League of Nations, and Professor R.H. Tawney invited him to apply for a new post in international affairs at the London School of Economics.

Toynbee was strongly attracted to the position in Constantinople, where he would be able to observe firsthand what still seemed to him the pivotal interaction of Moslem and western peoples in the postwar world. Yet he was reluctant to cut himself and his family off from English society and civilization. As it turned out, he did not have to choose right away. Instead, Toynbee's wartime chief, J.W. Headlam-Morley, offered him an interim appointment with a struggling new institution, the British Institute of International Affairs. Toynbee's job was to write a book-length survey of international affairs since the Peace Conference.

This was a task close to Toynbee's heart, and it allowed him to put off any new academic commitment for a year. He much preferred writing to teaching anyway; and his awkwardness with colleagues, whether at Balliol, in the wartime Foreign Office, or at King's College, London, confirmed his distrust of collegiality. He had been a "loner" in school, and the job he took in 1924 allowed him to work alone again. He was responsible only to a review committee, headed by Headlam-Morley, who proposed to vet his work before publication to assure accuracy and scientific detachment. It was an arrangement that proved mutually satisfactory, and what had begun as a temporary one-year appointment was, in due season, converted into a career, lasting until Toynbee's retirement in 1954.

VI

Chatham House and Ganthorpe: A New Equilibrium

1924–1930

Twelve days after *The Times* published Toynbee's letter announcing his resignation from the Koraes Chair at the University of London, J.W. Headlam-Morley invited him to "discuss a matter of some importance"[1] over lunch. Headlam-Morley had been Toynbee's immediate superior in the Political Intelligence Department during World War I. Afterward he stayed on as historical adviser to the Foreign Office and once had asked Toynbee to write a background paper about Ottoman "Capitulations" in connection with a dispute with France about the rights of British subjects in Tunis.[2] He was also one of the founders and leading spirits of the British (later Royal) Institute of International Affairs, and it was in his capacity as Chairman of the Institute's Publications Committee that he invited Toynbee to lunch in January 1924.

The institute had originated in a meeting of British and American representatives at the Peace Conference on 30 May 1919 at the Hotel Majestic in Paris. The reigning notion among those present was that an informed public opinion about international affairs was the only way to make sure that future wars would not break out because of secret diplomacy and the irresponsibility of officials dealing with matters about which ordinary persons before the war knew next to nothing. Initially, the group projected an Anglo-American institute, whose first task would be to prepare and publish an authoritative history of the peace conference. An American banker, Thomas W. Lamont, provided £2000 to get that project under way, and editorial supervision was entrusted to the well-known English historian H.W.V. Temperley.

Long before Temperley's project came to completion, however, the Americans decided that they preferred a separate organization and rejected the British idea of a binational institute. Delegates returning from Paris in 1920 decided instead to join the already existing Council on Foreign Relations in New York; whereupon the British contingent, having lost their American colleagues, organized a matching British Institute of International Affairs. On 5 July 1920 the new British Institute opened for business by renting two rooms in Mallet Street, London from the Institute for Historical Research.

A private society, it depended entirely on membership subscriptions and gifts from donors. Lionel Curtis, a Liberal imperialist, and G.W. Gathorne-Hardy, a Conservative, became joint secretaries and nursed the fledgling institute through its first years. Lionel Curtis (1872–1955) had become a lecturer on imperial history at Oxford in 1912. As such, he was largely responsible for inventing the term and popularizing the idea of the British Commonwealth of Nations. He believed passionately that Britain could remain great only by nurturing its special ties with the Commonwealth countries and with the United States. Inspired by that idea, he set about raising funds throughout the English-speaking world on behalf of the new institute and met with significant success when a Canadian, R.W. Leonard, bought Chatham House,[3] on St. James's Square, London and gave it to the institute as a new and more adequate home. Substantial contributions also came from John D. Rockefeller in the United States, from Abe Bailey in England, and from corporate donors such as the Bank of England. By 1922 the institute had 714 subscribing members and had begun to offer weekly off-the-record talks by various British and foreign experts on matters of current international concern.[4]

In 1924, as the sixth and final volume of *The History of the Peace Conference of Paris*[5] neared completion, the question of how to follow up that substantial scholarly enterprise arose. The council guiding the institute proposed the publication of annual surveys of international affairs and by January 1924 saw its way to financing that enterprise by hiring someone to write a volume that would link events up to 1924 with what had been reported on in *The History of the Peace Conference* and follow that with a survey of the year 1924 itself. (Because the peace treaties were not ratified all at once, the volumes dealing with the Peace Conference broke off at different times for different countries, terminating the narrative of events anywhere between 1920 and 1923.) The council had £1000 to offer to the person who would undertake this task and was prepared also to supply secretarial assistance. In return it expected 700 pages of narrative and 300 pages of documents, and it prescribed urgent deadlines: 1 December 1924 for the "liaison" volume and 1 February 1925 for the survey of international affairs in 1924.[6] Headlam-Morley became chairman of a new Publications Sub-Committee, charged with the task of getting the proposed survey volumes into print.

This was the project Headlam-Morley broached to Toynbee over lunch in mid-January 1924. From Toynbee's point of view, such a job offered him a chance to renew his wartime concern with foreign policy and its proper management. The greatest drawback he foresaw was that the task of keeping up with international affairs in all parts of the world and recording what happened as authoritatively as available information permitted would leave him little or no time to earn extra income from journalism or to work on his philosophy of history. Yet Toynbee was determined to carry through that undertaking just as quickly as he could. Another obvious drawback was that after completing the assignment, he would again have to look for gainful employment; but from the beginning everyone hoped that the council might be able to make the enterprise permanent. All depended on how valuable the annual surveys might prove, and on how productive the council's efforts at fundraising might turn out to be.

Given the fact that his salary from the University of London was about to stop, Toynbee was eager to accept the offer Headlam-Morley made to him on behalf of the institute. He bargained for an extra £200, arguing that the assignment would require him to give up his accustomed journalism. He also asked that the deadline for the second volume be set back to 1 March 1925. Headlam-Morley agreed to these terms on 12 February,[7] and three days later Toynbee began work in the new job, even though his appointment as Koraes Professor did not expire until 30 June.[8]

Initially, it was unclear whether Toynbee's assignment was to solicit and then edit the contributions of others to the projected surveys (in which case he was to pay the actual writers from the £1200 the institute had allocated to the task), or whether he would write it all himself. When it became clear that he proposed to dispense with outside aid, the Publications Sub-Committee of the institute decided that the volumes would carry his name as author, but to avoid possible political complications two of the sub-committee members undertook to read through his manuscript so as to be able to iron out any delicate issues with Toynbee before publication.[9]

In due course, after a very strenuous ten months, on 11 December 1924 Toynbee was able to declare to his principal assistant: "Here is the Preface corrected";[10] a stout volume of 526 pages entitled *Survey of International Affairs, 1920–1923* duly appeared in the early summer of 1925, published by Oxford University Press. Simultaneously, *The World after the Peace Conference, being an Epilogue to the "History of the Peace Conference of Paris" and a Preface to the "Survey of International Affairs, 1920–1923"* was published separately because Toynbee's introduction had become so lengthy that it could not fit into the prescribed dimensions for the survey volume.

Successful completion of the first volume satisfied only half of the assignment. Toynbee therefore continued to write at a furious pace, with the result that the institute was able to publish a 528-page *Survey of International Affairs, 1924* early in 1926. The two main volumes were only slightly longer than the 500 pages prescribed by the institute's agreement with Oxford University Press, but they differed from the council's initial expectation inasmuch as the 300 pages reserved for documents shrank to a mere fourteen pages in the first volume and a slightly more substantial thirty-five pages in the second. Nonetheless, as an anonymous reviewer in *The Times* remarked: "The B.I.I.A. is to be congratulated on finding a writer capable and industrious enough to perform the task single handed."[11]

The task did indeed require phenomenal industry and rapidity of composition. Few writers can gather information and then produce over 1100 printed pages in fifteen months, as Toynbee did in this instance. Fewer still are capable of arranging the tumult of current events into perspicacious and coherent narrative. And no one but he could have adorned such a very contemporary narrative with historical comparisons ranging globally across time and space.

Yet as he was at pains to acknowledge in prefatory notes to both volumes, he owed a great deal to "the constant assistance and collaboration of the members and staff" of the institute itself, and had "been able to draw upon the fine collection of official documents and carefully classified press cuttings in the library. If this great fund of information had not been at his disposal it would have been impossible to undertake the work, but for this very reason is it equally impossible to make individual acknowledgements." [12]

This last phrase was not really true. Toynbee had wished to thank Veronica Boulter by name. She was the member of the institute staff initially assigned to provide him with secretarial assistance. As it turned out, she did far more than any ordinary secretary and acted instead as a full-fledged collaborator almost from the start. But Miss Boulter was modest and shy, and she may have felt presumptuous in exceeding her assigned secretarial role. At any rate, she demurred emphatically from Toynbee's plan to thank her in the preface, and he compromised with the sentence quoted above. [13] Nevertheless, their collaboration was vital to the success of the undertaking, and once commenced in this fashion it continued across an ever-expanding portion of Toynbee's activity until his death. Veronica Boulter soon ranked second only to Rosalind on Toynbee's horizon, and after his divorce from Rosalind, she became his wife in 1946. Toynbee's young and unpretentious assistant therefore deserves rather more of our attention that she herself would feel comfortable with.

After graduating with first class honours from Newnham College, Cambridge, Veronica Boulter became one of the very first employees of the new British Institute of International Affairs in 1920. Born on 23 March 1894, she grew up in rural Wiltshire, where her father was a Church of England rector. Plain living and high thinking presumably prevailed in the Reverend Sidney Boulter's rectory. At any rate, his daughter emerged with an unusual indifference to worldly things, a capacity for long hours at a desk that rivaled Toynbee's own extraordinary *Sitzfleisch,* and a quiet efficiency in attending to details, whether in reading proofs, compiling indices (one of her specialties), or getting information down on paper in a straightforward and accurate, if somewhat pedestrian, fashion. Her readiness to look things up, and persistence in getting them right, nicely complemented Toynbee's habit of work, for he often rushed ahead, organizing what he had to say in the light of an overarching vision, leaving gaps for details which he did not have accessible as he wrote. With Veronica to assist him, Toynbee could concentrate on writing rapidly and feel sure that any information his own remarkable memory did not provide would be supplied by her.

As their collaboration developed, Veronica came to occupy an emotional place in Toynbee's life resembling the role his mother and younger sister, Margaret, had occupied in his youth. For Veronica soon learned to admire and look up to Toynbee, respecting his learning, his powers of imagination and synthesis (far superior to hers), and his prodigious productivity. She competed with him in the sense of working just as hard or even harder than he, taking work home in the evenings, and minding the shop when he went traveling or took holidays with his family in the country. She always managed to keep up with the pace he set, by looking things up and then editing, correcting, and adjusting his text to get details right. Her work soon became a service of love, sexless, but infused with a powerful affection all the same.

One may surmise that Veronica, like his mother, began to find vicarious realization for her own ambitions in Toynbee's success. He gallantly and repeatedly sought to give her public credit for their joint accomplishment, but Veronica always held back, preferring anonymity and a private glow of satisfaction when any compliment came Toynbee's way. For his part, Toynbee soon became rather more dependent on Veronica than she was on him. He needed her help to achieve the goals his job and vaulting private ambition set before him, and he basked in her admiration. Toynbee very much needed to be admired—needed, in fact, female admiration. His mother and sisters had accustomed him to that condition in his youth; Rosalind's increasing alienation from him, less and less disguised in their daily encounters,[14] meant that he needed Veronica's support and companionship as a supplement and emollient to his intense, fiery relationship with Rosalind.

At first, Toynbee was eager to have Rosalind meet and appreciate Veronica, but Veronica turned down his invitations to visit them at Castle Howard in summertime,[15] realizing, perhaps, how out of place she would be in such aristocratic surroundings, and aware also, no doubt, of the ambiguity inherent in her complementarity to, and rivalry with, Rosalind in Toynbee's affections. The two women must have met in London, but presumably saw as little as possible of each other, and Toynbee soon learned to acquiesce in that arrangement. He needed and depended on both of them, achieving a precarious equilibrium by dividing his energies between home and office, punctuating duty in London with prolonged Easter, Christmas, and summer holidays at one or another of the Howard or Murray country establishments that remained available to Rosalind for visits.

The first *Survey* was well received and achieved a second printing in 1927. Headlam-Morley and a colleague on the Publications Sub-Committee had vetted Toynbee's manuscript to make sure he maintained the level of accuracy, detachment, and coverage they expected. There was some nervousness about possible charges of partisanship, since Toynbee's campaign against Lloyd George's Turkish policy was well known. G.M. Gathorne-Hardy explained in the preface to the first volume: "The Institute includes representatives of every school of thought and is therefore precluded under its constitution from expressing an opinion on any aspect of international affairs. These volumes are confined to facts. But facts cannot be stated within manageable space except by means of a careful selection, which may of course be influenced by the personal views of those by whom the selection is made. In issuing these volumes the responsibility of the Institute is therefore limited to the appointment of the writer and the provision of the necessary funds. The final responsibility for what they include or omit is assumed by the writer. In this respect the position of the Institute is analogous to that of the publishers." He also explained the council's idea in commissioning the surveys as follows: "Speeches and articles are the main factors moulding public opinion on foreign affairs.

They are usually prepared at short notice by political leaders and publicists, who thus discharge, under heavy pressure, a function of vital importance. Without such aids as the Institute is now trying to give them they are left to collect or verify their facts by reference to the files of newspapers. . . . The primary object of these publications is to enable speakers and writers to gather in the time available for their task the factual material, carefully checked, upon which to base the advice they offer the public.''[16]

Toynbee, of course, had been carefully instructed in this sense before he undertook the job, and the first volume therefore resembles a reference work, with occasional resort to tables listing times and places of diplomatic meetings and careful summary even of quite trivial international agreements. Commitment to the League of Nations and what may loosely be called a liberal policy is nonetheless apparent, if only by the prominence given to the "Organs of International Authority and their Proceedings" in the first *Survey* and to "Security and Disarmament" in the second.

On the other hand, the *Survey of International Affairs, 1924* bears a much more personal impress, being less encyclopedic and more analytical than its predecessor. Toynbee decided to organize his second volume around four themes and deliberately omitted other matters, reserving them for future treatment. In this way he loosened the constraint of his confinement to a single twelvemonth, and was able to offer his readers a far more comprehensible (but also more subjective) portrait of the times. After recording the year's deadlock over "Security and Disarmament" he gave pride of place to the United States and the Soviet Union, with a discussion of "The Movement of Population" since 1920 that featured the impact of new American immigration laws, followed by an even lengthier report on the Third International and the Soviet government's efforts to reestablish diplomatic relations with other states. Such distribution of emphasis meant leaving some things out, and in a prefatory note Toynbee promised to deal with some of the themes omitted from the 1924 *Survey* in a subsequent volume.

Yet at the time he wrote that prefatory note, in the spring or early summer of 1925, Toynbee was not sure he would be able to continue at the institute. After he had shown his mettle by completing the first volume on time, Lionel Curtis and other members of the council decided to try to keep him on by appointing him Director of Studies for the institute.[17] But there were no funds, and until a donor could be found the council could make no firm commitment.

Throughout the first months of 1925, therefore, Toynbee again canvassed alternatives. As before, he was drawn to Turkey and renewed conversations with the Turkish Ambassador in London about a possible professorial appointment in that country. In the end, family considerations kept him from accepting a job in Turkey, but, as he wrote to Lady Mary, he regretted it. "Turkey is the particular point at which I can handle this business of the relations between East and West. . . . Since I was about Tony's age I have very slowly been building up a knowledge of the Islamic world and Oriental languages,'' and refusing the Turks' invitation meant losing the chance of becoming "a professional instead of an amateur orientalist.''[18] Gilbert Murray tried again to find a post for Toynbee in Paris in connection with the League of Nations Institute of Intellectual Cooperation, but negotiations dragged on and financing depended on a French government grant, which was not forthcoming. America also beckoned, and Toynbee did in fact accept an invitation to take part in a summer conference at Williamstown, Massachusetts, devoted to putting American businessmen in touch with up-to-date thinking about international affairs. The conference was scheduled for August 1925. Toynbee planned to use this opportunity to visit his friend Rob Darbishire, and, as he wrote to Gilbert Murray: "I shall try, while I am in America, to get journalism—either for a journey in Turkey or in some wider field of international affairs which I can follow in England." [19]

Uncertainty about his future preyed on him, and in early March he had another, relatively mild, nervous collapse that confined him to bed for ten days. Offers of financial help from the Murrays were at best an ambivalent relief to Toynbee's worries, since, however enticing such unearned income might be, financial dependence on his in-laws was incompatible with Toynbee's sense of what a man should be. Something of his tangled state of mind emerges from the following letter.

9th March, 1925

Dear Murray,

I am grateful—more than I can say—for what you have done and the way you have done it. Rosalind has told you that . . . we have decided to accept your making up the deficit for the time between the end of the B.I.I.A. payments for these two volumes and the beginning of the Williamstown lectures—though we shall make the deficit as small as possible by letting the house as soon as the work for the B.I.I.A. is finished. She tells me, too, that you are going to give us Ganthorpe permanently. I had meant to take nothing more from you . . . but, again, on thinking it over—which I have had leisure to do in bed—I see that, while I have been making my budget balance, I have done it at a strain which has made me difficult to live with for Rosalind and the children, so that in a sense I have been passing the strain to them. Do you remember the Irishman who wanted to make his blanket larger, so he cut off one end and sewed it onto the other? I find I am often doing that. I think I am adding something by an extra effort, and then it turns out that I am taking it out of another side of my life or another person. I suppose that the law of the constant quantity of energy is true of life as well as of physics, but it is disconcerting, for the demands of life are not constant, but have to be met as they come. However, what I mean about Ganthorpe is that I shall try to use the very great relief of strain which this will give me in order to put back my piece of blanket at Rosalind's end again and make things easier and happier for her. . . .

Yours ever
Arnold [20]

Accepting income and gifts of property from the Murrays was no substitute for a salaried job, but Toynbee was reluctant to sign on with another employer as long as the chance of continuing at the British Institute of International Affairs remained open. He therefore refused to stand for a chair in Greek at

University College, even though Gilbert Murray had drawn his attention to the possibility there. "I don't want you to think that I mean to go on accepting your help indefinitely and still refuse openings because they are not exactly in my line," he wrote. "What I really want is to shoot my bolt and see if anything comes of it before I definitely lower my expectations. . . . The two first B.I.I.A. volumes will be published by about May, and by September I shall know whether they are going to be of immediate practical value or not. They may bring an endowment to the Institute. . . . I know I may be exaggerating the possible effect of the volumes. When one has put a lot of work into something, one is apt to measure possible results by one's own efforts—generally quite wrongly. Will you tell me whether you feel it reasonable to hold on until the autumn, or whether you think that the line I propose to take then [to seek other, less congenial work] I ought to take now?"[21] Murray presumably replied encouragingly, and a week later Toynbee wrote to him: "A small windfall came yesterday. President Lowell [of Harvard] has asked me to give eight lectures at the Lowell Institute for $800. I shall do this in October."[22]

When Toynbee sailed for New York in July, his future was still unsettled, but when he got to Williamstown, he found a telegram from Rosalind awaiting him with reassuring news. A Glasgow businessman and coal magnate, Sir Daniel Stevenson, had agreed to fund a chair at the University of London for Toynbee, with the understanding that he would divide his time between teaching international relations and preparing annual surveys for the institute. Toynbee lost no time in accepting the proffered appointment, reporting to his mother: "Well my luck has turned too. Sir Daniel Stevenson has made a double foundation of £500 a year for a professorship of international history plus £500 for a directorship of studies at the B.I.I.A. It was all in negotiation the last month before I left, but when I arrived here I found a cablegram from Rosalind to tell me that the University part of it had passed the Senate.[23] . . . So I am settled again, and in London. I still find it hard to realise, but that weight of uncertainty about the future, which has been hanging over me for two years, has gone. . . . I can look ahead again, and plan to write my philosophy of history at leisure, over a number of years."[24]

His future thus assured, Toynbee remained in the United States until October 1925 to give the Lowell Lectures. This was his first visit to America, and he found much to absorb. At the Institute of Politics in Williamstown he met a mix of American businessmen and academics selected for their interest in foreign affairs. "It is interesting to study the future lords of the world," he wrote to Gilbert Murray. "The United Fruit Company interested me most. He addressed us on American policy in the Caribbean, and was Nordic man at his best—or at least, his most typical: the noble savage."[25] From Williamstown he went to Kentucky to visit his friend Robert S. Darbishire, and during the ensuing month Darbishire introduced him to Appalachian hill folk, whose speech and mode of life still echoed the eighteenth century. "Have you ever eaten gray squirrels?" he wrote back. "They were the *pièce de résistance* at a feast given us . . . in the knobs, as the hill country is called here. While we ate them, the women waved branches over our heads to keep the flies off—quite like an Assyrian bas-relief."[26] In addition, Darbishire's collection of arrowheads picked up on the farm and a visit to salt licks where herds of buffalo had formerly relieved their craving for salt fired Toynbee's imagination with visions of the Indian hunters who had once occupied the land.

A stopover in New York, where he stayed with the Lamonts—the partner in the Morgan Bank who had helped finance the preparation of *The History of the Peace Conference of Paris*—allowed him to experience a very different style of American life; and while delivering the Lowell Lectures in Boston in October he stayed with Harvard's President Abbott Lawrence Lowell, who, as a Boston Brahman, represented yet another, contrasting, sort of American. By the time he got back home, ready to resume work on the surveys and to start his new duties at the University of London, Toynbee had built up a picture of America's variety that his later travels extended and confirmed. He had also made a number of personal connections with key figures of the Anglophile East Coast establishment who played an important part in his subsequent career. It was thus a very good summer, not least because it had insulated him from the family frictions that increasingly burdened his life.

In outward things, family circumstances improved. Shortly before Lawrence's birth in 1922, the Toynbees had bought a house at 3 Melina Place in St. Johns Wood, a very fashionable part of London. The two older boys, Tony and Philip, were away at school most of the time, and Rosalind was able to take pleasure in her youngest, Lawrence, because the main burden of looking after him fell upon a faithful nurse named Bridget Reddin. She was a pious Irishwoman who, after Lawrence went off to school in 1932, became Toynbee's typist at Chatham House. By then she had become a trusted family retainer, and sometimes even dared to mediate between Arnold and Rosalind, as we shall see.

Yet beneath the surface, family relations were increasingly strained, Rosalind must have been disappointed by the all too obvious failure of her literary career. Novels that had to be subsidized to achieve publication and when published sold only a few hundred copies[27] were not what she had imagined for herself when she began writing at age eighteen. Success as a hostess and keeper of an intellectual salon was an alternative career for which she was eminently well suited, but her husband's anxiety about extravagant spending and his reluctance to interrupt his work for parties were obstacles to social success that she could not overcome.

Yet Rosalind, having emphatically abandoned her postwar flirtation with the Labour Party, enjoyed aristocratic company and was both witty and vivacious. It is not surprising, therefore, that when Toynbee was away in America in the summer of 1925, her letters to him reflect a very busy social life. Soon after he departed, she reported gleefuly that the German ambassador had asked whether she were the wife of the "celebrated Mr. Toynbee";[28] and when the Queen and Princess Mary visited Castle Howard in August, she was on hand to act as assistant hostess. "I wish you could have seen the Toms [Tony and Philip] being presented—standing very proper and tidy and clean in the garden room . . . making excellent little bows as the Royalities shook hands with them."

The royal visitors planted trees to commemorate the occasion, a ceremony witnessed by "the farmer tenants, the clergy and a few other special people who
were lined up along the hedge." Afterward, the royal party withdrew to the
privacy of Castle Howard's drawing rooms, where, Rosalind reported, "I have
found that I was not at all shy and didn't feel awkward or wondering how to
behave. They [the Queen and Princess] were very friendly and simple and I
thoroughly enjoyed the experience."[29] But receiving royalty, going to the races,
and engaging in other social events left her little time for other pursuits. "I am
so sorry you have felt neglected in letters," she wrote to Toynbee just a week
after her account of the royal visit; she explained in the same letter: "I have
not had any time to do my novel and haven't quite finished your Turkish documents. . . . I shall have to try at the novel at Boothby."[30]

Rosalind sometimes found her older sons difficult. This was especially true
of nine year old Philip, who, for example, swallowed a pin shortly before
Toynbee left for America and had to be X-rayed. "It seems so typical of him
to make the most inconvenience and upset to everyone that we can't help feeling very much annoyed with him," was Rosalind's reaction.[31] And at the end
of the summer: "There has been a rather hectic end to our happy holiday. Poor
Philip of course—has developed impetigo, a skin disease commonly found in
the slums I believe—and very contagious—and I have had to keep him at home
an extra week. . . . I am feeling rather down at the moment, trying to pack
up, with Philip on top of me, and hating myself for being cross with him."[32]

On the other hand, she could also write: "America certainly sounds interesting and more attractive than I had supposed. Well, I expect I shall go with you
sometime—and in the meantime the boys are being very dear and good, and I
am enjoying being with them very much and feel sure it was right to stay."[33]
And she rejoiced in signs of Toynbee's success: "Look what a tremendous
review [of the *Survey* for 1924] in today's *Times*. It was on the middle page
with all the leaders and I never spotted it at all because it was so grand and
huge."[34]

By the time Toynbee returned to London in mid-November 1925 the older
boys were safely packed off to school. Rosalind was glad to have him back
with an assured career ahead. His new dignity as Director of Studies at the
British Institute of International Affairs (rechristened Royal in 1926) carried
with it an annual travel allowance of £153 and a salary of £500. The professorship at the University of London also paid £500, according to the settlement
Sir Daniel Stevenson had made. The aim of his endowment was "to foster a
spirit of international cooperation, peace and goodwill," but nothing was said
about how the holder of the two posts would divide his time between the two
institutions. Toynbee duly gave an inaugural lecture at the University of London early in 1926, claiming that by cultivating an international point of view
one could hope to escape from national bias, and declaring that his dual position with the university and the institute would keep him in touch with both
academic and world affairs—a combination, he declared, that was "ideal for
historical work and rare in our times."[35]

Yet as he explained defensively long afterwards, his first duty as Director of

Studies was to produce the annual surveys and, although "initially teaching was also expected at the University of London, experience quickly showed . . . that a single chairholder could not do both jobs simultaneously."[36] Nothing preserved in his papers explains how Toynbee managed to shed responsibility for teaching at the University of London. It was surely his own doing, and was formally ratified only in December 1928, when terms of agreement among the Royal Institute, London University, and Sir Daniel Stevenson were renegotiated.

Toynbee disliked teaching. His encounters with students at Balliol and London had not turned out well. Since the cutting edge of his own thought was always reaching beyond the bounds of what he already knew, having to give tutorials and lectures on familiar topics simply distracted him from what he wished to pursue—that is, the completion of his own education so as to be able to write the great book he had been projecting in one form or another ever since his undergraduate days. Students of international affairs at the University of London who wanted to work with Toynbee could always seek him out at Chatham House, and a very few did so.[37] But Toynbee never gave a course of lectures at the university. Instead, as he explained in 1953, he preferred to use the scant margin of leisure left after work on the surveys for his *Study of History*. This "has been resented by some of my colleagues in the University, and I have always been sorry about this, without ever being convinced that my personal choice was wrong."[38]

What seems to have happened was that on getting back into harness in November 1925 Toynbee faced the task of catching up with the surveys, since the hiatus in his employment at the institute had allowed most of the year 1925 to pass while he did other things. Starting so late, he could not expect to publish a book about events of 1925 on anything like the scale of the first two volumes before some time in 1927. Yet closing the gap so as to bring out each annual survey within a calendar year of the affairs it recorded was what the council of the Royal Institute wanted him to achieve. That, at any rate, was what Toynbee set out to do, and by July 1928 he had caught up. "I feel rather battered" he wrote to Veronica. "However there is 1927 gloriously finished as far as I am concerned, though not, I am afraid, for you." She, of course, had still to make the index. He went on: "I feel quite bewildered at having no arrears on my mind, and feeling that anything I deign to write during the next three months will be a work of supererogation. In a leisurely way I shall get something in hand, all the same."[39]

That something was, of course, what his wife had started to call the "Nonsense Book," and eventually became his monumental *A Study of History*. Writing to Gilbert Murray from their summer lodgement at Castle Howard, Toynbee recorded the household's hectic literary pace. "Rosalind this day posted her MS [a new novel titled *Hard Liberty*] to Chatto. . . . I finished the *Survey* for 1927 the day after you left and am still resting my bones and pleasing my fancy in considering the Philosophy of History, before taking up the unpleasant subject of [the *Survey* for] 1928."[40]

Toynbee did not actually begin to write a publishable manuscript of his great work in 1928. Instead he devoted his summer to a task he had begun in his spare moments during 1927, to wit, expanding the small entry about comparison of civilizations that appeared in the outline he had made in 1921 into an extended summary of what he now wanted to say. The task was substantial, since it called upon him to order his thoughts about just how many civilizations had arisen and disintegrated in the course of recorded history, and to fit what he knew about each of them into the pattern of rise and fall he presumed to be universal. By the time he finished writing this expanded outline in the summer of 1929, it totaled no less than 750 manuscript pages, with a structure and sequence that conformed in the main to the published work.[41]

By 1928, when Toynbee's surveys had caught up with the calendar, the University of London had, perforce, become accustomed to doing without teaching from the new professor of international affairs. No one questioned his right to use spare time for the book he so much wanted to write, but the university authorities clearly felt that they were being shortchanged. Terms of appointment were therefore renegotiated. Toynbee was formally relieved of teaching duties and given the title of research professor of international history. The university's contribution to his income was reduced to £200 per annum and a new appointee undertook the teaching of international affairs. But reduction of income from the university was more than balanced by an increase in Toynbee's salary from the Royal Institute, which agreed to pay him £1200 a year and also contracted to supply a typist in addition to Miss Boulter's full salary. In return, Toynbee promised to produce a 500 page survey volume each year, completing the next manuscript by 31 October 1928 and by 31 August each year thereafter. He also agreed to refuse outside journalistic and lecturing income unless the council of the Royal Institute gave prior approval, though an exception was made for weekly articles on international affairs which he contributed to the *Economist*.[42] Toynbee was also formally authorized to leave London any time after 1 June each year, so that he could and did do much of the actual writing of the surveys in the country. Unlike ordinary academic appointments, the new contract was subject to termination on three months' notice by either party.[43]

In this way Toynbee became able to devote almost all of his time to writing, or to the gathering of information needed for his writing, escaping all the ordinary chores of an academic career. He also evaded nearly all administrative tasks within the Royal Institute of International Affairs by the simple device of letting a vigorous and senior staff member, Miss Margaret Cleeve, take over the management of all the research projects and publications that the Royal Institute added to the surveys in the course of the 1920s and 1930s. As Director of Studies, Toynbee sat on the Research Committee that approved everything the institute undertook, but carrying out committee decisions rested entirely with Miss Cleeve. His title therefore became somewhat misleading, for the only studies he directed were those he executed himself. He was by temperament a loner, and found collaboration difficult.[44] Committee work simply squandered time that he preferred to spend at his own research and writing.

But, as we saw, the Royal Institute of International Affairs proved very accommodating. The council valued the surveys Toynbee produced so rapidly[45] and did not care whether he worked in London or in the country, so long as the volumes appeared punctually. He was therefore able to move back and forth between the house in London and any of the Howard–Murray country places that Rosalind arranged for them to occupy over extended Christmas, Easter, and summer holidays. Toynbee always took his work with him, for even after he had caught up on the annual surveys and could count on a summer lull between completion of one volume and commencement of the next, the vast enterprise of his ''Nonsense Book'' took over. It occupied every spare moment he could find, since he had to carry through a vast program of reading in East Asian, Amerindian, and other exotic histories, before his knowledge could come anywhere near catching up with his ambition of understanding the whole human past. Only so could he hope to anatomize the tragic pattern of rising and falling civilizations with the necessary authority.

One must admire the fierce energy he devoted to his self-appointed tasks, regardless, or almost regardless, of competing claims on his attention. His ultimate accomplishment rested on prolonged, private, privileged study. No one can say that he did not use his time effectively, nor can one doubt that the ruthless pursuit of his ambition made ordinary family life impossible and cost him dearly from the mid-1930s onward.

But for the time being, the career and life pattern Toynbee defined for himself in the late 1920s was, or seemed to be, eminently successful. The surveys came out on time and commanded respectful attention from reviewers. They became fixtures on the library shelves of universities throughout the English-speaking world and beyond. Several hundred copies were also distributed by the Royal Institute of International Affairs to its members at a price below what Oxford University Press charged the public. Sales covered editorial and manufacturing costs satisfactorily, and when the sale of a particular volume exceeded one thousand, as happened frequently, royalties accrued to the Royal Institute.

This satisfactory economic base rested on the success with which Toynbee created, year after year, a convincing vision of what really mattered from the confusion of daily events as recorded in newspapers. His information came in fact mainly from the daily press, but he did not have to search through the world's newspapers himself. Instead, Chatham House maintained a staff of remarkably competent women who clipped the leading newspapers of the western world for stories pertaining to international affairs, and then arranged them by subject in cartons which were immediately available to Toynbee and other researchers.

Official papers, speeches, and other public documents supplemented newspaper reports, but official publication often lagged further behind events than the Royal Institute's surveys. The surveys actually attained a quasi-official status, since Toynbee adopted the practice of submitting his text to experts in the Foreign Office and other specially knowledgeable officials who were asked to comment. He was never an official spokesman for the British government, however, reserving the right to decide for himself exactly how to adjust his text in the light of what his official critics might choose to tell him and wish him to say. Frictions with officialdom were trivial before 1932 but became increasingly significant after that time, as we shall see.

Successive volumes emphasized different parts of the world, partly in response to the course of events and partly as a reflection of Toynbee's effort to master the historical past of distant parts of the earth like China and Japan, Africa and the Americas. Everywhere outside of Europe he viewed current events as the manifestation of a worldwide collision between Western civilization and other, broken-down and therefore no longer autonomous civilizations and cultures. To understand contemporary manifestations of this clash, he felt he had to master the earlier histories of each people and civilization concerned. Consequently, as he treated events in Morocco, Syria, and Arabia, or in China, Africa, and the Americas, he played contemporary newspaper accounts off against his ever-increasing knowledge of the deeper past of every part of the earth—a knowledge fed by unceasing reading and note-taking for his great book.

In this fashion, the surveys contributed to the "Nonsense Book" and vice versa. Toynbee actually achieved what he had promised in his Stevenson Inaugural Lecture, combining study of current affairs with historical research to the benefit of both. Thucydides and Polybius had done the same in their time, and Toynbee rather liked to compare himself with these distinguished predecessors, but he did so in private, without quite claiming to be their equal, or referring to the likeness very often in public.

Successive volumes reflected Toynbee's personal point of view in varying degree. The survey for 1925, for example, got entirely out of hand because Toynbee treated "The Islamic World since the Peace Settlement" at such length that it required a separate volume all by itself.[46] In his writing, he always exceeded planned and projected limits. Words came tumbling from his pen, as fast as his fingers could move, and as he wrote his imagination provided a series of wide-ranging historical comparisons, together with striking metaphors and other embellishments. The effect was to expand the text far beyond what was needed for straightforward narrative. Yet much of the value of his surveys depended on the flashes of insight and surprising comparisons with which his pages abounded. Nevertheless, the effervescence of his account of the Moslem countries in the 1925 survey was not repeated. In succeeding years he stayed more nearly within the 500 page limit prescribed by the contract with Oxford University Press, usually by omitting some themes for more perspicacious treatment in subsequent years.

He got away with the anomaly of 1925, at least partly because his account of conflict in the Islamic world between westernizing and autochthonous impulses was in fact masterly. This volume took up again the themes he had concentrated on while serving in the Foreign Office during the war, when he had argued that the future of the British empire depended on achieving a satisfactory *modus vivendi* with the "east." He now recognized that what he had called the "east" was only a part of the cultural diversity with which Great Britain and other westerners had to deal; but he still saw the Moslem role vis-à-vis the West as paradigmatic. The Moslems had a longer experience of encounter with the West. More clearly than the Chinese, Africans, or Amerindians, they had arrived at an acute and painful historic dilemma, because they wished passionately to throw off the ascendancy of Europeans yet simultaneously sought "to adopt the military technique, the political institutions, the economic organization, and the spiritual culture of the West."[47]

Toynbee supposed that Moslem fundamentalists, who wished to reject the West root and branch and reaffirm the pristine purity of one or another sectarian version of Islam, were no more than a rear guard, destined to historical futility. This makes his pages seem a little old-fashioned in the 1980s,[48] but as an analysis of what was happening in the 1920s, when westernizing dictators like Mustapha Kemal in Turkey and Riza Shah in Iran were in the ascendant, this volume of the survey stands up very well in the perspective of sixty years. A truly remarkable understanding of topographical and geographical details informs his narrative of the Rif war, which pitted Berber mountaineers against both Spanish and French armies, 1925–1926; and Toynbee's account of French policy in Syria and of British policy in Palestine explains a great deal about struggles of the 1980s in Lebanon and Israel, which are aimed, largely, at undoing what the two Mandatory Powers of the League of Nations achieved in those years.

The survey of 1926 devoted half its pages to an account of affairs in the Far East and Pacific area. The basic theme was the same. Toynbee saw China's confused upheavals, Japan's foreign policy, and Indonesian revolts against the Dutch as further symptoms of the clash of civilizations, in which the West was in the ascendant while other civilizations had all broken down. China's agony was complicated, of course, by the fact that Japan was one of the imperialist powers acting upon that country, so that China's encounter with the West was also an encounter with Japan, whose civilization seemed to have "broken down" far more successfully than Toynbee expected. Here was a puzzle that needed closer study, and when a chance to visit the Far East came along in 1929, Toynbee eagerly embraced the opportunity to investigate the matter at firsthand.

The survey for 1926 is further distinguished by the fact that for the first time Toynbee's political narratives were supplemented by separate essays on economic and legal topics. Being little concerned with such matters, Toynbee gladly farmed out "Inter-Allied Debts," "The International Steel Cartel," and "The United States and the Permanent Court of International Justice" to others. He made this a normal practice thereafter.

For 1927 Toynbee extended his self-education to the New World by devoting nearly half the survey to the Americas, meaning Latin and Amerindian America as well as the United States and Canada. Successive survey volumes for the next three years were less eccentric in their distribution of attention, putting European and League of Nations affairs in the forefront and updating discussion of Moslem, Chinese, African, and American international developments every year or two so as to produce a more or less complete record of whatever western journalists judged newsworthy from those parts of the world. Disarmament, reparations, and mandates, plus various boundary settlements in the colonial world and fascist Italy's intrigues in the Balkans bulked very large in that record. But for all the failures, postponements, and half-measures recorded in his surveys, on the whole Toynbee clearly judged that the League of Nations' ideal of peaceful settlement of disputes among nations was more or less living up to expectations. After Lloyd George had been overthrown in 1922, partly because of his failures in Turkey, Toynbee found himself in general sympathy with British foreign policy, and was, by now, thoroughly at home at Chatham House. In addition, he contrived to devote a major part of his attention each year to the "Nonsense Book." Outwardly, therefore, he had attained solid success and could look forward to still greater recognition when his great book would finally come out.

He made time for the Nonsense Book in part by relying more and more heavily on Veronica Boulter to write portions of the survey. She concentrated on what Toynbee found least interesting, preparing accounts of League of Nations debates and other formal diplomatic meetings. "I have got into your League of Nations draft and am going ahead like a house on fire," he wrote to her in 1927, and declared that what she had written was "turning out to be exactly what I wanted."[49] As Miss Boulter became more practiced and self-confident, her drafts became final copy. By 1931 Toynbee could tell her: "I certainly do congratulate you on your account of the London Conference. You have marshalled that vast mass of stuff and given it shape, so that the narrative carries the reader's mind along with it, bringing out the real crises as they arrive. . . . I am really delighted with your chapter."[50]

She long resisted any sort of formal acknowledgment of her role in preparing the surveys, but the title page of the volume for 1929 reads "By Arnold J. Toynbee" and in smaller print beneath, "Assisted by V.M. Boulter." This remained the pattern thereafter, faithfully reflecting the reality. Toynbee planned each volume and wrote most of the text himself. But he could never have met the council's rigorous deadlines (more or less, for he sometimes lagged by a few weeks) without incorporating what she wrote into the finished text, or without her devoted assistance in bringing his own manuscript into final form by checking facts, correcting details, and preparing copy for the printer.

Eventually he relied on her judgment for more substantial matters. "Will you consider whether I have been fair to France?" he asked. "I want to be sure of not doing them injustice—the more so as I feel Gallophobia rising in myself at an alarming rate."[51] Or more playfully: "I am such a monster that the one thing I never forgive in my underlings is when they are impudent enough not to say what they think of my work, so go on saying it or your head will be in danger of falling. I have taken up almost all of your points, I think, partly by changes and partly by additions."[52]

But if Chatham House and the human relations supporting production of the annual surveys went well, the same cannot be said of Toynbee's relation with his wife and children. Tony, the eldest, wished to go to Eton,[53] but his parents persuaded him to try the scholarship exams at Winchester instead. He failed, but was admitted in the autumn of 1928 as an ordinary fee-paying student, with Lady Mary undertaking to meet all his school costs. Toynbee said he was not disappointed, but admitted: "I had a twinge of that quite irrational wish that one's children should do like oneself."[54] And when Tony had just started at his old school, Toynbee told Darbishire: "I get a curious and delightful happiness out of Tony being at Winchester. He is entering into it a great deal more than I did."[55]

Living up to his father's expectations, however, was more than Tony could or would endure. He reacted to the onset of puberty not by redoubling his academic work, as Toynbee had done, but by retreating sullenly into himself. Philip got into trouble at school by the opposite sort of behavior, going in for "dirty talk" that shocked his teachers.[56] By contrast, Lawrence, Rosalind's favorite, remained at home and gave his parents no cause for concern.

Clearly, Tony's and Phillip's bad behavior was provoked, at least partly, by Rosalind's conspicuous preference for her youngest son, combined with Toynbee's no less conspicuous withdrawal from ordinary family responsibilities. Rosalind's imperious temper drove her to dominate everyone around her, as her grandmother had done. Toynbee acquiesced, fretting only when expenses outran income, or when some new example of their financial dependency on gifts from the Murrays galled him anew. He always had desk work to do and habitually retreated from family frictions by burying himself in his books. But his solid professional success in the late 1920s made such a policy look like a more or less satisfactory *modus vivendi* for everyone concerned. Rosalind, the boys, the Murrays, and Toynbee's now quite distant mother and sisters.

In 1929 Toynbee was invited to go to Japan to take part in a meeting of the Institute of Pacific Relations in Kyoto. This was a chance too good to pass up, and he accepted gladly. Such a trip would allow him to see important parts of Asia and get a feel for landscapes and people as he had done so successfully in Greece in 1911–1912. It also meant parting from his family for what turned out to be five months, but Rosalind conceived the daring idea of accompanying her husband on the first lap of the journey by driving across Europe as far as Constantinople and then returning without him. She decided to take the two elder boys (aged fifteen and thirteen) along on the adventure, and arranged to have Gwen Raverat, an old friend from the "Cambridge crowd" of her youth, meet her in Constantinople and take Toynbee's place in the car for the return trip.

Rosalind had acquired a car in 1927 and quickly became a proficient driver. In May 1929 she bought a new Ford for the trip across Europe, and Toynbee took lessons so as to be able to help out with the driving.[57] Rosalind put her heart into the enterprise, gathering lists of Ford dealers along the proposed route, alerting British diplomats in the Balkans to their intended schedule, and getting expert advice on just what spare parts to carry along in case of breakdown. She also solicited advice from other motorists who had sampled Balkan highways. One of them reported: "The road from Temesvar to Calaphat [sic] is quite beyond description. . . . I was frightfully disappointed, because it certainly means that you can't bring the infants that way." Crossing the Danube on "an antediluvian, ramshackle barge had to climb out in reverse on planks at an incline of 1/6. Nearly toppled into the river."[58]

Risks merely whetted Rosalind's appetite for the adventure, and she was very much in charge of the little party that duly took off from London on 23 July 1929 and arrived in Constantinople twenty-three days later, having clocked 2044 miles on the Ford's odometer. Their longest day's travel was 138.5 miles from Arras to Verdun; but roads got worse as they traveled east. As she had been warned, crossing the Danube between Rumania and Bulgaria meant getting on and off a small barge by driving along precariously positioned planks. In Bulgaria, washouts in the road over the Balkan mountain range also challenged Rosalind's skill as a driver; but she and the Ford triumphed over every difficulty until the last lap of the journey, where the Toynbees encountered insurmountable bureaucratic obstacles and had to board a train, car and all, to assure the government they would not spy on border fortifications between Bulgaria and Turkey.[59]

The American College for Women in Constantinople offered the family its hospitality, and the boys stayed there while the parents went by train to Ankara, where they parted. Toynbee continued eastward, heading first for Damascus while Rosalind returned to Constantinople, collected the boys, Gwen Raverat, and the car, and then started the long drive back on 6 September. She returned by a more southerly route, traveling through Yugoslavia instead of Rumania, and arrived safely three weeks later. The major crisis on the way back was malfunctioning of the distributor. This produced loud backfiring, which "must have suggested machine gun fire."[60] But Rosalind got the British legation in Yugoslavia to mobilize local Ford representatives to adjust the timing and so solved a problem that had threatened to immobilize her car at the Bulgarian–Yugoslav border. Her management of this crisis quite charmed the border guards, who found it incredible that two women, accompanied only by young boys, dared to drive across Europe.

The rest of the return trip turned out to be a bit of an anticlimax. As a traveling companion, Gwen Raverat was disappointing, for she complained too much of risks and discomfort along the way. "You would have been pleased," Rosalind wrote to Toynbee "to hear the boys decide that you were the only right person to travel with."[61] There were satisfactions for Rosalind too. "I have received £28/8/0 from the *Evening News* for the articles I sent," she told Toynbee, and provided him with an exact accounting of all her expenditures on the return trip showing that it had cost exactly £101, despite having to buy new tires along the way.[62] Altogether it had been a fine adventure for the family. Rosalind renewed something of her earlier collaboration with her husband over the Greek massacres in Turkey, and Toynbee was able to share some of his learning with the boys by commenting on all the places and landscapes through which they passed.

In the ensuing months, Rosalind devoted more attention than usual to domestic tasks. "I have been doing a great deal of turning out of cupboards and drawers, throwing away masses of rubbish, including my old MSS. I came on our wedding photographs (I have not thrown them away) and was amazed at how long ago they looked—the long skirts and funny hats—and how young we looked—just children. I suppose in fact we were—rather nice and rather queer looking at them." Moreover, she signed off by using the pet name of their early marriage: "Much love from your own Puss."[63] Rosalind also visited Toynbee's mother at Oxford and found his younger sister, Margaret, "much more prosperous and lively and friendly than a year or two ago."[64]

Prolonged separation from his wife and from the familiar round of work at Chatham House had a rather different effect on Toynbee. He had approached the trip with a mixture of anxiety and eagerness. Crossing from Dover, he wrote back to his mother: "I am looking forward to this journey every bit as much as I did to it [Italy and Greece] eighteen years ago."[65] Yet he had a strong foreboding that he was destined to die in some distant land and feared that he might never be able to finish his great work. "I have got the skeleton of 11 out of 13 parts of my Philosophy of History on paper now," he wrote to Darbishire on the eve of departure. "I shall get the last two done before I go, so that if I leave my bones in the Syrian desert or Gobi, I shall at least have the satisfaction of knowing that I have left the skeleton of the book in a tin box in Maida Vale."[66]

In anticipation of such a possibility, he prepared a long memorandum before he left, headed "Directions for Publication." It began by declaring that there should be no tampering with the structure he had worked out for his book. "All the divisions and headings are to stand." Then followed a list of thirty-two experts to be consulted in connection with the "historical illustrations" of his main ideas. Five critics were to be enlisted as well: Gilbert Murray, the historians Lawrence and Barbara Hammond, F.J. Teggart (of California), and an old school friend named Hamish Paton. "The specialists are to be consulted on points of fact, the critics not only on points of fact but on the whole thing." For the conversion of the raw materials he was leaving behind into a finished work, he left the following instructions:

> I should like best that Rosalind should do the writing out of the book—though it will seem funny stuff to her. I should like to leave the question of length and language entirely to her discretion, so long as the exact structure of the book (as given in the Table of Contents) is preserved.
> I should like Miss Boulter, if she would, to do the kind of editorial work which she has done for me so splendidly on the survey.
> My Mother and Margaret might also like to have a hand in producing the book.[67]

An odder last will and testament would be hard to find, and the version of *A Study of History* that might have emerged from such a collaboration defies the imagination. Toynbee clearly had little sense of the tensions he aroused among the womenfolk surrounding him, being so wrapped up in his own ideas and consumed by his extraordinary labors as to lose sight of everyday human reality.

But however oblivious he was to the dynamics of personal relations, Toynbee had a keen eye for landscapes and delighted to exercise his imagination by interpreting them historically. Seeing so many of the lands of Asia, about which he had previously only read, was therefore a great excitement for him, and whenever the schedule of his journey allowed, he renewed the exploits of his

youth. Thus when, on an excursion from Aleppo, he first visited the banks of
the Euphrates, he swam across, and boasted: "I claim credit, as there is quite
a current, and it is about a quarter of a mile wide even in this low-water
season."[68] Later, in China, he walked to a haunted temple somewhere near
Peking, and was nearly marooned for the night because he had ventured so far
from the beaten track.[69]

But most of the time such adventures were ruled out by the dictates of his
itinerary. He usually had to content himself with what he could see from train
windows, punctuating his travels with official and semiofficial calls on the great
and mighty of the lands he traversed. In Iraq, for example, he called on King
Faisal and wrote back to Veronica asking her to send copies of his survey
volumes to the King.[70] In India he visited the Maharaja of Jodhpur, in China
he met and admired Chiang Kai-shek,[71] and in Japan he even attended an Imperial Chrysanthemum party. "Raising a frock coat and top hat for it was my
greatest achievement yet," he told Veronica.[72] British diplomatic representatives received him all along the way, often making him their personal guest.

What he saw in the Middle East, India, and China all more or less conformed to Toynbee's expectation, in the sense that it confirmed the view of
current affairs that he had formulated in 1921 when studying the Anatolian war.
His experiences, shaped mainly by encounters near the top of the social pyra-
mid, seemed to attest to the disintegration of all the Asian civilizations as a
result of collision with an intrusive West. Efforts to imitate and incorporate
western techniques and ideas were everywhere in train. Yet the aim of such
efforts from the point of view of local leaders was to throw off political and
economic dependency as quickly as possible.[73] All this matched the recipe for
the world that Toynbee had worked out for himself during and after World War
I and confirmed him in the view that all civilizations went through the same
cycle of rise and fall.

Only Japan did not fit. Important elements of Japan's indigenous culture
clearly remained intact. On his first arrival in Japan, Toynbee made a point of
exposing himself to this side of things, spending a night at a Buddhist monastery "far up in the mountains among immense pines"[74] and going to see Nō
plays, which reminded him of the classical Greek theater. Indeed, he concluded
that what he saw in Japan was "a classical culture still alive. I can't exactly
explain what I mean by traditional Japanese culture being classical, but you
feel it very strongly indeed when you go into the temples. They feel like Greek
temples, retranslated into wood."[75] A classical culture, still alive, was quite
an anomaly for Toynbee to fit into his philosophy of history, and for the time
being he did not really try.[76] Instead, he stored up his experiences and continued to wonder about Japan's uniqueness.

Toynbee's papers contain nothing about his participation in the Third Meeting of the Institute of Pacific Relations in Kyoto, which took place between 31
October and 8 November 1929. For one thing, he met a number of Americans
who were professionally interested in international affairs in Kyoto. Chief among
them were Quincy Wright and Owen Lattimore, with whom he maintained
friendly contact subsequently. But in general the conference for which he had

traveled so far seems to have made little impression on him, and his contribution was, presumably, marginal.

One reason for this result was that the formal proceedings of the Institute of Pacific Relations were in practice quite eclipsed for Toynbee by a very different sort of encounter he had with a fellow historian and professor at the University of London named Eileen Power.[77] She attended the I.P.R. meeting rather by chance, having gone to the Far East to prepare herself for writing a book about the distinguished company of Europeans who traveled to China at the time of the Mongol Empire. Being already on the scene, more of less, she was somehow invited to participate in the Institute of Pacific Relations Conference at Kyoto.

Toynbee, of course, was already acquainted with her, and warmed to the company of a fellow historian from home. But his relation to Eileen Power intensified into unmistakable sexual attraction as he traveled through Manchuria and northern China in her company afterward, visiting such places as Mukden, Harbin, the Great Wall, and Peking. Miss Power was both elegant and charming. She combined a lively mind and wide-ranging historical interests with a strikingly handsome appearance, emphasized by a studied stylishness of dress.[78] As Toynbee explored China in her company, he found her delightfully responsive to his ideas about the clash of civilizations in the Far East. At the very end of the trip, after visits to Nanking and Shanghai, Eileen Power told him, under seal of secrecy, that she was engaged to be married. Probably she told him of her attachment to Rushton Coulborn, a married man and much her junior with whom she lived for several years in the 1930s before marrying another historian, Michael Posten.

The news upset Toynbee profoundly. He felt that the match was improper, yet was spellbound by her charms. Rosalind, like her mother, paid little attention to dress, and Toynbee was not accustomed to consorting with elegant women. More importantly, Eileen Power was keenly interested in things long ago and far away just as he was. This made her a far more sympathetic companion than Rosalind had become with all her condescending raillery against his obsession with the "Nonsense Book."

After a sleepless night, Toynbee burst into Eileen Power's room, implored her not to marry the man she was engaged to, and declared his love for her.[79] Embarrassed and surprised by his behavior, she asked him to leave, and he did so. They parted soon thereafter. Subsequently, on 6 January 1930, Toynbee wrote her a rather confused letter, half renewing his plea that she break her engagement and half apologizing for his intrusion. The letter is odd enough to quote *in extenso*.

> After my first reaction to the news of your engagement it is still a bit on my mind that I shall be one of a rather small number of people whom you will have told about it before making your final decision, for the way your picture fitted into my frame obviously threw out my judgement, besides going through my guard.
> . . . . "The unreal world" [of foreigners in China] *is* to the point . . . (That is easily tested when you come home.) Not so that other thing that was on my mind à propos of myself. This strange and unexpected thing that has happened to you

in the unreal world not only may vanish, but certainly will vanish, as soon as you leave the unreal world behind—and thank God for that, for it is a bad thing which leads nowhere. This, while true for me, has, I am sure, *no* application to you, and it was just an irrational and most devastating association of ideas in my mind, at first thoughts, when taken by surprise.

I dare say I am labouring this point for nothing and might have taken it for granted all along that there was not one [chance, stricken out] risk in a hundred of your final decision being influenced, even in the slightest degree, by the "reaction" of somebody whom you had taken into your confidence in the interval. So perhaps I might have spared myself a night of wondering what to do and the experience of being shown the door by an angry lady who thought I was going to forget—or had forgotten—my manners—though I hadn't and wasn't, but was only desperately embarrassed and tongue tied.

Do you know the game of describing a person well known to the company, by some true, but un-suggestive fact about him, and then making the others guess who it is? For instance: "A married man of forty who was shown the door by a maiden lady (same age) for telling her he was in love with her the morning after she had told him she was engaged."

By the way, I found my status quo restored when I woke up on the morning we sailed from Shanghai,[80] so you may take it that that nuisance is now abated. . . . It is rather horrifying for poor human beings to find the elemental forces inside them; but I suppose that ultimately is what makes all our wheels go round, and a price has always to be paid.

Now I shan't write or speak any more about your affairs unless you choose at any time to reopen the subject. I look forward to seeing you in the summer "according to plan".

Yours ever
Arnold Toynbee[81]

Eileen Power replied with a brief, noncommittal note, but added as a postscript:

> I meant, my dear Arnold, to say nothing about that unlucky episode; but a passage in your letter (which I have just re-read) causes me to take up my pen again. I really can't have you thinking that I tried to eject you lest you should "forget your manners." . . . You gave me a sudden and violent shock, for which I was totally unprepared, and what really animated me was a frantic and quite irrational desire to stop you from putting into words what I didn't want to hear. . . . It was silly of me, for the damage had been done. . . . And that's the last I shall speak of it. I am glad that you came out of it at once, for I should hate to have anything spoil our being friends, and I did *so* much enjoy wandering about with you.
>
>
>
> Yours,
> EEP[82]
>
>

The powerful upwelling of feeling that his encounter with Eileen Power provoked in Toynbee marked a watershed in his life. He went through considerable agony of soul both before and after the scene in her room and, as was his wont when his deeper feelings were concerned, recorded the upshot in Greek verse, published long afterward as "Grammatikos Minotauros" in his book

Experiences. This poem is sufficiently self-revelatory to deserve translation and comment.

The title of the poem translates literally as "Man of Letters Bull-Man." His use of the word Minotaur recalls the myth of how Theseus liberated Athens by slaying the Minotaur of Crete. The whole poem is, indeed, an artful recapitulation and transformation of that myth. The first lines set the scene by describing the man of letters in whom a "heavenly spirit" that is "equal to the gods" mingles with a "beast," born of the "primaeval Bull." The remaining lines of the poem may be translated as follows:

When years had passed, the man ruled both the Bull and the "heavenly spirit" in
     their grown state
and always restrained the arrogant beast with a stronger hand,
and without care, not anticipating disaster, his own ruler, proceeded apace.

Listen God what do you suffer in your mind? Or do you hate
your cunning work, long in the making, me?
Or have you formed something mortal better than yourself?
Or do you nonetheless envy me ruling the evil beast?
Or, having been maddened, have you struck me? This I know clearly:
Having been struck a terrible blow by Aphrodite, I myself am mad;
my limbs fight banefully against one another; no counsel from the inner heart

has appeared; naked and orphaned I array myself against
the earthy, primaeval Bull coming against me.

Rejoice and greet me, saved beyond hope; rejoice comrades.
Having conquered I return. The deadly rival lies slain.
For all the weak hereditary gods alike abandoned me to the
welcoming force of the beast; but an unknown, unhoped-for One
stood before me and gave me ten-fold war-strength
And then I strike with the naked hand, and the beast fell face downward
at one daring blow—a miraculous thing.
And I proceed swiftly, my own master, joyful in my heart,
as formerly, but knowing the shining deeds of an unknown God in the contest:
very much unknown in his form. Appear then
face to face, for those praying to see you.[83]

The echo of Theseus' feat of killing the Minotaur is clear; so is Toynbee's appraisal of his personal feat of subduing his sexual emotions in order to concentrate on his book.[84] That feat had invited the envy of the gods. "Have you formed something better than yourself?" he asked the jealous god. A heaven-provoking boast indeed! And Toynbee's act of hubris was duly succeeded by Ate: the destroying madness of his passion for Eileen Power. All this fitted the classical idiom Toynbee had absorbed from his Oxford education. Resort to the familiar literary mode salved his distress and distanced him from the actual experience.

But a new and different note is struck in the strophe celebrating victory over his sexual passions through the help of an unknown God, a God whom he

invites in the last line of the poem to appear to his votaries in some less ambiguous way. This refers to a mystic experience that somehow freed him from
his emotional turmoil. Thereafter he could no longer believe that "religion
itself was an unimportant illusion," which was the view he had carried away
from his classical education.[85] Instead, he became convinced that a supernatural reality had communicated with him in some quasi-personal way. Such an
entity was far closer to the Christian God of his childhood than to any of the
pagan gods of Greece. His use of the phrase "the unknown God" for that
which intervened when "weak hereditary gods" had fled shows that he realized
this, for it artfully echoed words St. Paul used when preaching to the Atheni-
ans.[86]

Yet Toynbee did not in the end return to the religion of his childhood. He
found too much of Christian doctrine incredible for that to be possible. Instead
he began a long spiritual travail, seeking to define a faith of his own in the
light of this and a second similar encounter with a spiritual reality that communicated its presence to him in a way he could not understand intellectually,
yet could no longer doubt or deny. The effort to come to terms with these
mystical experiences bulked large in his subsequent thought and writing, and it
is clear that his mature views did not achieve definition in 1930, or for years
thereafter. Still, it is worth quoting the words he used, some thirty-six years
afterward, to describe his encounter with an unknown God in 1930.

> How are we to picture to ourselves a god who is spiritually higher than the
> highest human spiritual flights, yet is at the same time, in one of his aspects, a
> person like enough to a human person for communication, person to person, to be
> possible between God and a human being? I find this incomprehensible intellec-
> tually. The nearest I have come to understanding the mystery has been in two
> experiences that were not acts of thought but that felt as if they were flashes of
> insight or revelation. Each experience came to me at a moment of very great
> spiritual stress. The earlier one came when I was in a moral conflict between the
> better and the worse side of myself, and this at a moment when the better side
> was fighting with its back to the wall. The second experience came at the moment
> of death—a tragic death—of a fellow human being with whose life mine was
> intimately bound up. [This refers to the suicide of his son Tony in 1939.] On the
> first occasion it felt as if a transcendant spiritual presence, standing for righteous-
> ness beyond my reach, had come down to my rescue and had given to my inade-
> quate human righteousness the aid without which it could not have won its des-
> perate battle.[87]

This testimony is all we have to go on in trying to figure out what happened.
Whatever the character or psychological springs of the experience, it carried
with it an emotional force that changed Toynbee's outlook on the world permanently and profoundly. It, with the subsequent similar experience in 1939,
made him over from a satisfied more or less Stoical neo-pagan into a seeker
after fuller and less ambiguous revelation of the transcendental spiritual reality
whose existence he could now neither understand nor doubt.

Oddly enough, his turbulent encounter with Eileen Power had the effect of reknitting his marriage ties with Rosalind. When he began to feel attracted to Miss Power, he wrote to Rosalind, telling her about it. This letter does not survive, but two letters she wrote in response reveal a remarkable detachment and insight. On 11 December 1929, Rosalind wrote to an address in Peking, hoping to intercept him there, as follows:

Dearest

Your letter from Moukden[88] about Eileen reached me here [at Castle Howard] this morning. Naturally, to a certain degree, I am disturbed by it. . . . I am not afraid that you will want to leave me and go off with her. If I was there too I would not be afraid at all (though perhaps that is hubris—and I shall find when you come home that my "mana" is all gone. That does happen to people.) . . . I expect it was unnatural the way you never felt the least inclined to fall in love with people, and it is natural that you should do so now, for I expect it made me feel something lacking in you—though I know that sounds paradoxical. But it was in a way perhaps because you were not interested enough in people, and after all I want you to be that, even if more risk goes with it. I think possibly your feeling like this now is a sign of your having grown up or wakened up more—and if it makes me feel less sure of you and take you less for granted (as it will!) that will be good too in the end. . . . Now, probably, I shall take more trouble to keep you and please you, for people are like that, I am sure, even sophisticated people like you and me. . . .

I don't think she would have been a better wife for you than me, though she would have been quite a good one. There would have been less ups and downs— she would have been an easier wife, I dare say, but not better really. . . .

Tomorrow I go home, and not more than a month till you are home too.

Your own

Puss[89]

Three days later, after getting another letter from her husband, announcing a change in his itinerary and a postponement of his return by two weeks beyond the original time, Rosalind wrote again, since she figured that her first letter would not reach him after all. The tone is almost the same, as this excerpt shows:

I am very glad that you wrote all about what you were feeling, like that. Being kept out and not told would be the worst thing that could happen. I should mind that more than anything, so do go on telling me, even if, at Peking, it has all got much worse and more serious. I *do* mind, badly as it is. Having contemplated the possibility before does not make much difference, for one does contemplate, intellectually, so many things. I knew I would mind if you did fall in love with Eileen, but I don't think I realized how badly I should mind. Perhaps it is a good thing for me, for I think it makes me realise how much I do love you—and I haven't always realised that enough. . . .

Another six weeks[90]—it seems interminable. I wonder if it seems as long to you as it does to me, or if you are beginning to dread coming back to me?

Yours ever

Puss[91]

Return he did, braving hunger and other severe discomforts during a ten day trip on the Trans-Siberian railway.[92] When he got back to London, on 29 January 1930, his infatuation for Eileen Power was over and he was greeted by a wife resolved to be more agreeable than she had sometimes been before the abortive affair. Work at Chatham House resumed, where Veronica was as efficient and helpful as ever. The older boys were safely away at school, and seven-year-old Lawrence, who lived at home and went to day school, continued to give his parents no trouble. As for Toynbee's mother and sisters, Margaret had a satisfactory job in Oxford, working for the *Dictionary of National Biography,* and could therefore live with Edith Toynbee and keep her company. Even Jocelyn was more relaxed, having found a niche for herself as a historian of classical art at Cambridge. Toynbee and everyone around him could therefore settle down to a renewed and more self-conscious domesticity.

His emotional storm had indeed helped to clear the air, and its residue kept his marriage together for the next decade, not least because Toynbee's new belief in a transcendental spiritual reality brought him intellectually closer to Rosalind, who had begun to develop a fervent yearning for Catholicism, partly, perhaps, from association with Bridget Reddin, Lawrence's one-time nurse, who brought a vigorous and mildly assertive form of Irish piety into the household.

In addition, Ganthorpe Hall had at last become available to Rosalind. When Lady Mary transferred Ganthorpe to her daughter in 1925, the place was leased to tenants, but the contract expired in 1930. Rosalind could therefore establish a country home of her own at Ganthorpe, instead of spending her treasured vacations as a somewhat awkward guest at Castle Howard or (less often) at Boothby. Accordingly, the Toynbees moved in during the Easter vacation of 1930, furnishing the place partly from their London house. Rosalind went ahead to attend to arranging furniture, and wrote back to her husband: "It will be nice when you come, for having the boys without you makes it seem like last Christmas, which I don't like! And you must be at the beginning of this lovely house. Only one and a half days now til you come.[93] After he arrived, Toynbee reported to Lady Mary: "We are all supremely happy at being here. It is better even than our seven years' anticipation, and that is saying a great deal."[94]

He did, indeed, come to love Ganthorpe, and it was there that he began to realize his life ambition by writing the first six volumes of *A Study of History.* However, Rosalind's attachment to the place reflected a rather different ambition. The project of making her husband into a country gentleman, while she played the role of chatelaine and landowner, even if only on a modest scale, appealed to her increasingly reactionary imagination. She had come to value social hierarchy and expected the deference of her inferiors, much as her grandmother had done. Owning and occupying Ganthorpe therefore offered her a chance for a new start—one in which her literary disappointments could be left behind, and where she could find a new, more satisfactory role both for herself and for her husband.

Rosalind set vigorously about that task, and Toynbee set vigorously about his task of combining responsibility for the surveys with writing his great book.

For a while, the incompatibility of their respective visions of how life at Ganthorpe should be led was obscured by the very intensity with which they set out to realize their disparate dreams. How those efforts fared, and eventually clashed, will be the subject of the next chapter.

Toynbee at the height of his powers
Bodleian Library, Toynbee Papers

Ganthorpe Hall, where volumes I–VI of *A Study of History* were composed between 1930 and 1938
Lawrence Toynbee photo

VII

Triumph and Defeat

1930-1939

Toynbee's achievement in the 1930s was extraordinary, for on top of the annual *Surveys of International Affairs* which continued to come out each year more or less on time, he also published the first six volumes of *A Study of History* (I–III, 1934; IV–VI, 1939). The *Study* attracted instant acclaim and attained surprising sales even in its unabridged form. At age fifty, Toynbee was well on his way to realizing his great ambition when, in 1939, the onset of World War II interrupted the pattern of his life. In addition, Toynbee maintained a furious pace of outside work throughout the 1930s. He attended a conference on Commonwealth affairs in 1933 and edited its report; and as his reputation mounted, he started to make extra money by giving BBC talks, many of which were printed as pamphlets. An even more lucrative sideline was lecturing on university campuses in America and Britain, which sometimes also resulted in publication. On top of all this, he continued to write a weekly column on international affairs for the *Economist* and engaged in other casual journalism as well.

Among historians, his reputation was as yet unmarred by the harsh criticism that *A Study of History* attracted after the war. As far as the surveys were concerned, Toynbee's denunciation of British failure to live up to League of Nations rules for collective security in the years after 1932 roused some animosity in the Foreign Office and in Parliament, but his advocacy of the League's cause attracted support from others.

Toynbee's professional successes were, however, accompanied by a crescendo of private failures. In 1932 or 1933[1] Rosalind joined the Roman Catholic church, and her subsequent espousal of an aggressive Catholic piety allowed her to assert intellectual and personal independence against her father's emphatic agnosticism on the one hand and her husband's hesitations in matters of faith on the other. The result was a gradual wearing out of the Toynbees' marriage. But deterioration of his relation with Rosalind was both obscured and

149

exacerbated for Toynbee by prolonged and acute frictions with Tony and Philip, beginning in the early 1930s and climaxing on 15 March 1939, when Tony, at age twenty-four, died of a gunshot wound he had inflicted on himself a few days earlier.

The timing was extraordinary, for 15 March 1939 also marked another kind of defeat for Toynbee: it was the day on which Hitler seized control of the rump of Czechoslovakia, thereby discarding the Munich agreements of the year before and making mockery of British and French efforts at appeasement. After this, World War II became inevitable and foreseeable. By 15 March 1939, therefore, Toynbee's hope of making another war impossible by helping to guide public opinion aright was in shambles; his family was in disarray; his great book was unfinished.

Religion offered relief of a sort. But firm belief in the existence and importance of the transcendental spiritual reality that he had encountered in 1929 and met again at Tony's deathbed in 1939 did not allow Toynbee to become comfortably Christian. Roman Catholicism beckoned as a way of resolving doubts. and rejoining Rosalind, but he continued to disbelieve the truth of key Christian doctrines, and clung, a little desperately, to the dictates of his intellect despite Rosalind's imperious faith and his own emotional inclinations.

Measured against such private agony, Toynbee's professional triumphs weighed lightly indeed. Yet in retrospect it is hard to doubt that his private failures stemmed very largely from his single-minded dedication to work. Writing, always writing, he never had time to pay attention to his sons, or to others around him. This was a far cry from the role of country gentleman that Rosalind had projected for him when they first moved into Ganthorpe. To be sure, Toynbee did make some initial efforts to live up to her expectations. In 1931 he wrote to Veronica, "I have been leading a peaceful existence building a bit of lawn and trying to show by example that the optimum strength of a challenge for evolving civilisation is a mean between two extremes."[2] Rosalind appreciated his effort and at the end of the year confided: "I wonder what 1932 will be like? 1931 has not been a bad year for us personally, however depressing for the world at large." And she signed off as of old: "Very much love from your Puss."[3]

But Toynbee was not to be deterred from his main goal. In January 1930, on getting back safely from Asia, he knew that the time had come to start the great book he had been planning and preparing for ever since childhood. His problem, as always, was to find time for the task. Initially he thought he could complete the book in two summers.[4] But summers were never entirely free, since each survey volume, even after it had all been written, still had to be proofread and corrected, and incidental undertakings likewise encroached on disposable time. "I have been digging away at a bathing pool," he wrote to Veronica, "and having polished off the BBC talks, have just got down to the great book of nonsense."[5] By 11 August 1930 he reported having 150 manuscript pages of the big book in hand; by the end of the month the sheaf of completed pages was up to 187; by mid-September he recorded, triumphantly, that Part I was finished.[6] When typed copies of what he had written came back from Chatham House in October, Toynbee exulted: "I am astonished at the amount that has dripped off the end of my pen. I have now done two thirds of the Book of Genesis."[7] And in the next Easter vacation, when he was again at Ganthorpe, he remarked: "The worst of getting into my nonsense book is that I find it so hard to take my mind off it again."[8]

But his duties at Chatham House required him to do so for something like half the year. The pattern he devised for writing his book while simultaneously producing the annual *Survey of International Affairs* required him to work daily at Chatham House from January to June, with the exception of an Easter visit to Ganthorpe in the spring. As each calendar year drew toward its close, Veronica sorted through the Chatham House press clippings to eliminate duplications; then Toynbee looked them over and drafted a scheme for the next survey, deciding what to emphasize and assigning responsibility for each section to himself, to Veronica, or to some third party.

Long afterward, in a memorandum written after Toynbee's death, Veronica explained how they worked. "Having allocated the chapters, Arnold would settle down to write the most important chapters himself, and I would do the same with my chapters. By about the end of June, Arnold and Rosalind would leave London for Ganthorpe, where Arnold would concentrate on writing the next batch of the *Study of History,* but would deal with any correspondence relating to the *Survey,* check the typescript, and finally see the current volume through the press. I used to prepare the index for each volume, and also managed in my spare time, to index proofs of the *Study* as they came from the O.U.P. [Oxford University Press]. I generally had to forego a summer holiday, but for several years I used to stay for a few weeks at Ganthorpe with Arnold and Rosalind."[9]

As time passed, Veronica tended to take a larger share in the actual composition of the survey, "though I think there was only one year when my parts exceeded Arnold's in length, and he always wrote the most important parts himself. We did, however, after the first few years, always have a regular writer on economic affairs, and particular topics were often allocated to other writers."[10] Outside authors' names were listed with the chapters they wrote in the Table of Contents of each volume, but Veronica's work so blended with Toynbee's that no distinction was made between their respective chapters. Toynbee, indeed, once declared that he could not always tell which of them had been responsible for the initial draft of a particular passage. The reason was that she revised his drafts more and more freely, and he revised hers, making the final text into a genuinely collaborative product. Her name, subordinated to his on the title page of successive volumes, recorded their professional relationship.

In spite of a conscientious effort at impartiality, successive volumes of the survey inevitably recorded something of the author's reactions to the course of international affairs. Once Lloyd George's Near Eastern policy of backing the Greeks against the Turks had met defeat in 1922, Toynbee found himself in general sympathy with the course of British foreign policy until 1932. His attraction to the Labour Party had dimmed after 1922 almost as swiftly as it had arisen, and Toynbee retreated from political activism toward a nonparty, vaguely liberal point of view in domestic and foreign affairs.

Like the British government, he blamed the French for obstructing disarmament and opposing treaty revision that might have repaired some of the inequities of the Versailles settlement. "I do feel intensely bitter against the French," he told Gilbert Murray. "Such clever people have no right to be so stupid." [11] And to Miss Cleeve at Chatham House he remarked: "I suppose my feelings against them go back to the Peace Conference. . . . In my anti-gallicanism I am almost in danger of becoming patriotic." [12]

His patriotism manifested itself also in the way he interpreted the crash of 1929 and the depression that followed in its wake. The survey for 1931 begins with a remarkable essay on "States of Mind" during the "Annus Terribilis 1931." It begins: "The year 1931 was distinguished from previous years . . . by one outstanding feature. In 1931 men and women all over the world were seriously contemplating and frankly discussing the possibility that the Western system of Society might break down and cease to work." He went on to assert: "And this order was British in the sense that British enterprise and technique and initiative and responsibility had played the leading part in building it up. Moreover, "The Germans and the English were the chief sufferers in 1931 from the disorganization of the economic order of which the British had been the principal artificers. The French and the Americans, on their side, were open to the imputation of being mainly responsible for the disorganization." Comparison with the fall of Rome pervaded the essay, but Toynbee concluded that French and Russian "disbelief in the survival power of the British economic order"—worldwide, based on free trade—posed a question as yet "quite impossible to answer." [13]

The year 1931 was an "annus terribilis" politically as well as economically, for on 18 September Japanese troops attacked China and began the conquest of Manchuria. Japan's attack defied the United States as well as Britain and the other League powers, for it violated not only the Covenant of the League of Nations, but also the Kellogg Pact, which in 1928 had outlawed war, and the Washington Treaty of 1922, to both of which the United States was signatory. Toynbee therefore called Japan's aggression "an 'acid test'—to borrow President Wilson's famous phrase—of the whole system of 'collective security.' " [14] But in the course of 1933 it became clear that neither the American nor the British government was prepared to defend the principle of collective security in the Far East. The United States contented itself with registering disapproval by refusing to recognize Japan's conquests and, in the absence of a firmer lead from America, Great Britain shrugged off responsibility for the Far East by declining to activate the League machinery of sanctions against Japan.

Toynbee viewed such behavior as craven and a betrayal of Great Britain's legal obligations. Privately he expressed his feelings vigorously enough. "This means abdication of any attempt at organized law and order," [15] he wrote to Gilbert Murray. "I get the impression that the British public are inclined to let not only the League but the China trade and even the Commonwealth go hang— not seeing, in their blindness, that this is 'finis Britanniae.' " The Tory policy of "peace at any price" he found "a complete breech with the traditions of English imperialism. How odd it is." [16] Four days later Toynbee proposed that the League of Nations Union (a private association headed by Gilbert Murray) should organize a boycott of Japanese goods; [17] but when his father-in-law invited him to join the executive of the League of Nations Union to do something about it, Toynbee refused, saying he wanted all his time for the Nonsense Book. "I think I have never sat on a committee without wishing all the time that I was otherwise employed. . . . I am always put off by the kind of approach that one has to make to a subject when the object is not just to study the facts but to get something done and to promote a particular cause—even a cause with which I am sympathetic. . . . And though League propaganda and national propaganda are as different as white and black, the intellectual approach . . . is, I think, the same in either case." [18] His personal conduct thus mirrored that of the British government. Like H.M.G., Toynbee refused to engage in Far Eastern struggles.

Chatham House had acquired a quasi-official status by the early 1930s. As a result, those who directed its affairs were loath to offend the government by open opposition to official policy. Toynbee therefore remained very circumspect in reporting the huffing and puffing at Geneva over Japan's aggression and Great Britain's refusal to press for sanctions. [19] "How utterly scandalous Simon's [Sir John Simon, British Foreign Minister] Far Eastern Policy is," he wrote to Veronica. "I find it really hard to stick to the *Survey* style in dealing with him—though of course it is the most effective style, really, for being thoroughly nasty." [20]

Yet however much he disapproved of the way Great Britain backed away from its obligations to collective security, Toynbee was very much aware of contervailing considerations. For one thing, any active British (or American) policy in the Far East would intensify the violent collision of civilizations he so deplored. He recognized, too, that unless the United States participated to the full in applying League sanctions, Britain and the other members of the League of Nations could not expect to deter Japan by mere threats and trade boycotts. And since no one in Europe was ready to fight a war in the Far East for an abstract principle, however important the ideal of collective security might be, practical politics required Great Britain to back away from armed confrontation with Japan.

The Far Eastern imbroglio in the early 1930s illustrated the fact that Britain was not merely a European power. Inspired by the apparent reconciliation between Britishers and Boers in the Union of South Africa after 1907, Lionel Curtis and some other key figures at Chatham House had long cherished the hope that all or most of the British Empire could be transmuted into a cohesive, active political commonwealth. Only in this way, they believed, could Great Britain become Greater Britain and maintain the leading role on the world stage it had enjoyed in the nineteenth century.

That was not the way things were going, of course. In India agitation for independence was gathering force, while Canadians, Australians, and South Africans were far more inclined to assert local sovereignty than to accept the lead of Great Britain. But the economic depression and Japan's aggression in China did present acute new problems for all the Commonwealth countries, and imperial spirits of the Royal Institute of International Affairs decided that an unofficial conference to discuss how to cope with these circumstances might help officials and political leaders unite the Commonwealth more effectively. Accordingly, the first unofficial Conference on British Commonwealth Relations met in Toronto, Canada, 11–21 September 1933, sponsored jointly by the Royal Institute of International Affairs and by a newly founded sister institute in Canada. Toynbee of course attended and edited the conference proceedings afterward.[21]

The major question at the meeting, according to Toynbee, was whether the Commonwealth could hope to prosper as a new sort of great power, transcending oceanic barriers by virtue of shared political heritages, or whether the Commonwealth and its separate national parts ought instead to rely on the League and the framework of collective security enshrined in the Covenant. As he reported to a colleague: "The fact that, all the same, they were drawn back, in spite of themselves, upon the Collective Security System is . . . one of the outstanding results of the Conference." Indeed, without the League, the Commonwealth itself had a "rather poor chance of surviving." As a result everyone ended up paying "lip service to the Collective System."[22] Thus the meeting was a crippling defeat for Lionel Curtis' vision of the future, but no real disappointment for Toynbee, who pinned his hopes on the League and peaceable change and had never really believed in a revivified, politically consolidated British Empire.

After the conference ended, Toynbee visited Harvard, where he gave a series of lectures at President Lowell's invitation, treating some of the themes of his as yet unpublished *A Study of History*. "The Bostonians anyway, seem to lap the nonsense up," he reported.[23] He was simultaneously reading proof of the first volumes. "It is certainly pleasant to see the Nonsense Book getting into print," he wrote to Veronica. "I have just sent back corrected proofs up to page 208 this morning. Print gives things a specious air of convincingness. I think I have been very cunning in quoting Scripture and other authorities, to my purpose, as I see the extracts in the footnotes."[24]

From Harvard, Toynbee traveled west, heading first for Chicago, with stops along the way for lectures at Buffalo and other places. His principal host in Chicago—which he called a "false Babylon"—was Northwestern University, whose president asked him to make " 'contacts' with the faculty, rather as if I were an Old Master or a Parisian creation temporarily on exhibit. It is flattering to have such a rarity value, but, though they are kindness itself, it is no joke. By the time I reach Texas, I shall be a sort of dodo or great auk."[25] He liked his westernmost stopping place at Denton, Texas, seat of North Texas State Teachers' College, far more than he did Chicago. From there he turned back eastward, lecturing on current affairs at every stop, and traveling via St. Louis and Atlanta to Washington and New York. Thence he returned to London and Chatham House, having been away for nearly three months.

He paused in his travels to see old friends, including both Robert Darbishire and "Will Westermann, whose opposite number I was at the Peace Conference."[26] In Kingston, Ontario he stayed with the Canadian missionary Dr. MacLachlan, who had introduced him to Turkish circles in Smyrna in 1921. But such visits were brief and scarcely relieved the hectic pace Toynbee set himself. He was intent on maximizing his lecture fees and eagerly accepted everything that came his way. From the financial point of view his tour was indeed a success. :"I am glad to hear that you are bringing back more money than you expected,"[27] his young son Lawrence wrote.

But the price in terms of nervous exhaustion was heavy. "In many ways I have enjoyed this expedition," he reported, "but all the same, if I do go to hell, I now know what I shall be doing there: and that will be to carry out a time table like this that doesn't ever come to an end." Yet he was impressed by American hospitality, and enjoyed being lionized. "I once more find myself liking them better on closer acquaintance," he told Veronica.[28]

Toynbee was, in fact, under specially heavy pressure throughout the three months he spent in North America. In the last days of the Commonwealth Conference, when controversy about how to shape the report flared up, he suffered what he called "a mild collapse," becoming completely unable to sleep.[29] Sleeping draughts provided some relief, but Toynbee's dependence on them, which became almost habitual for him in the ensuing years, was worrisome to Rosalind,[30] who was suffering a bout of ill health herself. Her wisdom teeth were judged at fault, and in November she went to a nursing home to have them removed. She then spent a period of recovery with her parents. "Mother has been a dear . . . and quite endlessly kind," she wrote. Yet, "I am afraid I have felt rather up against the extreme arbitrariness and rigidity of her rule. It brings back with rather startling vividness the experiences of my youth. . . . It is awfully pathetic rally, for it does go so far to frustrate the effects of the endless kindnesses."[31] On the other hand, "My relations with Dad have been quite happy and easy, and I think he is pretty well reconciled to my 'apostacy.'[32]

The apostacy to which Rosalind referred was her conversion to Roman Catholicism. Her own account, published in 1949, dates her interest in Catholicism to early childhood, when she heard stories about persecutions her Murray ancestors had suffered for their Catholic faith in Ireland. But definite commitment came, she claims, in adolescence, when at age sixteen she had been sent to Italy for her health and first read the *Little Flowers of St. Francis*.[33]

Rosalind's adolescent attraction to Catholicism was a way of exploring prohibited things, for her religious upbringing had been governed by Gilbert Murray's aggressive agnosticism, which viewed the rise of Christianity and the other mystery religions as a deplorable "failure of nerve,"[34] that is, an abandonment of the rationality that made the ancient Greeks worthy of study and emulation. But, as we saw, Rosalind experimented with other forms of rebellion too. Her adolescent flirtations with anarchism and the pursuit of blood sports were more important than religion, to judge from surviving letters; and her attraction to Catholicism did not become important again until the late 1920s. By then, the literary failures of the novels in which Rosalind's personal frustrations had found thinly disguised portrayal may have impelled her to give a more overt expression to the rebelliousness she had dutifully suppressed during the first years of her marriage. At any rate, the earliest surviving record of her revived religiosity dates from 1929, when she went on holiday with her parents and found herself debating "the religious hypothesis" against her father.[35] Opposing her father did not yet mean opposing her husband, however, for in the same letter she continued: "It makes me appreciate more how curiously and blessedly you and I do agree about many important things."

The next stage carried her rebellion from words to acts. When Toynbee was traveling in Asia (and, while wrestling with sexual temptation in Shanghai, encountered the spiritual reality that so altered his outlook), Rosalind became a Christian by dint of an Anglican baptism. This took her parents aback and may have surprised Toynbee as well. He explained to her mother: "As for Rosalind's baptism, it happened while I was away, and she rather wondered if I should mind about it. I didn't, for I don't believe she will alter her dear self by putting on any label which she fancies; so I don't worry—though I dare say I shouldn't be quite logical in taking any extreme calmly—for instance, I should be worried if she thought of becoming a Catholic! But happily that doesn't seem to be on the horizon. For myself, I steadily become more undogmatic. I simply can't conceive of myself belonging to any religious institution."[36]

When Rosalind did express a wish to receive instruction in the Catholic faith, Toynbee's initial reaction was to oppose her new eccentricity. He suggested she should see a psychiatrist instead; and being a self-consciously sophisticated, modern woman, she obliged him by consulting Dr. Sylvia May Payne of the Institute of Psycho-Analysis. Dr. Payne declared her entirely sane, but for a while Rosalind held back. As she wrote in 1949: "It seemed to me that I must do no more, that it could not be God's will for me to persist against my husband's wishes. . . . Then suddenly, quite unexpectedly, he relented: without my saying anything more about it, he came in one day and said he had been wrong, that I should be free to see a priest if I wanted to, and that he had in fact arranged an interview for me, through his secretary who was a Catholic."[37]

The secretary in question was, of course, Bridget Reddin, Lawrence's former nurse, whose role as family retainer gave her a specially privileged status by virtue of her years of loyal and humble service. Her part in Rosalind's conversion was probably far greater than written records show. Bridget Reddin's ready acceptance of the status of pious and faithful family retainer exactly fitted Rosalind's aristocratic aspiration and expectation of deference from inferiors. A religion that sustained such behavior, and induced Bridget willingly to shield her from all the disagreeable tasks involved in raising her youngest son, attracted Rosalind as much for its social implications as on any theological ground. This is plain in the polemic she wrote against her father, *The Good Pagan's Failure,* which attributed the triumph of barbarism and egalitarianism in the late 1930s to the abandonment of the Catholic view of human and celestial hierarchy.[38]

Whatever her motives, after her conversion Rosalind had finally emerged from her parents' shadow[39] and launched upon a decade-long struggle to convince her husband that he, too, should accept Catholicism as the only cure for the world's ills. Her arguments did not persuade him, and she may have resented her husband's intellectual independence. Her son Philip, who had just turned seventeen at the time of her conversion, disdained her efforts at intellectual debate on matters of religion.[40] In their published form her arguments often remain inchoate and seem more shrill than convincing.[41] Nonetheless, Rosalind had a strong will and a high opinion of her own abilities. This must have made her son's scorn and her husband's unwillingness to agree with her theological convictions very hard to take.

Gilbert Murray kept his feelings about his daughter's conversion very much to himself. Lady Mary reported: "There really isn't much to say. Her father hasn't talked to me about it, and has gone on the glacier. [They were on holiday in Switzerland.] I think it is a pity that he has minded this . . . but I cannot think it will really separate them speculatively much more than has been the case for some time. I think, except for her happiness (and that's a big except!) I should be sorry if you followed her, because it might put your historical genius in blinkers."[42]

Toynbee replied a few days later: "Very happily, as it has turned out, her becoming a Catholic has made no rift between her and me. In fact, I am sure the result has been to bring us closer together." Still, he continued: "There is no likelihood at all, so far as I can foresee, of my becoming a Catholic too." He went on to point out that Rosalind's Catholicism was a more constructive response to the loss of belief in "an apparently reasonable world before the crash came" than were its Marxist and fascist competitors. "Personally, I don't propose to take any of these ways, but to stay outside, unattached, and go on looking at things, though, in the world as it is, that is perhaps a bit like sitting in the observatory on Vesuvius during an eruption. I think the practical thing to keep in mind about Rosalind is that she has a very strong and definite character, and the Catholic church isn't likely to be able to turn her into a different person, even if it wants to."[43]

Toynbee was thus able to brush off the significance of Rosalind's conversion, for himself and for her. What he could not dismiss so lightly was the change in his relation with Tony, his eldest son. Beginning in 1930 Tony reacted to the onset of his adolescence by refusing to compete with his father's dazzling record at Winchester. At home in summertime he refused to make up for slipping grades by extra work and withdrew into sullen silence. Not knowing what to do, his parents sought medical advice from a doctor who informed Toynbee that Tony suffered from "slowness in reaching puberty and consequent tension, taking the form of opposition to me and determination to be unlike me in all respects. This seems to fit the facts."[44]

Tony's determination to differ from his father persisted, with the result that in 1932 he was withdrawn from Winchester for academic failure. Though he expressed a wish for a military career in India,[45] his parents decided to send him to Germany instead, with the idea that a year of private instruction might prepare him for admission to a German university. The scheme meant escape from his father's shadow, and Tony accepted the new arrangement cheerfully enough. "I do believe we may, with luck, have launched him on the right path," Toynbee remarked.[46]

But when Tony came back from Germany to Ganthorpe for the summer of 1933, relations with his parents reached a new low. He persisted in sullen withdrawal, idling his time away by playing with guns, and then quarreling with Philip over the affections of Laura Bonham Carter, "to whom," Rosalind reported, "they have both, it seems, fallen victim."[47]

Toynbee was profoundly puzzled and distressed by Tony's behavior, so contrary to his own adolescent response to the high expectations his parents had had for him. He had neither time nor patience for fruitless discussion, since various writing tasks always weighed upon him. Yet Tony's withdrawal made family relations painful. "I am afraid you went away with a heavy heart on account of the boys being so silly and unresponsive," Edith Toynbee wrote soon after his departure for North America.[48] He replied: "Yes, I have been worried, not so much about the boys in general as about Tony. . . . There is something all tied up and unhappy inside him, but one can't get at what it is. . . . He has been doing no work at all this summer and showing no interest in anything else, (e.g., not in tennis, which he was keen on last year)."[49] Rosalind tried to put a cheerful face on family distresses by telling her husband: "I believe it is going to be for the best. I don't think it is either your fault or the boys' fault, but the two of them and you is altogether not going to work out for the present."[50]

But Rosalind, too, found Tony hard to take. "He says himself that he is not interested in anything, and that revival of the fire arms mania is only an attempt to fill a vacuum and not a genuine interest. . . . What can one do for him?"[51] She decided to try Dr. Payne, the same psychiatrist who had counseled her in connection with her wish to become a Catholic. Tony "leaped at the suggestion."[52] The result, Rosalind wrote to Toynbee, was that Dr. Payne "considers the jealousy with Philip extremely important and is emphatic that the policy of keeping them as much apart as possible is right. She also thinks him extremely fond of me—though somehow inhibited in expressing it."[53]

By the time Toynbee returned from North America, the older boys were safely out of the way and family relations relaxed correspondingly. Philip, who was starting his last year at Rugby, actually throve on Tony's disgrace. He did well academically and also became a prefect and member of the school's rugby team. "You may be right about Tony being the 'element of discord,' " he wrote to his mother, "though personally I am inclined to blame myself just as much. I can't help feeling that next year will be all right. . . . I am beginning to feel a real affection for this place."[54] As for Tony, he was duly admitted to the University of Bonn, where he promptly joined a student dueling fraternity and reported home: "I took an active part in a demonstration against the treaty of Versailles with the rest of the Rheins [his fraternity]."[55]

On the other hand, Lawrence, who was just eleven years old, attracted his mother's undiminished affection by being "his nicest and happiest little self."[56] He accepted instruction in the Catholic faith without protest, and in 1935 left home to enroll in the school maintained by the monks of Ampleforth Abbey, not far from Ganthorpe. This established a link with Ampleforth that became increasingly important for Toynbee in the following years.

However painful he found them, Toynbee never allowed his family difficulties to interfere with his work, and the work that always mattered most was his big book. When in the summer of 1930, at age forty-one, Toynbee first began to busy himself with the monumental task of writing up his notes for *A Study of History*, he was sure in his own mind of the significance of the enterprise but profoundly uncertain about its reception by others. As we have seen, he gave up referring to his book as a philosophy of history and adopted Rosalind's derogatory nickname, the Nonsense Book, as a sort of camouflage. Toynbee nonetheless maintained an unwavering commitment to his self-appointed task of working out the full scope and implications of his vision of history. His reading convinced him that the record of events in China, India, and even in the tangled Middle East could be structured into a series of tragic civilizational cycles, each conforming to the archetypical career of Hellenism that he had sketched so elegantly in his Oxford lecture "The Tragedy of Greece" in 1920.

As he read, he looked for correspondences with what he already knew about the ancient Mediterranean world, and his eager imagination usually succeeded in finding what he sought. Sometimes he failed. "In a day or two I shall be sending back Furlani's book on Babylonian religion. He has helped me solve my equation—X:Babylonian religion = Xity:Greek religion, X proving to be naught, as I had suspected. This is odd, but these exceptions tell me something."[57]

A few awkward anomalies, like the sudden emergence of the Ummayad Caliphate in the seventh century A.D.—a universal state without any obvious preexisting civilization for it to unite—provoked Toynbee into making rather implausible assertions to save his system. Thus he declared that the Arab conquerors, inspired by Mohammed's newly minted revelation, were "unconscious and unintended champions"[58] of a "Syriac" civilization that had gone underground a thousand years before at the time of Alexander's conquest. No one before Toynbee had conceived of Syriac civilization, and it seems safe to assume that he invented the entire concept in order to be able to treat the Ummayad Califate as a universal state with a civilization of its own.[59]

Despite a few implausible inventions like this, Toynbee's synthetic imagination and extraordinary erudition combined to create a powerful and entrancing vision of the human condition. He knew its power, believed in its truth, and labored furiously to give intelligible form to the whole human past by spelling out the recurrent patterns and overarching cycles he had discerned. States of mind and feeling were always the critical factor, according to Toynbee. In different places and times, human beings had accepted the restraints and benefits of civilization for a while, but only as long as "creative minorities" succeeded in meeting successive challenges so successfully as to be able to attract the willing assent of their fellows to whatever innovative actions they advocated.

But persistent flaws in human nature meant that, sooner or later, the growth of every recorded civilization had been checked by the unleashed brutality of war. Repair of war's ravages in turn required or at least permitted a now "dominant minority" to achieve peace by establishing a universal state. But this involved resort to wholesale compulsion, because the minority was no longer creative, and could no longer count on the willing assent of subjects and followers. Compulsion in turn bred alienation, creating internal and external proletariats that eventually combined to supplant and then destroy the civilization in question by transferring loyalty to a new religion and allowing barbarian war bands to break up the weakened universal state.

Such a vision of human affairs gave enormous importance to war as the axis and occasion of the breakdown of civilizations. In the early 1930s, when Toynbee was writing the first three volumes, World War I still dominated the public mind, and the obvious question raised by his vision of the human past was whether Western civilization had already broken down, like all the others, or whether the challenge of creating a viable international order could be creatively resolved, giving the West another chance for growth. The question still seemed open when the first volumes came out in 1934. The first installment of *A Study of History* therefore served as a grandiose background argument for the advocacy of collective security—advocacy that more and more emphatically infused Toynbee's annual *Surveys of International Affairs* as the Fascist and Nazi challenges to Europe's peace intensified after 1934.

By December 1931 Toynbee had written enough of his book to approach Humphrey Milford, publisher to Oxford University Press. "I wonder," he wrote, "whether you would consider publishing a large work of mine. . . . I am calling it 'A Study of History.' "[60] The letter went on to compare the work with Frazer's *Golden Bough,* projecting a total of six volumes to be delivered and printed in three batches, two volumes at a time. Four weeks later Milford replied: "I can't resist your *Study of History,* vast though it be,"[61] and went on to propose a draft agreement. A contract, dated 18 February 1932, soon followed. It prescribed a total of six volumes, each between 500 and 600 pages, and offered a royalty of 10% on the first 750 copies sold, $12\frac{1}{2}$% thereafter, to be divided between the author (75%) and the Royal Institute of International Affairs (25%). Toynbee felt grateful to Humphrey Milford for accepting his book so quickly,[62] though in fact at least two other publishers showed lively interest in a work whose qualities were already a matter of gossip in English academic and literary circles.[63]

The enterprise became known because, after his first summer's work on the book in 1930, Toynbee started to send parts of the text to various friends and colleagues for criticism. At first he was hesitant and humble, perhaps genuinely unsure of what reception his grandiose new approach to human history would provoke. "I enclose the first batch of my Nonsense Book—sheets 1–187 inclusive," he told Veronica, "I haven't a notion whether there is something in it or whether it is awful rot. The critics will give me light."[64] To his father-in-law he wrote:

27 October 1930

Dear Murray,

Here is the first part of my book of nonsense. If you are able to look at it sometime I shall be immensely interested to know how it strikes you. I don't know myself whether it is really nonsense or not.[65]

Gilbert Murray's reply seems not to exist; but, though he took exception to some details, his general tone was very warm, as Toynbee's response makes clear:

Dear Murray,

It really was good of you to read my stuff so quick. I wanted your comments more than anybody's, and yours are the first to come in.

I gather that it hasn't struck you, at first glance, as being all my eye—and this is a great relief: it is so difficult to judge for oneself whether one's private thoughts are sense or nonsense.

The besetting danger of work like this is crankiness, and I see that, to your mind, I came nearest to this in my animus against Western civilisation. I will correct this in the two places where you point it out, when I revise, and keep looking for it all through. . . .

I am conscious of having a certain "down" on Western civilisation and have often tried to think out why I have it, in order to get it into proportion. I expect it is largely personal "uncorporateness," which makes me rather hostile, in the same sort of way, to Winchester and Balliol and the British empire; partly it is the effect of the War, which for anyone of my age, is bound to seem the chief expression of Western civilisation, so far, in one's own lifetime, and partly it is the effect of a classical education. I think I have a tinge of Renaissance feeling that the Ancient World is the real home of the human spirit, and that what came after is rather a pity—like dog latin.

The letter takes up a few specific points Gilbert Murray had queried and then goes on:

I am very glad you like my metaphors. More and more, I see how right Plato was in trying to convey the really important ideas in myth.

This whole work of mine is really a myth about the meaning of history and I suppose (so far as I can see it through my small spectacles) the meaning of life.

Do you find that things which you read or thought twenty or thirty years ago come into place in your thoughts now, as if you had been unconsciously planning, all the time back, what you are making today? This keeps on happening to me in this work. For instance, when I was thinking about the geneses of civilisations and had rejected race and environment as explanations, the "Prologue in Heaven" of Faust, which I hadn't read since I was in school, came into my head, and I saw that the wager between God and the Devil was a far better explanation of the dynamic force which brings a civilisation into existence. Now I have been re-reading Faust, and find a dozen things in it which make my own ideas clearer.

Well, thank you very much indeed for giving your time to my stuff. Sometime,

I should like to have a talk about it, and I hope to send you a second installment
next summer, when I have finished the Survey for 1930. . . .

Yours ever,
Arnold[66]

Of all the persons to whom he submitted his first volumes for advance criticism—Norman Baynes, Alfred Zimmern, G.F. Hudson, J.L. Hammond, George
Gooch, H.J. Paton, H.A.R. Gibb, his mother Edith Toynbee, and others[67]—
Toynbee felt most at ease with his father-in-law, and this account of how he
tapped schoolboy memories and drew on hidden recesses of his mind in writing
the early volumes of A Study of History is therefore unusually frank and revelatory of Toynbee's method. Long afterward, he claimed to be empirical, in
contrast to Spengler's dogmatism;[68] but a better term would be unabashedly
intuitive or poetic. To be sure, historical writing is always poetic. Toynbee's
work was unusual only in making the poetic dimension of his achievement
more transparent than the conventions of academic history, as defined in German seminars of the late nineteenth century, had allowed.

In 1934 the power and sweep of Toynbee's portrait of the past impressed his
English contemporaries profoundly. This is attested by the extraordinary reception accorded to his first three volumes. The *Times Literary Supplement* devoted an entire front page and two inside columns to a review which hailed the
book as a "nobly conceived and assiduously executed work" and only modu-
lated its praise by observing that its "mark of greatness is a question which
posterity must be left to answer." Leonard Woolf, writing in the *New Statesman and Nation,* declared that "The sweep and scope of this work is so magnificent that the reviewer must feel that it is an impertinence . . . to attempt
to deal with it in a few hundred words. . . . There has been no book published since the war which anyone with a healthy appetite for print and speculation will read in greater excitement and enjoyment." J.L. Hammond in the
Manchester Guardian called it "a large, measured, tranquil and philosophical
examination of history," and concluded: "There can be few people for whom
the reading of Mr. Toynbee's history will not be a deeply significant event."
Gilbert Murray was among the very first to acclaim his son-in-law's work,
declaring in *The Observer,* "It is without doubt a great book.[69]

The only shadow of the sort of academic attack that Toynbee was later to
suffer came from the pen of E.L. Woodward, who, after calling the book an
"extremely interesting and ingenious study" went on to ask: "One may wonder whether Professor Toynbee (or any single man in the present state of learning) can really be sure that his selection from the accumulated data about the
past is not open to attack."[70]

Critics' praise was matched by surprisingly vigorous sales for such a learned
and lengthy a work. By January 1935, a mere six months after publication, the
first edition was exhausted, and royalties of £333 had accrued to Toynbee and
Chatham House.[71] A second printing allowed correction of small errors of the
sort his mother excelled at discovering.[72] Toynbee could not have hoped for a
warmer reception than his first three volumes received. On the strength of his

achievement, Oxford awarded him an honorary degree, which pleased him so much that for the next few years the title pages of his books identified him as Arnold J. Toynbee, Hon. D. Litt. (Oxon). His later career as a public personality even got a small start when a popular magazine called *Everyman* hailed him as "Personality of the Week" on 4 January 1935.

Private letters echoed the public praise. Alfred Zimmern wrote: "The more I turn the pages the more I admire the 'Nonsense' (forsooth!).[73] D.C. Somervell, a schoolmaster quite unknown to Toynbee personally, but who subsequently linked his name with Toynbee's by condensing the first six volumes to a few hundred pages, wrote to say he had read the book twice and found it "enthrallingly interesting."[74] Rosalind's reaction to her husband's spectacular success is not recorded; but she can scarcely have felt completely comfortable when a work she had derided for years as the Nonsense Book won such praise. Remembrance of her own literary failures may well have made Toynbee's success bittersweet for her.

Two observations about the initial reception of Toynbee's first volumes seem worth making. First, volumes I–III of *A Study of History* made comparatively little popular impression in the United States, where Toynbee's post-World War II fame was to center. To be sure, a handful of distinguished professors reviewed the work quite sympathetically in American learned journals, but New York's literary circles remained indifferent, perhaps because reviewers were put off by the way Toynbee sprinkled his pages with untranslated quotations from Greek, Latin, French, and German authors. Americans habitually dismissed works of scholarship as unreadable anyway and must have been daunted by a work whose first three volumes treated only two out of thirteen projected headings. Outside London, therefore, initial reactions to Toynbee's book were professional and polite, and usually also perfunctory.[75]

Second, from the perspective of more than half a century that has now elapsed since the volumes first came out, it is easy to recognize some of the special circumstances that contributed to the warmth of their initial reception. Timing was important, for 1934 came exactly at the tip point between the postwar and prewar eras in Great Britain. Hitler's rise to power in Germany in 1933 acquired new menace for Great Britain when he allied with Fascist Italy in 1936—a realignment that was provoked by the ineffective British and French reaction to Italy's assault on Ethiopia in 1935. A work that made warfare the key to civilizations' breakdown therefore spoke to the public anxieties of the late 1930s with special power and poignancy. Whether Western civilization has already broken down, or was about to break down, hinged, according to Toynbee, on whether the Great War of 1914–1918 would actually prove to be, as advertised, "a war to end war," or would instead turn out to be only World War I, as we have since learned to call it. As a result, all the far-ranging investigations of the distant past that Toynbee incorporated into the first volumes of *A Study of History* had an extraordinary topicality for British readers in the late 1930s.

Another secret of Toynbee's success was, ironically, the affirmation of some familiar ideas despite his "Copernican"[76] denial of the centrality of national and West European history for the human past as a whole. Toynbee's fundamental iconoclasm did not prevent echoes of nineteenth century notions from cropping up in his pages. For example, he avoided the word progress, but still spoke of a "law of civilisations moving forward" (I, 176) toward "the goal of human endeavors: the mutation of Man into Superman" (I, 194; cf. also III, 383). In another passage he referred to an "omnipresent power which manifests itself in the performances and achievements of all Mankind and of all Life," which may be conceived as transcendent and called God or declared instead to be merely an *élan vital,* immanent in things (I, 249). Such remarks harked back to his Victorian and Edwardian youth, and softened the impact of his principal message—the demotion of Western civilization to the status of one specimen among twenty-one "philosophically equivalent" civilizations. Such a mix of old and new fitted the time. His book was radical and revolutionary with just enough whiff of older faiths to allow its readers to feel that the future was not foreclosed, that Western civilization might yet be saved, and that God or His secularized equivalent, *élan vital,* was still in charge.

The path-breaking quality of Toynbee's work is, however, its principal claim to enduring significance. The greatness of these volumes rests, above all, on the simple fact that they expanded the range of historical consciousness beyond anything conceived by historians before him. By surveying the past for plausible equivalents to the Roman Empire (universal states) and for similarly plausible equivalents to the combination of Christianity and barbarian invasions (which he took as indicative of a relation between successive civilizations that he dubbed "apparentation and affiliation"), Toynbee discovered no fewer than twenty-one civilizations. His roster of twenty-one fully developed civilizations, together with extended discussions of an oddly assorted group of "abortive" and "arrested" civilizations—Eskimo, Ottoman, Spartan, and others—made the scope of his researches and the range of his analysis literally worldwide.

Toynbee's pen thus made an intellectually comprehensible vision of all human history accessible to his readers. His history was simple in its grand lines, however subtle and extended in its exempla. The scheme had the elementary virtue of being complete, with room for all of humanity's known and knowable past. Nothing of the sort had been available to the Western world since sometime in the seventeenth century, when the educated classes had been compelled to abandon the biblical account of humanity's common descent from Adam, due to its incompatibility with new knowledge flowing into Europe from the Americas and elsewhere.

One might argue, perhaps, that Johann Gottfried von Herder (1744–1803), with his vision of autonomous cultures, each expressing a distinct national spirit bodied forth primarily in language and literature, was potentially as comprehensive as Toynbee's parallel civilizations.[77] But Herder was only marginally concerned with the non-European world, being interested mainly in rebutting French pretentions to a monopoly of true civilization. Moreover, in the subsequent decades of the nineteenth century, European scholars narrowed their view by denying historical significance of eras and places that failed to contribute to "Progress"; and regardless of whether "Progress" was defined in terms of truth, freedom, power, or wealth, it remained always a European monopoly.

This turned out to be the meaning of Hegelianism for historians, whose parochialism (compared with their predecessors of the eighteenth century) was reinforced by the growing margin of superiority Europeans enjoyed over other peoples as a result of the industrial and democratic revolutions.

Toynbee challenged this smug view of history more effectively than any predecessor except Spengler. Between them, Spengler and Toynbee announced a new breadth of historical consciousness; and of the two, Toynbee was the more significant inasmuch as the English language had a wider reach than Spengler's German, and Toynbee's examples ranged more broadly and stayed closer to everyday notions about human affairs than Spengler's generalizations did. (Spengler explained history as the product of personified abstractions—"Dionysiac" and "Apollonian" Kulturen, for example. Such discourse is profoundly alien to the English language, and seems opaque and unconvincing to anyone not already attuned to German philosophical Idealism. Spengler's readership was therefore inherently more limited than the public that was prepared for Toynbee's less abstract mode of discourse.)

Toynbee achieved his *tour de force* with a dazzling array of poetic inventions. He gave old terms new meanings: "challenge and response," "withdrawal and return," "internal and external proletariat," "mimesis," and many more. Breathtaking new perspectives opened as he imagined some of the might-have-beens of history. The "Forfeited Birthright of the Abortive Far Eastern Christian Civilisation" (II, 446–452) is a particularly good example of the fertility and range of his imagination. In this passage, Toynbee treated Nestorian and Latin Christianity as equivalents, comparing their different fates in confronting the Arab Moslem assault of the early eighth century. But the Nestorian culture of Central Asia was something of which few of his readers had any notion, whereas Latin Christendom was their birthright—the center of medieval history as taught in schools. To treat it as equivalent to Nestorianism, and no more than a semibarbarous rival to the civilization of Islam, as Toynbee made it out to be, shocked and challenged all the familiar ethnocentric notions of the past his readers had grown up with. Such novel vistas, renewed repeatedly as he discoursed on other unfamiliar passages of history with a tone of confident mastery that not even Spengler had dared to aspire to,[78] made reading *A Study of History* an intellectual adventure.

After more than half a century, reading Toynbee's pages still remains an adventure. The dazzling range of his information, the boldness of his comparisons, the perspicacity of his reflections, together with odd bits of information, such as the capture in 1794 of the Dutch fleet by French cavalry, galloping across the ice in which the ships were frozen fast (II, 263), all combine to make his first three volumes worth anyone's attention, even if his twenty-one civilizations and their cycles no longer seem as convincing as they did when the book was new, and when the threat of another war that was widely expected to destroy European civilization loomed unmistakably ahead.

However satisfying the success of his first three volumes may have been, their publication did little to relieve Toynbee of the strain under which he had been working. His responsiblities for the annual *Surveys of International Af-*

fairs continued as before; journalistic writing ate into his time; and he still had eleven of the thirteen heads of *A Study of History* before him. A grant of £8000 from the Rockefeller Foundation of New York to support Toynbee's researches[79] helped Chatham House finances, but that only made him feel, more strongly than ever, an obligation to "show results as soon as I can by publishing the middle batch."[80] He therefore pressed strenuously ahead, even at the risk of his health. After working so hard at the book that his "bearings get over-heated,"[81] he had to rest. "I am now trying to get out of the net of sleepless-ness by stopping writing for a few days," he confessed to Veronica.[82] But three days later he was back at his desk "because . . . it seems like a race between finishing it and being overtaken by whatever may be going to happen in public affairs."[83]

Toynbee's mounting fear of another war led him to alter the original plan of the work substantially. "I have now decanted all but the dregs of VI, VII and VIII into V, and am within sight of the end of V, so I am not doing so badly," he confided to Veronica.[84] The Roman numerals referred to the main heads of his original plan, according to which Part V, "The Disintegration of Civilisa-tions," was to be followed by Part VI, "Universal States," Part VII, "Uni-versal Churches," and Part VIII, "Heroic Ages." By "decanting all but the dregs" of these parts into Part V, Toynbee greatly extended the scale of his treatment of the "Breakdowns of Civilisations" and "Disintegrations of Civil-isations" and when the next batch of volumes duly appeared in August 1939, a week before the outbreak of World War II, only those two sections of the original plan were listed in the Table of Contents. Nonetheless, with the ap-pearance of the second batch of volumes (IV–VI) most of what he had to say about the fundamental rhythms of the rise and fall of civilizations had been put onto paper. Only "Contacts between Civilisations in Time" and "in Space," together with projected remarks about the "Prospect of Western Civilisation," and his own inspiration remained untouched from among the themes he initially projected. By the time war came, Toynbee could therefore feel that regardless of what happened next, he had in fact been able to complete the basic task he had set himself in the 1920s.

He achieved this goal only by working at a pace that taxed his physical strength to the limit. Sleeplessness returned to haunt him as public and private anxieties intensified. Repeatedly, he had to use "sleeping draughts," though he feared and disliked dependency on drugs.[85] In 1937 he hit upon an alterna-tive, which helped him to relax and attain something like peace of mind, by asking permission to visit Ampleforth for a few days and participate in the routines of monastic life. "I am spending a week with my friends in the mon-astery here," he wrote to Veronica. "It is a particularly serene and cheerful place and is doing me a lot of good."[86] This visit inaugurated a close friend-ship with Columba Cary-Elwes, a monk at Ampleforth, who was serving as librarian when Toynbee first stayed at the monastery. The emotional relief he found in submission to monastic routine also helped to sustain the intensified religiosity that characterized Toynbee in these years of strain. But brief, occa-sional respite scarcely affected the furious pace of his labors. "I have revised

two-fifths of Part IV in eight days (including Christmas), which isn't bad going,"
he boasted to his ever-faithful helper and colleague. "Your answers to conun-
drums have been . . . speeding me on my way." [87]

It is scarcely surprising that a man who could spend Christmas day at his
desk continued to provoke rebelliousness in his sons. Family problems became
particularly acute when Philip joined Tony in defying his parents. He did so,
aged eighteen, by becoming a Communist in 1934. His flamboyant temperament compelled Philip to act on his new principles in extremely conspicuous
ways, thus distressing his parents even more than Tony's sulkiness had done.
"We have had a most awful crash with Philip," Toynbee reported to his mother,
Edith. "He ran away from Rugby on a moment's impulse last Wednesday
while we were in Berlin, went to a 'Young Communist cell' in London, took
part in the row at the Mosley meeting—though he came back of his own accord
to Rugby on Friday morning and thereby escaped being expelled—we have to
take him away at once: we did it yesterday." [88]

Faced with the problem of what to do with Philip, the Toynbees turned to
Ampleforth. The monks agreed to accept Philip as a visitor, hoping to induce
repentence and provide a setting in which he could prepare for entering Oxford
next fall in spite of his forced withdrawal from Rugby. Philip submitted more
gracefully than might have been expected, but in the following summer he once
again ran away to join his new Communist friends. "Alas, for your kind inquiry: we are in more trouble with Philip," Toynbee informed Veronica. "Passing
through London from Oxford to Dieppe, he broke his word, and went to see
the Communist boy Romilly again. . . . What is really serious is that Romilly's 'Young Communist Bookshop' is, as we have heard from Philip himself,
a nest of vice; and also that, now that we can't rely on Philip's word, the
possibilities open to one for dealing with him narrow themselves down enormously.

"I am, I confess, rather laid out, partly because I am a less stalwart man of
action than Rosalind, but even more because the Old Adam rises up in me and
resents the position furiously—which is certainly human but still more certainly
wrong and futile. . . . What a life!" [89]

Philip's exhibitionism and Tony's sulkiness thus succeeded in provoking their
father to anger. This only redoubled his trouble, since he believed that an angry
reaction to his sons' behavior was wrong. Toynbee was not accustomed to
being wrong, yet found no escape. No wonder sleeplessness haunted him! Or
that he buried himself so desperately in his work!

In a memoir written after Toynbee's death, Philip, looking back on his youth,
concluded: "He [Toynbee] simply had no understanding of children and young
people, and no great interest in them either. My two brothers and I attracted
his attention largely as nuisances. How clearly, even today, I can see his head
poking out of the window of his study, his face a mask of nervous irritation,
as he sternly reproved us for making too much noise." Moreover, "he kept
telling us how expensive it was to bring us up," and, more generally, "hadn't
the time or energy left to study and understand the people who were closest to
him. . . . His tendency was to put friends and members of his family into

certain pigeon holes, as it were: 'Frederick, kind but wasteful,' or 'Dorothy, very strong minded, but foolishly indulgent towards her own children.' Once somebody had been put into his or her pigeon hole, there was little hope that he would ever get out of it. . . . He had, as it were, dealt with this friend, once and for all, and felt no need to examine him any further.''[90]

Philip's recollection seems fair, as far as it goes, but it omits Toynbee's leading trait: his persistent effort to smooth over awkward relations with the people around him through a studied mildness of manner and by burying himself in his work. As we have seen, Toynbee lacked ordinary sensitivity in personal relations. He very much needed to be surrounded by admiring females, but his habits and outlook simply did not allow him to sympathize with his sons in their adolescence. They responded by engaging in opposite and extreme forms of rebellion against the fiercely high expectations he and Rosalind had for their offspring.

As for Rosalind, she found consolation in religion for the breakdown of her dominion over her sons. A letter she wrote to her mother in August 1938 reveals something of her state of mind. Lady Mary had just returned from a visit to Ganthorpe, and Rosalind felt a need to justify herself, as follows:

> I believe, too, that in a way you perhaps thought me hard in my attitude about the elder boys. . . . I don't know if I can explain it all properly, and present "my defence" so to speak, so that it seems less callous or indifferent than I am afraid you thought me. You see, I have minded so terribly . . . about the badness of the boys—who have been I suppose what I have cared about most in the world, and given most thought and effort to. . . . There was a time when I first realized these things and made myself face them—years ago now—when I almost was defeated; when I hated and feared life so much that I nearly went to pieces. But instead I did find God in a way I never had before, and that . . . infinitely outweighs all the evil and the suffering of this world. . . .
>
> I know that in Bun [Lawrence] I am still most vulnerable. I still do hope in him and for him, and mind almost as much as ever how he grows up and lives; but I try, so far as I can, to be prepared in him also for disappointment and sorrow, and if it comes I shall not be less sure than now of the absolute goodness of God.[91]

This, then, was the painful family background against which Toynbee ground out the second batch of A Study of History in the attic at Ganthorpe. Moreover, his familial distresses were simultaneously reinforced by the paradox of professional failure looming behind his remarkable personal success. As we saw, he had sought to justify his decision not to enlist in the army during World War I by devoting himself instead to the really important task of making a secure and lasting peace. But peaceful settlement of international disputes became more and more of a will-o'-the-wisp after 1934, when rearmament got under way and the actuality of war in Ethiopia (1935–1937) and Spain (1936–1939) presaged renewed combat in the heartlands of Europe. The precipitous collapse of the League of Nations' ideal of collective security after 1934 therefore undermined Toynbee's self-esteem in hidden yet agonizing ways.

His hidden, personal hurt made his reaction to Mussolini's attack on Ethiopia almost apocalyptic. When the League of Nations first took up the question of Italian aggression, Toynbee thought that the Italians would be easy to stop. "Personally, I should like to close the [Suez] Canal and I would dare Italy to go to war with us. I find it very hard to stomach allowing a very horrible war in East Africa when we could stop it in this way in a moment."[92] He saw the confrontation in eschatological terms. "It is the real thing all right—all the forces of good and evil are on their hind legs now. I think we shall manage."[93] And again: "I think there may be war; but then, if we don't take the risk of a lesser war, in a good cause, against the least formidable of the predatory Powers, I feel perfectly certain that we shall soon be fighting for our lives—just to save our skins. . . . The British Empire is the big prize."[94] Three months later, his mood was still high. "I believe that we are now going to have a showdown, whether Mussolini attacks us or submits to defeat without hitting out; and I believe that, if once the Covenant is enforced successfully against a Great Power, the trick is done—provided that at the same time we open the safety valve of peaceful change at our own expense. All bluffs have now been called and the issue is clear."[95]

Then in April 1936 Toynbee's hopes and calculations were all abruptly shattered. Britain and France refused to back effective sanctions against Italy after all. In private, Toynbee exclaimed: "The whole thing is so infantile, as well as so evil, that it makes me sick to think about it."[96] Publicly, he made his distress known by writing another poem in Greek—this time an ironical adaptation of verses by an obscure ancient author named Moschus—and submitting it to *The Times*. On 22 April 1936 Toynbee's Greek verses were published in the letters to the editor column, along with a translation into English provided by his long-time associate Chatham House, G.M. Gathorne-Hardy. The translation reads as follows:

Epitaph on Abyssinians and Europeans

Without our arms or art these men could dare
War's utmost frightfulness, since men they were;
And, in close fight, to death untrembling passed,
As free men, battling nobly to the last.
But we, whose science makes us strong and great,
Are doomed to share the tortures of their fate,
Yet not their soldiers' grave; the gods in scorn
Withhold that privilege from men foresworn.

Toynbee's outrage rested on his conviction that the British government had betrayed national and imperial interests, not to mention the interests of humanity at large and the principles of the League along with his own personal and professional *raison d'être*. How to handle such malfeasance in the *Surveys of International Affairs* presented him with a delicate problem, since the authorities at Chatham House were not accustomed to attacking the British government, and some members of the governing council of the Royal Institute of International Affairs entirely sympathized with what had been done.

Toynbee was clear in principle. "The treatment of Great Britain," he wrote
to the Vice Chancellor of Birmingham University, "when she appeared in an
unflattering light should be just the same as that of France or Germany or Japan
in previous volumes in cases in which those countries had played an invidious
part on the centre of the stage." [97] All the same, he anticipated friction. "What
a story—and what a wigging we shall get for telling it so wickedly!" he wrote
to Veronica. [98]

Since Toynbee's views were well known, the authorities at Chatham House
took the precaution of checking the text of the second volume of the *Survey*
for 1935, [99] which dealt with the Italo-Ethiopian war, and agreed to publication
on condition that Toynbee add to the preface a remark to the effect that "other
honest people thought otherwise." [100] Reviewers did indeed accuse him of
abandoning objectivity by passing "moral judgment" on British policy; but
Toynbee was unrepentant, deriding his critics for avoiding the question of whether
or not his reproaches were deserved. [101]

Britain's betrayal of the League in April 1936 constituted a kind of watershed in Toynbee's outlook. Thereafter, he abandoned hope of any this-worldly
solution to the ills besetting Western civilization. He convinced himself that
only a redirection of mind and spirit, substituting the worship of God for the
suicidal worship of parochial national states, could forestall the breakdown of
Western civilization. He never entirely despaired of the possibility of such a
conversion (or reconversion), though he believed it might require further suffering and travail before human minds awakened to the misdirection of loyalties
dominating the twentieth century. His vision of history suggested, indeed, that
nationalism and the collective self-worship it had turned into cried aloud for
the withdrawal of a creative minority from the hurly-burly of everyday affairs
in order that it might at some future date return with a healing pattern of belief
and faith. He aspired to belong to such a minority, earnestly seeking a fuller
knowledge of God. Yet despite resort to prayer, and the sporadic relief he
found by conforming in vacation times to the pattern of monastic life as practiced at Ampleforth, he never actually found what he sought. His personal
encounters with transcendent spiritual reality remained fleeting and ambiguous,
despite their importance for his private life.

Toynbee's abandonment in 1936 of liberal, this-worldly hope for peaceful
settlement of international disputes, and his enhanced religiosity, meant that
subsequent events, up to and including the outbreak of war in September 1939,
did not excite him nearly so much as the crisis over Ethiopia had done. As the
war scare of 1938 headed toward its climax at the Munich Conference, he
could therefore say: "I find that the bitterness is past: one had one's excruciat-
ing moments in 1935–1936, and this seems just the natural epilogue to that." [102]

All the same, the hope that peace might somehow still be preserved was
something Toynbee could never entirely abandon. In the circumstances of the
late 1930s, this meant, first and foremost, some kind of accommodation between Hitler's Germany and the victors of 1918. Toynbee had long felt that
the Versailles settlement penalized Germany unjustly and had faulted the French
for their refusal to make concessions. Even after Hitler came to power in 1933,

Toynbee continued to believe that offers of boundary revisions in the east and rearrangement of the colonial pattern of Africa might conciliate German public opinion and render Hitler's fanaticism harmless.[103] By 1934, after a visit to Germany during which he discussed Germany's "peaceful intentions" and need for colonies with Alfred Rosenberg, a leading Nazi ideologist, Toynbee concluded that Hitler was more subversive domestically than he had supposed, but he felt "less alarmed about the foreign policy of the Nazi regime . . . than I had been before."[104]

Two years later, in February 1936, he returned to Germany for what turned out to be by far the most significant of his encounters with National Socialism. Dr. Fritz Berber, an expert in foreign affairs and confidant of Ribbentrop, had invited Toynbee to address academic audiences in Bonn, Hamburg, and Berlin. In Berlin Toynbee spoke before the Akademie für deutsches Recht on "Peaceful Change." What he had to say was very welcome to his German audience. "We in England," Toynbee declared, "are beginning to think very hard about possible ways and means of arriving at some peaceful adjustment between 'have-nots' and 'haves'—of whom we are the chief." What he had in mind was to return former German colonies to a German administration, subject, however, to the terms of a "deed of trust" and "international inspection." He went on to propose internationalization of administrative personnel for such technical services as communications and health throughout tropical Africa "on the model of the China Maritime Customs."[105] This was heady stuff for German nationalists. A British diplomat informed him afterward: "Your lecture was an eager topic of discussion everywhere . . . especially among business men with foreign connections."[106]

His public performance was, however, trivial by comparison with the use Hitler tried to make of Toynbee's visit to Berlin. At the end of February 1936 Europe's balance of power was on the verge of decisive change. Italy, harassed by the League of Nations for attacking Ethiopia, was moving toward alignment with Germany. The French had just ratified an alliance with the Soviet Union. Hitler's program of rearmament had begun to make the German army and air force formidable. Accordingly, he decided to take advantage of Italy's breach with Britain and France by reoccupying the west bank of the Rhine, demilitarized by the Versailles (1919) and Locarno (1925) treaties. The advance of German troops on 7 March 1936 toward the French border caught the French and British by surprise. They referred this new treaty infraction to the League, where weeks and then months of debate eventually degenerated into passive acquiescence. His successful coup strengthened Hitler enormously, both at home and abroad, since, after March 1936, the French army could no longer count on fighting on German soil in case of another war, as had been the case before.

When Toynbee arrived in Berlin at the end of February 1936 Hitler was, of course, acutely aware of the risk he was about to take. His program for rearmament was far from complete and when the German army marched in it actually had secret orders to withdraw from the demilitarized zone if the French mobilized and began a counterinvasion. To make the humiliation less likely,

Hitler therefore took special steps to conciliate (and confuse) French and British opinion on the eve of this, his first overt military move to restore German power. A first extraordinary gesture to influence opinion in his favor came on 20 February 1936, when Hitler granted an extended private interview to a French publicist, Bertrand de Jouvenel, and made an impassioned appeal for an end to Franco-German enmity.[107] A week later, Hitler (or someone in his entourage) selected Toynbee as a suitably influential shaper of British opinion for a parallel private interview.[108]

For nearly two hours, therefore, the Führer discoursed to Toynbee about "his personal mission to be the saviour from Communism," on the importance of an understanding with Great Britain, and on the limited character of his aims in Europe and overseas. Unlike de Jouvenal, whose report of his interview appeared in *Paris-Midi* on 28 February 1936, Toynbee did not rush into print with Hitler's message. Instead, soon after returning to England, on 8 March 1936 he prepared a confidential memorandum for Anthony Eden, British Foreign Secretary, and the Prime Minister, Stanley Baldwin.[109] The document exists among Toynbee's papers, with an acknowledgment of its receipt from Eden, dated 11 March 1936. Since the reaction of the British government had already been defined on 7 March, the day of the invasion,[110] Toynbee's report of his interview with Hitler cannot have made much difference, though it may have reinforced the readiness of Baldwin and Eden to acquiesce in Germany's military reoccupation of the Rhineland.

Nevertheless, Toynbee's memorandum is interesting as an indication of his own viewpoint. In summarizing the lengthy speech he had heard in the German Chancery, Toynbee emphasized that Hitler wanted the former German colonies returned but had no ambition to conquer Europe. Toynbee paraphrased the Führer as follows: "The fundamental principle of National Socialism was to build a Reich on an exclusively national basis—reuniting the whole German nation, but not including anyone else." And again: "I want England's friendship," he said, "And if you English will make friends with us, you may name your conditions—including, if you like, conditions about eastern Europe." Toynbee then interjected in his own voice: "I have the very strong conviction that in this rather vital point, Hitler was quite sincere in what he said to me."

Nonetheless, Hitler did not entirely disguise his intention of changing the balance of power in Europe and the world. "Austria is bound to fall to us sooner or later," he remarked, and then went on to say, "If we [Great Britain] wanted a friend in need against Japan, why should that friend be Russia? Why should it not be Germany?"

Much of the Führer's discourse was historical, directed against the Russian menace. Hitler, in fact, impressed Toynbee with the precision of his knowledge of the past and with the sweep of his argument, which treated Stalin's communism as the latest in a long series of assaults upon Europe by Asiatic barbarians, reaching back to the age of the Huns and Avars.[111] More importantly, Hitler convinced him that the Nazi government really wanted and needed a peaceable understanding with Britain and France. Toynbee summed up his estimate of Hitler's needs and intentions as follows:

The weakness of Hitler's position is that he has always played a dramatic role in the sight of his German audience. Up to now, his role has been that of champion against Bolshevism. . . . If that still continues to be his principal role, it is hard to see how he can avoid coming into military collision with the Russians sooner or later. . . . My impression is that he has begun to realise the danger . . . and that he is eager to change his role and to appear as "the good European" and "the associate of England"—allowing his anti-Russian role to fall into the background. This would be an alternative way, for him, of getting the prestige and justification on the home front that he simply must have, in some form or other. If he can get it in a way that might lead to peace instead of to war, I believe he would be vastly relieved.

I therefore believe that any response from the English side . . . would produce an enormous counter-response from Hitler.[112]

Toynbee's optimism did not evaporate at once. The long drawn out debate in Geneva about sanctions against Italy was still in progress; when, in March, the League also took cognizance of the Rhinelands issue, Toynbee saw a chance for a general revision of the Versailles settlement. "Things seem to be turning out in a more promising way than could have been expected ten days ago," he wrote on 20 March 1936. "I hope and believe that the result may be something in the nature of a new peace conference . . . at which a negotiated settlement will be substituted throughout for a dictated one."[113]

When these extravagant hopes collapsed in April, his disillusionment was correspondingly profound. Privately, he began to forecast the imminence of a new world empire, believing that the competition lay between Russians and Germans for the role of twentieth century Romans. "Personally, I am inclined to think that the Germans rather than the Russians will play the Roman part," he wrote to his American friend Quincy Wright, "and it is not inconceivable that England and France may accept this greatest of all *faits accomplis* without making another world war about it."[114] Two years later, at the time of the Munich crisis, he had changed his mind about British readiness to submit to the Germans: "My belief is that Germany is going to do, within the next few days, exactly what she did in 1914: attack a small country in the conviction that England won't fight, and then find herself mistaken."[115]

Face to face with the prospect of war, his mood fluctuated wildly, and almost from day to day. His immediate reaction to Chamberlain's return from Munich on 29 September 1938 with the promise of "peace for our time" was ambivalent. "It is of course possible," he wrote, "that Chamberlain's policy is a delusion. . . . But I still have the feeling that something very big has happened, and that [the Germans'] mischief making power will turn out to have been much clipped."[116] In a paper titled "First Thoughts on September 1938 and After" he declared: "The principle of self-determination of nations has now at last been applied equally for the benefit of the nations that happened to be on the losing side in 1919–21."[117]

Yet two weeks later his tone had altered. This became clear when he wrote an essay entitled "After Munich" for publication by Chatham House in hope of clarifying public opinion about the new posture of international affairs. Toynbee argued that France and Britain had passed up their "last chance at remaining Great Powers," since the principle of national self-determination in Europe was "bound to produce a Mitteleuropa under German hegemony." Further resistance to Germany, he averred, had become impossible. These views were sufficiently shocking that his old friend G.M. Gathorne-Hardy was assigned the task of informing Toynbee that for now at least Chatham House would not publish his paper, lest it become "a most dangerous encouragement to Hitler." [118]

A thread underlying Toynbee's shifting judgments of international affairs was his conviction that the plural sovereignties of what he liked to call "parochial states" was an evil that had, somehow, to be transcended. As he wrote to Veronica, "The only constructive thing to work for, I feel, would be to get beyond national sovereignty—and I should follow that thread a long way, even if it led over some very rough country." [119] This could well mean surrender to Hitler, as he admitted shortly after the war had begun. "It would be possible to argue," he said, "that the world is in such desperate need of political unification . . . that it is worth paying the price of falling under the worst tyranny. . . . I will not try to suggest what one's decision should be. . . . We are all groping our way and finding it difficult to see clear." [120] Yet Toynbee's old liberal hope of somehow solving international problems by reasonable rearrangements continued to haunt his fertile imagination. As Munich loomed, he mused to Veronica: "The results of a war are unimaginable: a permanent United States of Europe to hold Germany down in the hope of admitting bits of her into the federation piecemeal, is the nearest to anything constructive I am able to make of it." [121]

Such a hope seemed the merest pipe dream in 1938, yet craven submission to Nazi domination was unacceptable. Toynbee simply did not know what to think or what public policy to recommend. Instead, he devoted all the time he could to the task of finishing the next instalment of A Study of History, which dealt with what certainly seemed the all-too-apposite themes of "The Breakdown of Civilisations" and "The Disintegration of Civilisations."

He continued to be responsible for preparing annual Surveys of International Affairs, of course, and these volumes grew in size as events moved faster and faster toward the outbreak of World War II. But the Rockefeller grant and other income accruing to Chatham House allowed Toynbee to hire others to write substantial parts of the surveys. Two new members of the Royal Institute staff took on regular assignments, and some chapters were contracted to outsiders as well. But above all Toynbee relied increasingly on the indefatigable competence of his longtime associate, Veronica Boulter. She got the surveys out, more or less on time, all properly edited, indexed, and checked for accuracy. She as much or more than he deserves credit for the impressive row of solid and substantial volumes that Chatham House published year after year until the outbreak of war in 1939.

All the same, Toynbee's heart was not in his work on the surveys any more. After his passionately restrained denunciation of British policy toward Italy and the Ethiopian war, the surveys reverted to chronicling events in careful detail, with little or no overt appraisal of acts of policy. As we have seen, Toynbee was profoundly ambivalent about the course of public events, and he had no wish to clash again with the authorities in Chatham House or at Whitehall. By assigning most of the writing to others, and concentrating mainly on planning and editing, Toynbee could minimize his commitment to the surveys and free himself for more attention to what he really cared about. He had always felt that his real calling was to write a great history and the warm reception of the first volumes of *A Study of History* confirmed him in thinking that this was what really mattered. Accordingly, he drove ahead as fast as he could and by the end of 1938 had readied the bulky manuscript of what became volumes IV, V, and VI for the printer.

While these volumes were in press, two personal catastrophes struck Toynbee in rapid succession. In February 1939 his mother died. This provoked very poignant memories and powerful regrets. Exactly what he felt cannot be reconstructed because his sister Margaret destroyed a letter of condolence in which, among other things, he explored the break with his mother at the time of his father's mental collapse. Nevertheless, enigmatic references in her reply make it clear that Toynbee deeply regretted the distance that had grown up between himself and his mother. "I am glad," Margaret answered, "that you felt able to write to me quite frankly, as you did, about your relationship with Mother. But it has left me feeling much sadder for you than for myself. For while I believe with all my heart that, as you say, your early intimate relation with Mother is the one that matters most and is somehow eternal, it is very sad for you to have lost that relation partly through something that was unnatural. Mercifully, I don't believe that Mother guessed this. . . . But she also felt at times as if you'd gone very far away—rather extra far—especially in view of your earlier relations to her, and . . . it puzzled her." [122] It seems unprofitable to speculate on what was "unnatural" about Toynbee's break with his mother in 1909. That he blamed himself seems sure.

Nonetheless, Toynbee did not go to see his father and tell him of Edith's death. Instead, his two sisters undertook that task. Jocelyn wrote of their visit: "Margie and I have just got back from Northampton. We can't tell you how glad we are that we went. We told Father straight away, quite simply, about Mother; it really was just as if he weren't ill at all. He realised what had happened absolutely at once and cried in just the same way that Margie and I can't help crying when it comes over us. He never said one word of all the usual things about himself and his own unhappiness. . . . He was absolutely gentle and let us kiss him ever so often, and never shrank away as he sometimes does. . . . His love for Mother and us is there all the time, under that kind of cloud which descended upon him; and today, at least, it did come right through that cloud. It just shows how infinitely worth while were all those sad and painful visits Mother paid him, when it all seemed useless and no good and she just patiently went on going." [123]

Toynbee scarcely had time to absorb the trauma of his mother's death when an even greater blow descended. His eldest son, Tony, shot himself in a fit of pique after quarreling with a girl to whom he was engaged, and died a few days later on 15 March 1939. Three years before, Tony had come back from a year at the University of Bonn as a flaming anti-Nazi. Continuation of his studies in Germany was therefore out of the question, and he was not qualified for entry into a British university, having dropped out of Winchester without completing his schooling. Initially Tony proposed to enlist in the Red Army so as to play a genuinely active part in opposing Hitler. Toynbee disapproved but went along with his headstrong son by arranging an interview for him with Ivan Maisky, the Russian ambassador to Great Britain, who made it clear that foreign volunteers were not wanted in the spy-conscious Red Army. Comparing notes with her husband afterward, Rosalind declared: "It is odd about Tony. I too get the impression that he was in a way relieved that the Red Army was off."[124]

Tony had, nonetheless, exhibited very considerable linguistic gifts while in Germany, studying various Slavic tongues and even trying Old Persian. A career in the foreign service was foreclosed by his fragmentary university training, but the British consular corps seemed a suitable substitute. Tony duly took and passed the exams, and, as a first post, was assigned for three years to Peking in order to master the Chinese language. His parents reacted joyously. "I can't quite believe it has happened," Toynbee wrote to Veronica. Tony was showing "businesslikeness which no one has been able to wring out of him yet—except the country police when he was taking out a license for a rifle or revolver.[125]

In Peking Tony contracted a fever, which affected his heart. After months of bed rest, he was sent home, discharged from the consular service, and reassigned to the home office of the Department of Overseas Trade.[126] But a bureaucratic career in London definitely did not appeal to Tony and he decided to resign. This distressed his parents. "Having just heard from Philip that Tony apparently really does mean to resign, we have sent him an ultimatum. It is rather a crisis, as . . . he would be practically unemployable. I hope we shall be able to head him off—as we did (with Maisky's help) from the Red Army two years ago."[127] Apparently, in this instance Tony did cave in to his parents, but eight months later he took revenge by shooting himself. An unhappy love affair was the immediate occasion for his suicide, but the fact that his parents were so disapproving of his behavior aimed the shot at them as well as at the young woman with whom he had quarreled.

The further fact that he did not succeed in killing himself outright made their pain even more exquisite. Rosalind was prostrated, and before his son died, Toynbee had plenty of time at the bedside to regret his failure to play a proper role as father. He again underwent a mystic experience, which he recorded thirty years later as follows: "It felt as if the same transcendent spiritual presence, standing for love beyond my, or my dying fellow human being's, capacity had pulled aside, at that awful moment, the veil that ordinarily makes us unaware of God's perpetual closeness to us. God had revealed himself for an instant to give an unmistakable assurance of his mercy and forgiveness."[128] One can understand why Toynbee felt need of God's mercy and forgiveness at that moment. His distress was made even more intense by the fact that on the day Tony died Hitler marched into the rump of Bohemia, which had been left to the Czechs by the Munich agreement, thereby discrediting forever the idea that Nazi ambition was limited to Germany's national self-determination.

Philip was devastated by his brother's suicide and toyed with the idea of killing himself, too. Although he had done well at Oxford, despite or even because of his Communist commitment, reconciliation with his parents remained impossible.[129] But Lawrence, who was still a schoolboy at Ampleforth, came to the rescue, so far as anyone could. "We both of us felt," Toynbee confessed to Veronica, "as we saw him get out of the train, that he was being given to us for the second time over, as though he were just being born."[130] The three of them went off to France to recover as best they could; ineluctably, however, final proofs and the indices for volumes IV–VI of the *Study* arrived while they were abroad, and Toynbee dutifully sat down to check them through a last time before publication.[131]

The circumstances under which the second batch of volumes of his great work were prepared explains the exalted, prophetic tone with which Toynbee addressed the breakdown and dissolution of civilizations. Biblical phrases abounded, and Toynbee chose to surround his analysis of how civilizations die with exhortation to spiritual, religious reform. In one passage he clearly accepted Christ's Incarnation, a point of doctrine that he subsequently shied away from affirming.[132] He flatly predicted the coming world empire as well—something he afterward always backed away from in his public statements.[133]

Reviews were less unanimous than in 1934. The *Times Literary Supplement* compared Toynbee with Gibbon, concluding that Toynbee had shown how Gibbon erred in ascribing the fall of the Roman empire to the triumph of barbarism and religion. "In a work of much vaster compass [than Gibbon's] he has shown that the triumphs of barbarism are illusory and that the Galilean has indeed conquered." A.J.P. Taylor, a rising star among English historians, hailed "the stupendous learning of this great work," but found it "difficult to accept the main argument" and deplored Toynbee's "pontifical determination to force every event into a rigid scheme. The historian and philosopher R.C. Collingwood wrote privately: "Like everybody else who is at all interested in history (I suppose) I have been reading your last three volumes, and I must write you a word of congratulation. It is astonishing to me that anyone should possess such a body of sheer historical learning, . . . hardly less so that anybody who does possess it should be able to wield it instead of merely lying down under it."[134]

On the other hand, Gilbert Murray found it hard to sustain his earlier enthusiasm. "I am, I confess, seriously troubled about this chapter," he wrote after reading part of the manuscript. "It makes on me the impression . . . that you are becoming a propagandist for Rosalind! . . . I do not think that our religious emancipation is the sole or main danger. . . . Forgive my outspokenness. You know much more than I do, but I do feel that you ought to shake yourself and get for a bit into a different and more objective atmosphere." And he signed off: "Yours ever, The Old Rationalist."[135]

As it turned out, the swift and cataclysmic onset of war overshadowed publication of the second set of volumes coming out, as they did, on the very eve of hostilities. The war brought sharp changes in routine, including the conversion of Ganthorpe, where Toynbee had written all six volumes of his monumental book, into an army billet. "I wonder if we shall ever live here again," Rosalind lamented as she evacuated. "I don't suppose Castle Howard will ever get going again. It is just an epoch ending—only Ampleforth will go on. . . . It is so sad, and ending a chapter here."[136] As things turned out, her prophecies were wrong, but Toynbee's life with Rosalind did not long survive their severance from Ganthorpe; and the war changed much else for them both, as we shall see in the next chapter.

VIII

World War II

1939–1946

When Britain and France declared war on Germany on 3 September 1939 Toynbee's life took a sharp new turn. Professionally, he moved to Oxford where he became head of a quasi-governmental agency, the Foreign Research and Press Service, billeted in his old college, Balliol. Privately, the wound of Tony's suicide in March 1939 followed by the abandonment of Ganthorpe and the move to Oxford in September undermined the *modus vivendi* he had established with Rosalind. After a year of restlessness, she moved to London and in 1942 decided to separate herself from a man whom she no longer felt to be her husband.

The blow to Toynbee's self-esteem was almost mortal. He contemplated suicide and teetered close to madness, but in the end his religious convictions, Veronica's support, and his habit of work prevailed. In the teeth of intense personal desolation, he therefore continued to meet with outward success. The research team he headed was first entrusted with the task of defining British postwar aims and then, in 1943, was incorporated fully into the Foreign Office. Increasingly weighty responsibilities did not counterbalance the fact of personal failure, and after many vain efforts to achieve a reconciliation with Rosalind, he divorced her in 1946 and married Veronica. World War II thus turned out to be even more difficult for Toynbee than World War I had been. Rosalind's betrayal hurt even more than his self-betrayal in 1914–1918, since it could not be explained away. He emerged from the war with a facial tic, which he carried as a mark of his agony for the rest of his life.

The outbreak of war in 1939 did not come as a real surprise to anyone, least of all to the experts at Chatham House, who had begun planning a wartime role for the Royal Institute of International Affairs more than a year before hostilities started. Toynbee was particularly emphatic in arguing that the institute ought not to degenerate into a propaganda agency, official or unofficial. He much preferred the advisory, expert role he had played in the last years of

World War I. As he explained to Gathorne-Hardy: "I do not know whether you remember the thing called the Political Intelligence Department which got going towards the end of the war and was run by Terrell and Headlam-Morley. I worked in it myself, and can remember the relief in getting out of propaganda and into work where one's business once more was to find out the truth. If, in the event of another war, the Institute will step into this position, it would be doing work of first class national importance, and . . . we should neither have sold our souls nor lost our reputation through having taken part in the war in this way." [1]

Getting the Foreign Office to recognize that it needed help from outsiders was not easy, but agreement became possible when Allan Leeper, a Foreign Office official and one of Toynbee's former colleagues in the Political Intelligence Department, was put in charge of negotiations with Chatham House. On 4 August 1939 the Foreign Office agreed to pay most of the costs of transferring the staff of Chatham House, together with its collection of press clippings, to Balliol College, Oxford, where the group would have the task of providing accurate information on foreign affairs to any branch of the government on demand. Oxford University was a party to the arrangement also, for the Foreign Research and Press Service, as the new outfit was officially called, was expected to add university experts to its staff. Finances were complicated. The government provided 80% of the initial budget, the rest coming partly from the university (which continued to pay regular salaries to dons who joined the staff) and partly from the Royal Institute of International Affairs, whose contribution covered salaries for Toynbee and some key members of his entourage. [2]

As director, Toynbee found himself suddenly plunged into a maelstrom of bureaucratic infighting for which he was temperamentally ill prepared. "I expected to be fighting Germans and restraining my younger colleagues from hating the Germans too much," he confided to one of his old schoolmates, "but I have had, so far, to spend three quarters of my energies in resisting aggression from other Englishmen and trying to damp down departmental hate, which is much more enormous than any feeling against the Germans." [3] And to another confidant he wrote:

> I lead a curious life, nowadays, such as I never dreamed of before: in snatches I deal direct with eminent men over quite important affairs, and manage, with my colleagues here, to do some good work for them. But most of my energy still goes in . . . defeating wasting intrigue and backbiting. . . . Whether the present state of interdepartmental warfare is normal or abnormal I don't know: in the last war I was only a junior, and this kind of thing passed over my head. I suspect that where political power is, there intrigue and perfidy are also always to be found— politics being one of the slum areas of human affairs that has never been cleaned up; at least that is my reading of history, and this is now confirmed by what I find in my present job. . . . Ordinarily I lead a sheltered life, working among people all of whom I have respect and affection for, and this under conditions in which ambition and struggle for power don't come in at all. Now I have to swim in the great ocean of sewage and keep my head above the evil-smelling waves. If it were

for a personal career, I should sooner or later be "too proud to fight;" but luckily, as it is, I have to do the job as a duty to Chatham House and H.M.G. and my colleagues here, so I can do it with a certain amount of disinterestedness. Perhaps in this I partly delude myself. . . . As far as I can make out, I am not doing badly in administration; at any rate, I am holding my own.[4]

In a sense he did hold his own, for he presided over a tumultuous expansion of staff, and of services and roles for the experts gathered under him. He survived a bitter quarrel with Margaret Cleeve, who initially assumed that Toynbee would leave administration to her, as he had been accustomed to doing at Chatham House. His hit and miss style of management also survived unfavorable publicity, when letters to *The Times* complained of the high salaries Toynbee had arranged for some of his Chatham House associates.[5]

Even when, early in 1943, the growth of the F.R.P.S. brought a total of 177 professional staff under his direction, Toynbee sought to maintain a collegial, quasi-academic spirit. On Friday afternoons, for example, he organized an informal seminar, open to everyone, by asking senior members of the group to take turns at analyzing the important events of the week. This was intended to broaden perspectives and provoke a sense of camaraderie across specialists' lines, and it did so to some degree.[6]

Nevertheless, Toynbee's optimistic view of his administrative capabilities was not widely shared at the time. Efforts to unseat him never got very far, partly because his prestige was so great, partly because those who attacked him were unable to command much confidence in their own administrative capacities, and partly because Toynbee fought back, using personal ties arising from his Winchester and Balliol background and from his role at Chatham House, to rally support for him and for the organization he headed. Yet Veronica, his closest colleague and most sympathetic companion, felt that he was miscast as an administrator. In a memorandum written after his death, she remarked: "There is no doubt that this experience of acting as Director of a team of experts was not among Arnold's successes. . . . I remember, for instance, occasions when he was interviewing an expert about some difficulty that had arisen, and he would say "Yes, Yes," meaning merely that he was following the other's argument, whereas the expert would think that Arnold was agreeing to the course that he was proposing, and would be disappointed to find later that Arnold did not agree with him."[7]

Toynbee had always found it difficult to work smoothly with others. He expected a lot of himself and of those around him. Only Veronica fully lived up to his expectations and demands, and she did so by surrendering her life to her work even more fully than he did. But Toynbee also felt that any direct reproof of another person was insulting, and an intolerable breach of manners. When differences of opinion or policy arose, and when someone's performance fell short of his expectations, he therefore habitually backed away, hoping that conflicts would dissolve of their own accord. Yet, ironically, his perfectionism made him a surprisingly effective bureaucratic infighter. He was hypersensitive to criticism of his own performance and treated outsiders' challenges as unmannerly insults. Self-respect and injured pride then required energetic, emphatic, and prolonged rebuttal, and he was very skilled at that.

When new tasks, new personnel, and divergent outlooks generated frictions within the F.R.P.S. staff and with other branches of the government, Toynbee was always dismayed, but he was also stubborn and vain enough to persist in holding onto his directorship, repelling rivals by defending the efficiency and value of what his staff had produced, and endeavoring to cultivate the support of the Foreign Secretary, Anthony Eden, and other men of power and authority. He was successful in the attempt. "I think," he wrote to Veronica in June 1941, "we have definitely turned the corner: Eden has been convinced we are some use." In addition to weekly reviews of the foreign press, "we are to produce a new series of short weekly papers . . . on current topics (e.g., Arab federation)."[8] By the beginning of 1942 the Foreign Research and Press Service was also producing handbooks for various countries, modeled on similar collections of background information that had been compiled during World War I.[9]

Toynbee and others from the F.R.P.S. also assumed a leading part in British discussion of possible terms of peace, beginning as early as 1939.[10] Toynbee set up a Peace Aims section of the F.R.P.S. to help shape official British policy with respect to the postwar settlement. These efforts soon meshed with international consultations conducted by the World Council of Churches from its headquarters in Geneva. Under the aegis of the World Council of Churches, a delegation of American churchmen came to Balliol in October 1941 to explore peace aims with Toynbee and a group of Anglican clergy and laity. On this occasion Toynbee first met a man destined to play a critical role in promoting his postwar American reputation: the Reverend Dr. Henry Van Dusen, Chairman of the American Commission to Study the Basis for a Just and Durable Peace, and President of Union Theological Seminary in New York.[11] After the United States became an active belligerent, a second American delegation arrived at Balliol, including, this time, John Foster Dulles, the future American Secretary of State.[12]

Toynbee's role in these meetings had both official and unofficial aspects. On the one hand, he was privy to official British documents affecting peace terms, which he was forbidden to share with the World Council of Churches. Yet he was free as a private person to say what he wished to say. When he asked for official guidance about just how open he ought to be in talking with the Americans, he was told (a week after the meetings had taken place): "We certainly think that Toynbee could safely and with advantage talk freely to Dulles, although . . . some reticence on any subject on which the British and American governments are thought not to see eye to eye might not be out of place."[13]

Toynbee's role as go-between with Americans took a new turn in the summer of 1942 when the Rockefeller Foundation of New York invited him to visit the United States "to consult on post-war problems."[14] He was given leave to accept this invitation, and accordingly departed by air from Great Britain on 23 August and did not get back to London until 20 October 1942. This was Toynbee's first flight, and he found it "a most extraordinary way of travelling—

spending one day wandering about the back parts of a remote rural country in a charabanc [he took off from Ireland], and then next night being catapulted across an ocean." [15] Further air travel within the United States provoked a lyrical description of flying above the clouds. "We rose up through them, like a submarine coming to the surface, and then skimmed along above their upper surface, which was level and solid looking. It might have been the firmament in Genesis, or it might have been the west Siberian steppe." [16]

What he did in America was a good deal more important than merely to experience the excitement and novelty of air travel. First of all, he was admitted to the confidence of the section in the Department of State charged with postwar planning, headed by Leo Pasvolsky. As he told Veronica near the end of his stay: "I am staying with Lord Halifax . . . seeing quite a lot of Pasvolsky, discussing professional details." [17] He was warmly received by the Americans. "The Survey and the Nonsense Book have been my introduction—and they have served me well," he reported. [18] Moreover, Toynbee's ideas about a just and durable peace and those of the Americans he met seemed hearteningly congruent. "I do believe I have done a good job in making friends," he declared. "I hope it is going to lead quickly to regular cooperation and exchange of personnel; anyway it will help to grease the wheels. . . . They are a good lot, pretty much like us, with the same virtues and limitations." [19]

When Toynbee secured leave for the trip, the Foreign Office prohibited him from speaking publicly on current affairs but said he could nonetheless confer with "worthwhile groups." [20] The Rockefeller Foundation, which financed his trip, wanted him to circulate and sample opinion outside official Washington, and undertook, with help from the Council on Foreign Relations in New York, to assemble "worthwhile groups" for him to talk with all across the country. After spending two weeks in Washington with Pasvolsky and the State Department, Toynbee therefore started on a grand tour that took him to more than a dozen American cities, including Houston, Los Angeles, Chicago, and many smaller cities, like Louisville, Des Moines, and St. Paul. At each stop he discoursed on the conditions for a just and durable peace as he saw them to small groups of editors, lawyers, educators, and other professional people interested in international affairs. The burden of his message was that a durable peace would require the United States to take an active part in world affairs, repudiating the isolationism of the 1920s and 1930s and subordinating national sovereignty to some sort of world government.

The most important of these meetings occured at Princeton on 7 October, when Toynbee met with Dulles and other members of the Commission on a Just and Durable Peace—the group affiliated with the World Council of Churches whose representatives had already called on him at Oxford. Van Dusen summarized what Toynbee had to say on this occasion in a letter to the chairman of the British committee on a Just and Durable Peace: "Toynbee followed, to both Dulles' and my surprise, by insisting that there were no adequate solutions of the main problems of world order apart from world government. His specific proposal was a reconstruction of a World Association of Nations, to which all the United Nations would initially belong, and the Axis powers as soon as

possible; and the allocation to the four major powers of responsibility for police." This was strong meat for an American audience in 1942, as Van Dusen
made clear, explaining that the group at Princeton finally agreed that "there is
no reasonable likelihood of avoiding either a third world war or chaos, except
through organized and effective world government. Dulles was reluctantly
compelled to concede the validity of this conclusion."[21]

A typescript of the conclusions reached at this meeting exists among Toynbee's papers. It confirms Van Dusen's report, declaring, for example, that "the
new [world] government must thenceforth be independent of any nation or group
of nations, and those who determine from time to time its powers and personnel
must come to include those who are now neutrals and enemies.

"Some may feel that our proposals are overbold, others that they are ill-
timed. We have not acted without profound reflection on both these matters.

"As Christians we must proclaim the moral consequences of the factual in-
terdependence to which the world has come. The world has become a community and its constituent members no longer have the moral right to exercise
'sovereignty' or 'independence' which is now no more than a legal right to act
without regard to the harm which is done to others. The time has come when
nations must surrender the right to do that which is immoral."

Clearly, Toynbee had made some important converts by his unexpected challenge to traditional American isolationism. Nor did his triumphs end in Princeton. On the evening of that same day Van Dusen arranged a dinner party in
New York at which Toynbee met Henry Luce, publisher of *Time, Life,* and
Fortune magazines. Luce had become a powerful shaper of American opinion
through his publications. He had grown up in China, the son of a Presbyterian
missionary, and his extraordinary success as a publisher rested partly on the
fact that he remained accessible to missionary ideas such as the new world
order that Toynbee was preaching, and was ready to use his magazines to back
them. What he heard at Van Dusen's dinner table impressed him so much that
he asked Toynbee to spend an evening with "the Time–Life–Fortune group"
of senior editors and writers. This meeting occurred on 19 October, on the eve
of Toynbee's departure for London. No record of what transpired seems to
survive, either among Toynbee's papers or in the official archives of Time Inc.
But it is easy to infer that the 1947 cover story in *Time,* which established
Toynbee's American fame, had its genesis at the Van Dusen dinner party in
October 1942 when Henry Luce encountered him for the first time.

Toynbee of course had no inkling of how his future had been shaped by this
trip. A confidential report to the Foreign Office, which he wrote on his return,
listed India as the major stumbling block to good relations between Britain and
the United States. Toynbee prescribed the "liquidation of Imperialism" through
international administration of non–self-governing territories as the only way
to resolve American suspicions of the British Empire.[22] He had proposed much
the same thing in Berlin in 1936. This reflected the fact that Toynbee's political
ideas were not significantly altered by the onset of hostilities in 1939. He continued to think, as he had ever since World War I, that abolition of war was
the critical, overriding challenge confronting Western civilization, and that the

choice lay between a world-conquering empire, on the Roman model, and some sort of voluntary federation, for which the League of Nations and the Achaean and Aetolian leagues of ancient Greece offered historical (and, unfortunately, unsuccessful) precedents.

On the other hand, Toynbee's religious ideas did undergo perceptible change, partly in response to the failure of his hopes for the League, and partly in response to private family pressures. Toynbee interpreted the failure of the League as a failure of faith. People in Britain and elsewhere simply did not believe in the secular, liberal recipe for settling international quarrels by public discussion, law, and diplomacy. Hence politicians, even if they were so inclined, could not mobilize enough support for collective security and the rule of law to make the League system work. He further viewed the rise of communism and fascism as religious movements, whose all too obvious success in the 1930s rested on the fact that these faiths did succeed in commanding mass support. "I personally believe," he wrote in a letter to the editor of the *Manchester Guardian,* published on 9 April 1935,

> that the principal cause of war in our world today is the idolatrous worship which is paid by human beings to nations and communities or States. This tribe-worship is the oldest religion of mankind, and it has only been overcome in so far as human beings have been genuinely converted to Christianity or one of the other higher religions. . . . The spirit of man abhors a spiritual vacuum; and if it loses sight of God as He is revealed in Christianity it will inevitably relapse into the worship of Juggernaut and Moloch. . . .
>
> Indeed, the terrific hold which the institution of war manifestly possesses over the hearts of men lies, surely, in the opportunity for self-sacrifice which is offered to the individual by this cult of Moloch. An uncompromising demand for submission and self-abnegation and self-sacrifice is manifestly the strength of those militant State cults—whether Fascist or Communist—which have been most successful in gaining converts during these post-war years. People will sacrifice themselves for the "Third Reich" or whatever the Ersatz-Götzen may be, till they learn again to sacrifice themselves for the Kingdom of God.

Having formulated his view of public affairs in these terms, Toynbee never afterward wavered in affirming the bankruptcy of secular, rationalist efforts to solve serious political questions. His personal encounter with a spiritual presence in Shanghai in 1929, Rosalind's militant Catholicism (after 1933), and Philip's no less militant communism (1936–1939) had brought questions of religious and political faith into Toynbee's home with a vengeance. Tony's suicide in 1939, and the renewed experience of a transcendental presence that this crisis provoked, gave still further impetus to Toynbee's search for a faith worth fighting and dying for.

Concord with Rosalind required him to embrace Catholicism, yet Toynbee could never accept the doctrines of the Church, even when he most wanted to. But he could and did embark on a serious flirtation with Catholicism, initially for Rosalind's sake and subsequently because his visits to the monastery at

Ampleforth exposed him to an attractive regimen of life and to new friends, chief among them Columba Cary-Elwes.

Toynbee sketched the history of his attraction to Catholicism in a letter to Father Columba, written at a time when his plans to divorce Rosalind threatened to end their correspondence. It runs as follows:

> Here, as truthfully and accurately as I can put it, is my own account of my approach to the Catholic Church:—
>
> (i) My first approach was for Rosalind's and Lawrence's sakes after their conversion. But for Rosalind, it seems unlikely that I should have encountered the Church except in the most casual and superficial way. Her conversion, as you know, produced disquiet and foreboding in me. I resolved to overcome these feelings. . . .
>
> (ii) The second stage was that, through making this friendly approach to the Catholic Church . . . I came to admire and love it, (to be more precise) at least one Catholic institution (Ampleforth) and a number of individual Catholics (above all, yourself. . .). At Ampleforth I realized that I was at one of the windows through which God's light shines into this world. . . . It seemed to me that this revelation, through Ampleforth, of some of God's goodness and truth was an unsought and unexpected reward which had been given to me for having taken Rosalind's and Lawrence's conversions with love and sympathy and not with hostility.
>
> The longer I have been in this relation with Ampleforth . . . the more I have felt myself in communion with God, through, and thanks to, Ampleforth. . . . My movement has been one of steadily closer approach, but all the time on what I should call the institutional, as opposed to the doctrinal, plane. On . . . the distinctive belief of Catholicism . . .—the belief in transsubstantiation and the consequent belief in the power of the priest—there has been no shearing off because there had never been any approach. . . .
>
> If I had persuaded myself (and I have an ingenious mind) that I did believe this doctrine, I should have salvaged my relations with you, with Ampleforth, and perhaps even with Rosalind—and these are precious things in my life, not my intellectual prowess, which I use as a gift given to me by God to use, without finding in it my personal happiness. Don't you see that all my personal interests pointed the path which . . . you have prayed that I might follow, and that, in refusing to take it in spite of the enormous personal losses which this refusal has brought upon me, I have been resisting a severe temptation and have been doing my duty—a very difficult duty, I have found it—to God as that duty appears to me?[23]

Father Columba's duty as friend and confidant both to Rosalind and to Toynbee was difficult for him as well, which is why he reproached Toynbee for intellectual pride and broke off relations in 1944, thereby provoking the explanation quoted above. Fourteen years younger than Toynbee, Columba embodied a serene, unquestioning style of piety that stemmed from his birth into an English Catholic family, followed by a firmly Catholic education. At Oxford he specialized in modern languages, but by the time he met Toynbee in 1937 wide reading had made him a man of very considerable, though mainly theological, learning. He soon came to look up to Toynbee, accepted his guidance in current affairs and history with the enthusiasm of a convert, and hoped, when their friendship was new, that he might become God's instrument for converting to the Catholic faith a man whom he regarded almost with awe.

In that cause, Columba was not above flattery. "Please God you will, . . . like those giants S. Augustine and S. Thomas, . . . embrace all the knowledge of your time in the synthesis only possible to truth, natural and supernatural."[24] And again: "I wrote in my last letter of how you might be the philosopher to our kings. It amused me after to find (vol. VI [of the *Study*]) that philosophers never came off! Well then, you must be the saint."[25]

But as a monk and a Catholic, Father Columba also felt it his duty to offer spiritual instruction. In July 1939 he wrote: "You should . . . pray to God for half an hour a day. As this is unsolicited advice, treat it as it deserves." Toynbee responded: "I shall now try and take your advice about setting aside that half-hour a day for prayer."[26] A few months later, on 3 September, the day war was declared, Toynbee wrote: "Besides feeling you one of my closest and dearest friends, I also feel that you are my most direct door to God."[27] And Columba, in turn, declared: "The only authority we know or will know in our lifetime, I imagine, is the Catholic Church as it is now. I wait patiently for God to call you into it in his own good time, to be a leader of your generation, someone who will dare to face the fact of our ancestors' sin . . . and counteract it."[28]

But, as we saw, Toynbee found that his duty to God required him to refuse conversion, even though his attraction to Ampleforth and his attachment to Rosalind pulled him toward accepting Columba's faith. Yet from 1937 he visited the monastery for a week or two whenever he felt he could take a holiday, and he found such visits profoundly restorative. "I always feel in harmony here," he wrote to Veronica on one such visit, "because one can just look at the life that is being led and see that it is good, without having to think of religion in terms of the theology, with all its stumbling blocks."[29]

Daily prayer, private contemplation of a crucifix Columba provided for him, and the example of his wife all pushed Toynbee toward Christian modes of thought and expression. This drift toward conventional Christianity found public expression in a sermon, preached at St. Mary's University Church, Oxford in 1940. He used orthodox language throughout, closing with the peroration: "Membership of the Church links us with all the conscious and willing servants of God who have passed through this world before us. . . . We are compassed about with a cloud of witnesses who have fought and won before us the battle against the worship of false gods that we have to fight and win again now."[30]

Toynbee's Burge lecture, delivered at Oxford on 23 May 1940, just as France was collapsing before the German Blitzkrieg, was a rather more impressive affirmation of an almost wholly conventional and quasi-Catholic form of Christianity.[31] This lecture deserves comparison with his 1920 lecture "The Tragedy of Greece." In 1920 Toynbee set forth the underlying scheme for the first three volumes of *A Study of History,* suggesting that all civilizations followed the tragic pattern he read into Graeco-Roman history. In 1940 he set forth the theme that was to dominate the last volumes of his great work, by suggesting

instead that spiritual progress, achieved largely through suffering, gave meaning to human history on earth. In 1940, as France fell and the breakdown of
European civilization seemed surely at hand, Toynbee in effect simply reversed
his view of the historic relation between civilizations and religion. In 1920
religion served civilizations by acting as a chrysalis, passing on knowledge and
skills from one civilization to its successor. In 1940 civilizations served religion, since the suffering involved in their breakdowns provoked spiritual progress.

In the words of the Burge lecture: "If religion is a chariot, it looks as if the
wheels on which it moves towards Heaven may be the periodic downfalls of
civilizations on Earth. It looks as if the movement of civilizations may be
cyclic and recurrent, while the movement of religion may be on a single continuous upward line." Toynbee deprecated "our own Western post-Christian
secular civilization" as "at best a superfluous repetition of the pre-Christian
Graeco-Roman one, and at worst a pernicious back-sliding from the path of
spiritual progress." For "if our secular Western civilization perishes, Christianity may be expected not only to endure but to grow in wisdom and stature
as a result of a fresh experience of catastrophe." The Church, in "her traditional Catholic form," might then be expected to enter "into the inheritance of
the last civilizations and of all the other higher religions." [32]

Under such circumstances, spiritual progress would surely advance to new
heights, since "What Christ with the Prophets before Him and the Saints after
Him, has bequeathed to the Church, and what the Church . . . succeeds in
accumulating, preserving, and communicating to successive generations of
Christians, is a growing fund of illumination and grace—meaning by 'illumination' the discovery or revelation or revealed discovery of the true nature of
God and the true end of man, here and hereafter, and by 'grace,' the will or
inspiration or inspired will to aim at getting into closer communion with God
and becoming less unlike Him. In this matter of increasing spiritual opportunity
for souls in their passages through life on Earth, there is assuredly an inex-
haustible possibility of progress in this world." [33]

If the Oxford lecture of 1920 was a manifesto and preview of his *Study of
History* as first conceived, the Burge lecture of 1940 may be taken as a parallel
manifesto of Toynbee's revised outlook, which had appeared sporadically in
various passages of volumes IV and VI, published in 1939, and would dominate the final four volumes of the *Study*, published in 1954.

Two points seem worth making about this fundamental transformation of
Toynbee's thought. First and most obvious, his revised view was more nearly
conventional, treating Christianity as superior to all other faiths and asserting
the reality of progress, in spiritual if not in other matters. It represented a return
to the pattern of thought of his childhood as against the classicism of his
schooling.

Second, personal suffering provoked by deteriorating family relationships was
clearly the main catalyst that acted upon his mind, but private failures and
disappointments were powerfully supplemented by the course of public events
and the renewed outbreak of war. When, after the war, these personal and

public difficulties faded into the past, Toynbee's flirtation with Catholicism and his use of distinctively Christian language diminished, while his respect for Buddhism and other non-Western religions became more apparent. The year 1940, when it looked as though England would follow France to defeat, and when his relations with Rosalind were tottering toward an irremediable break, marked the high tide of his resort to conventional Christian language in public. By the time he got back to the task of writing his great book, Toynbee's faith had become more ecumenical, embracing all the world's higher religions, and, as we shall see, instead of allowing Father Columba to convert him, Toynbee in the end converted Columba from his rather narrow, inherited Catholicism into a very ecumenically minded Christian.

What checked Toynbee's approach to Catholicism, while simultaneously intensifying his religious quest, was the break with Rosalind. Outward events are clear enough, but Rosalind's inward evolution remains hidden, thanks to her own systematic destruction of personal letters and other papers, and to the suppression by others of what they felt were embarrassing records.[34] This makes it quite impossible to write anything like an adequate or sympathetic account of what she suffered, triumphed over, and succumbed to, though it seems likely that she experienced all three of these moods, and with very great intensity, before declining at the end of her life into deep despair.

Outwardly, then, the facts are these. After closing down Ganthorpe in September 1939, Rosalind joined her husband at Oxford. He gave her a job on the staff of the new Foreign Research and Press Service, assigning her responsibility for preparing a weekly summary of the Catholic press. This involved an almost uncanny return to 1914 for both of them. As Toynbee explained to Columba: "I find it very odd being installed here, for when war broke out in 1914, I was a fellow and tutor of Balliol, and the wave that washed me out of this college then has now washed me back again. . . . Today I turned aside from my daily round for an hour and gave a university lecture in Balliol Hall— standing on the spot where I gave my first lecture of all, more than 26 years ago. . . . My wife found this kind of work strange at first, but she is now getting used to it, and her particular job—which is to follow the *Osservatore Romano* and the Catholic press of the world—is congenial to her, more or less. Anyway, without such a job, she would feel rather lost without any household at Ganthorpe to look after. . . . We are in digs at No. 3 Ship Street . . . and with a benign landlady; so we are well housed."[35]

Toynbee's phrase "congenial to her, more or less" glided over the facts, for Rosalind found herself cooped up, deprived of her sons, deprived of her house, and in uncomfortably close proximity both to her parents and to her husband. Toynbee, on the contrary, liked the setup at Oxford, confiding to Columba: "I must say, in these times, I am glad to be here and not in London."[36] By September 1940, when the German air assault on Britain was moving toward its climax and when the reflected fires of London could be seen from Oxford each night, Rosalind decided that she must get into the thick of things and do something real, instead of just reading newspapers. Toynbee explained to Columba, on 10 October 1940: "Rosalind has started working in London for the

Sword of the Spirit, fitting offers of housing—room in the country—to homeless families in London. I am a bit sorry to have her in what is greater danger—
the danger has a certain amount of fascination for her—but she was getting
restive in sedentary work at Oxford, and wanted a change."[37]

Once again, in deprecating her taste for danger Toynbee only hinted at new
tensions with Rosalind. Between the time she left Oxford in September 1940
and the time he took off for America in August 1942 he visited her on weekends from time to time. These visits were sometimes stormy. On one occasion,
in fact, Rosalind accused him of cowardice for refusing to live with her in
London. As he admitted, years afterward: "Rosalind wanted me to throw up
my government work, which was then in Oxford, and join her in London. I
refused, because I felt I ought to stick to my job, and she taunted me with
being afraid of the Blitz. The barb stuck."[38]

Her taunt hurt because Toynbee thought she might be right. After all, his
behavior during World War I had exposed him to the same charge. In 1914
Rosalind had been among those urging him to stay at home, using his dysentery
as an excuse for escaping military service. Now she attacked him for having
obeyed her in 1914 and for refusing to accept her call to action in 1940. In this
connection, it is perhaps worth recalling an already quoted passage from her
novel *Unstable Ways,* published in 1914: "If only he would take her by force,
if only he would storm her citadel of criticism, if only his love were a less
timid, bashful thing. He was too considerate for her, too gentle. She was ashamed
of her own grossness, for she acknowledged it as such, but the hunger for a
less ethereal lover was upon her."[39] One may guess that Toynbee's habitual
submission to her will in practical and familial matters may actually have disappointed Rosalind. She may, in some moods at least, have wished for a mas-
terful husband, a hero in word and deed, and not, like Toynbee, merely of the
lamp. At any rate, calling her husband a coward was a supreme insult, both
for Rosalind to have issued and for Toynbee to receive. Good relations were
never restored.

Even more decisive was the fact that Rosalind discovered a new object of
affection in the person of a young Dominican named Richard Kehoe. He had
been born in the United States but was educated in England and became a
popular preacher and respected biblical scholar and theologian at Oxford. When
Rosalind and Richard first met is unclear; presumably it was while she was
working with Toynbee on the staff of the F.R.P.S. Nor is it clear when they
first became lovers. Whether Toynbee sensed a rival, who can say? His sensitivity to the feelings of those around him was slight, and the scandalous im-
probability of any sexual link between two such conspicuously pious Catholics
presumably closed off suspicion.

The two were, indeed, oddly matched. Born in 1905, Richard was fifteen
years younger than Rosalind, who must have been either forty-nine or fifty
when she first met him. Long after they had begun to live together openly,
Toynbee and others believed he was like a son to her. Very likely Richard was
an emotional substitute for her sons, whom Rosalind had loved with a quite
unusual intensity. But by 1942 Tony was dead, Philip remained rebellious and

unreconciled, and Lawrence, after finishing at Ampleforth and spending four terms at Oxford, had joined the army. In July he enrolled in a tank officers' training school, and both Toynbee and Rosalind expected him to be killed. Even short of that disaster, it was obvious that Lawrence, now age twenty, had grown up and was becoming independent of his parents. Rosalind had long fastened her affections more on Lawrence than on her other sons, and more on her sons than on her husband. Now she was bereft, and the young Dominican, with whom she shared an intense religious commitment, somehow drifted into the vacant place in her emotional life.

The redirection of her affections may well have been stormy, but evidence is lacking other than the fact that in May 1942 she suffered a collapse. "She is in the same nursing home in Highgate as I was," Toynbee wrote, "and it is the same complaint, namely simple tiredness. . . . Writing books and working in the East End, on top of doing one's own cooking and living through the Blitz, has been—not surprisingly—too much for her."[40] The books in question were religious tracts—one, published in 1941, entitled *Time and the Timeless*, and a second, which came out in the following year, called *The Life of Faith*. Neither made as much impression on the English literary world as her 1939 essay, *The Good Pagan's Failure*, did, and neither of them reveals anything significant about her feelings toward her family in general or Toynbee in particular.

Yet these feelings were very much in flux, for the next landmark was Rosalind's decision to leave him permanently. She came to this parting of the ways in the autumn of 1942, exactly at the time when Toynbee was laying the groundwork of his extraordinary postwar reputation in America. Veronica recalled, long afterward, how Toynbee discovered Rosalind's desertion, and what happened:

> On his return by flying-ship, to England, he went straight to Rosalind's flat, but found that she was not there—I think deliberately absent, since he had given warning of the time of his arrival. He rang me up in Oxford, and I did give him a warm welcome. When he and Rosalind did meet, she made conditions for the continuation of their marriage which he found unacceptable. He returned to Oxford, and just before Christmas 1942, he asked me if I would marry him if he could get a divorce on the grounds of desertion. Though I had become an agnostic by that time, I had a dislike of divorce, inherited from my father, an Anglican priest, Rector of a small parish in Wiltshire. But I did believe in Arnold's work, I was fond of him, and I saw that he badly needed help. So I said that if he did get a divorce from Rosalind I would marry him, and that if he and Rosalind could be reconciled, then I would go away.[41]

There was another side to the story, however, that put Toynbee's professional success and his private failure into exquisite and supremely painful juxtaposition. As he explained to Columba, apologizing for not having written:

> I have, it is true been particularly busy in my work. After more than three years out in the cold, my colleagues and I at Balliol have had what is quite a success. The Foreign Office has decided to incorporate us into their organisation as from

the beginning of April; they are amalgamating within another organisation of theirs; and they are almost certainly going to move us up to London. All this is giving me a lot of work, as you may imagine. It is welcome work, and also it isn't the cause of my long delay in writing.

The reason for the delay is that I couldn't write to one of my most intimate friends without telling him something which I can hardly bear to write and which it will grieve you terribly to read.

A breach which, in spite of efforts on both sides, has long been widening between Rosalind and me, has now come to a final break. There, now I have told you. I won't go into details at present, if you will forgive me, except to say that it was the prospect of my moving back to London that brought things to a head, so that the success of my work is bound up with a very awful calamity in Rosalind's and my lives.[42]

Rosalind's rejection of him hit Toynbee like an earthquake. He had not anticipated anything like this. Marriage and family were, for him, fundamental. Nothing he had done, or failed to do, justified Rosalind's repudiation of her wifely duties in his eyes. At first he found it all but impossible to understand what might have driven her to such an act. Yet criticism of Rosalind disrupted his emotional economy, since it meant destroying the exalted image of his wife as a "goddess" to be adored and obeyed which had governed his relation with her since before their marriage.[43]

The crisis of 1942 resembled the crisis Toynbee had faced when his mother got so wrought up over Harry Toynbee's madness as not to be able to pay her usual attention to him. In that instance, as we saw, his first reaction was to take refuge in flight by bicycling to Birmingham, presumably to visit a favorite aunt who lived there. Veronica obviously played the aunt's role in 1942, receiving him warmly on his return from America, and doing what she could to reassure him in the weeks that followed. But as her own account makes clear, she was not eager to take Rosalind's place as his wife. The working partnership they had established since 1925 was all she wanted or really felt comfortable with. She described her "many deficiencies as a wife" long afterward as follows: "I was still very shy, and though, in his company, I could do things like attending Embassy luncheons or cocktail parties, which I should never had tried to do by myself, I must sometimes have been an embarrassment to him."[44] But in the first weeks of his desolation, Toynbee desperately needed reassurance, so when he asked she halfheartedly agreed to marry him if and when a divorce became legal.

Yet Veronica as wife and Veronica at work would obviously remain one and the same faithful assistant. Her conditional promise of eventual marriage therefore did not really fill the gap created by Rosalind's repudiation of her bond to Toynbee. In matters of everyday living, Toynbee had conveniently transferred to his wife the demanding dependency that he had fastened on his mother in childhood. But Veronica could never play such a role, being his assistant not his superior. Toynbee felt this intensely, and, with Veronica's full approval, embarked on a long and tortuous effort to win back what he had lost by achieving a reconciliation with Rosalind.

Self-examination, and an earnest effort to correct the faults he found in himself, was the principal path he took in seeking this end, but every so often he blurted out his hurt anger and indignation at Rosalind's unbending resolve to break away from what she not unjustly called his "possessive dependence"[45] upon her. "She threw me away like an old glove," he once complained to his sister Margaret;[46] to his secretary, Bridget Reddin, he declared that Rosalind "had not done the right thing."[47] To Columba he remarked: "Our relation was, as I see it, part of her and my rightful scheme of things, and any change from that is a breaking of the scheme or framework within which her perfection should be built up."[48]

But for the most part he reproached himself. His first effort was "to pray to God more actively and fruitfully than I ever have since I was a child."[49] By 29 September 1943 he was ready to agree with Columba's suggestion that "the purpose of the calamity is to detach me from idolatrous worship of the creature and liberate me for making God himself my centre of attachment."[50] But five months later he met an intractable obstacle that prevented him from believing that his breakup with Rosalind was really part of God's plan for his good. "The estrangement between me and Rosalind and the break-up of our family is a wound that won't heal, and I am puzzled about how to take it. It is certainly an evil and therefore cannot, in itself, be God's will. . . . Does submitting to God's will mean acquiescing in something that cannot be God's will? This baffles me."[51]

From that conundrum the Christian mode of discourse which Toynbee had established with Columba offered no escape. As a matter of fact, Toynbee had already anticipated his theological impasse by seeking a different, literary solution—one he had habitually resorted to in time of emotional crisis as a young man. What he did was to write a lengthy poem, titled "Ganthorpe." As he explained to Gilbert Murray: "I hoped, while I was working on it, that it might help to lay the ghost. I have often found, as I daresay you have, that, in one's work, problems that have worried one for years cease to trouble when one got them into words, good or bad. But so far, unfortunately, this poem hasn't had the same effect."[52] Murray's reply does not seem to survive; but it must have been less than enthusiastic, since Toynbee's next letter explains: "What you say about its being too direct an expression is very true; I expect the real catharsis will come later, in the nonsense book. . . . I feel now, myself, that 'transmutation' is the way of salvation on the spiritual plane as well as in the artistic. . . . I have somehow to transmute what I have lost—this being the most precious thing I ever have had or shall have in my life (I can say this of Rosalind without disloyalty to Veronica or depreciation of her). I must transmute or die. So there it is—it is the one path I can see ahead of me as a possible way through."[53]

In 1969 Toynbee published a slightly revised version of this poem in his book *Experiences,* so he may subsequently have valued it more than he did initially. The first strophes proclaim his attachment to Ganthorpe, its familial and historical associations for him, and his hope of dying there. In the sixth strophe, he even boasts that the anguish of banishment will not sever the tie.

> Still, Ganthorpe, I possess you. . . .
> My joys and griefs have melted into your landscape,
> Your lineaments are written into my book,
> And readers who have never heard your name—
> In Pasadena, Birmingham, Bombay—
>   Will be touched unawares by your spell.
>
> Who else has possessed you as I do? . . .
> It is *my* ghost that will walk through Owlës Wood,
> It is *my* ghost that will sit at the attic window,
>   Holding you against all comers,
> Invincible in his demonic power.

But after this bold claim to a right of possession resting on his literary achievements, Toynbee goes on to assert that

> The power in which my ghost holds Ganthorpe
>   Is dark desire.

and

> I will not be bondsman to desire.

To escape that bondage, he then declares, in the final strophe,

> Desire, my tyrant—is also God's servant:
>   I will wear desire's yoke in the service of God,
> Feeling through stabs of memory the workings of God's will,
> Forging my pangs into acts in the furnace of God's love, . . .
> Teaching the demon desire to serve God as an angel of light.

The original text then invoked Rosalind, but in the published version Toynbee omitted the first two of the following lines, thus twisting the meaning of the poem's conclusion.

> Presence of Rosalind filling my outward and inward horizon—
> Presence entrancing, aloof, like a rainbow aloft over Raysmoor—
>   Flitting dryad, sleeping hero
>   Urgent muse, elusive goddess,
>   I have loved you beyond measure;
>   I shall love you always;
> I will dedicate my love for you to God.
>
> The beauty with which you enkindle my love
> Is the light of God's countenance shining through His creatures.
> Mistaking in you God's beauty for yours,
>   Loving you beyond measure,
> Worshipping you instead of your Creator,
>   I have shrouded God's light in my darkness,

I have changed the veils through which God reveals His face
Into palls eclipsing the Beatific Vision.
Beloved works of God, I shall love you always,
But henceforth I will love you for your Creator's sake
     Seeing you with Columba's eyes—
     Seeing God in you through a glass darkly,
     Seeing God with you face to face.[54]

The combination of classical with Christian themes in this poem is an interesting register of Toynbee's state of mind. But the fact was, as he confessed
to his father-in-law, that the incantation of this verse did not relieve the pain
and distress from which he suffered, any more than Columba's theology or his
own religious exercises had been able to do. He therefore turned to psychiatry,
accepting Gilbert Murray's suggestion of consultations with the same Dr. Sylvia Payne who had advised Rosalind, Tony, and Philip in past times of crisis.
To facilitate the consultation, he asked Father Columba to return a long, quasi-
confessional letter he had sent to him in May 1943 about his relations with
Rosalind, and gave it to Dr. Payne as a convenient way of providing background. The letter has since disappeared.[55]

"I am in the middle of my treatment by my psycho-analyst, Mrs. Payne,"
he wrote to Columba. "She is not giving me a full-dress analysis but what I
believe they call 'therapeutic talks'. . . . She brought me to see a syllogism
in my unconscious mind. 'My Father got my Mother's attention by going out
of his mind; if I go out of my mind now I shall get my wife's attention back.'
It is a great help to get this . . . part of oneself at its tricks."[56] By August
1944 Dr. Payne's ministrations pretty well convinced him that the principal
fault lay with Rosalind, and that Veronica's feelings ought also to be considered. He explained to Columba:

At present I can see only a part and not the whole. I can see that if Rosalind and
I could come together again, that would be right and should be done. But this
could—after what has happened—break up Veronica's life (though, in spite of
that, Veronica thinks it would be right). How can good come through the suffering
of an innocent person who, through her lovingness, has been involved in Rosalind's and my sin?

My sin was that, after Rosalind had refused me a home, I broke away from her
and turned to Veronica. The cause (though not excuse) was that being refused a
home where one has the right to have a home gives one an enormous shock. When
I recovered from that shock, I opened the door again, as you know, to Rosalind
with Veronica's blessing. . . . Except for that time, I haven't, I believe, gravely
sinned against Rosalind. What was awkward in me, as a husband, was not willful
sin but the psychological immaturity I am now trying to remedy by treatment. If,
however, . . . Rosalind and I came together again, I should have sinned most
grievously against Veronica by having broken down the barriers that formerly lay
between her and me in the good and affectionate but limited relationship which
we used to have.[57]

Columba found himself in a difficult position, for he was at least partially in Rosalind's confidence too. One of her letters to him survives and casts some light on Rosalind's state of mind. "I do most truly want his good," she declared, ". . . but I can't feel clear in my own mind about his projected new marriage." Yet a fundamental objection to the marriage—that it would prevent Toynbee from "coming into the Church at some later time"—was now removed, because "it now seems *almost* certain that our marriage was *not* valid, and that I shall get an annulment [*sic*] before long, without even having to go to Rome. The ground was one which you were the first to suggest—that I was not baptized when it took place, in 1913. . . . It has been an almost bewildering relief to me to find that, after all, the feeling I have had for so long of never having been really married to him, was, almost certainly, justified. . . . To me all this makes it seem much more clearly God's will that we should separate—that we were not meant to go on together . . . but I doubt if Arnold will see it so."[58]

Toynbee was, no doubt, taken aback to find that Rosalind now had reason to believe she had never been properly married to him; nonetheless, he offered to go through with a new, religious marriage if she would come back. But Rosalind was not interested and steadfastly refused to agree to live with him any more. He therefore reluctantly brought suit on the ground of desertion and on 27 May 1944 was granted a divorce that became final and legally binding only on 13 August 1946.

By the summer of 1944 Toynbee's attraction to Catholicism was wearing thin, thanks partly perhaps to Dr. Payne's secular counsels and partly to his sense that the Catholic Church had in fact taken Rosalind from him, and even given her grounds for supposing that they had never been legally married. He therefore chose not to visit Ampleforth as usual for his summer holiday, and after the divorce had been granted and the prospect of Toynbee's conversion faded, Columba decided, in September 1944, to break off their correspondence. It was not resumed until June 1946.

Yet Toynbee was nothing if not persistent, and his sense of irremediable and unjustified loss was not diminished by the prospect of legal divorce and remarriage. Veronica could not possibly replace Rosalind. Marriage would add little to the sympathy and admiration she already provided, and that little could not begin to fill the void Rosalind had created in his emotional life. Accordingly, he kept hankering after Rosalind, and before the divorce became final he arranged two lengthy interviews with her, one on 23 September 1944 and a second on 11 October 1945. He reported the result of the first effort at reconciliation to Gilbert Murray: "We talked for six or seven hours. It was calm and friendly. We really did, I feel, re-consider the whole thing. . . . Rosalind was kind but very clear: she has no affection for me, no pleasure in my company, no interest in my work. . . . We agreed that there weren't, here, the conditions for a successful restoration of the marriage." Toynbee was concerned for her future. "She has gone into a world of her own, by way of some kind of sealing off of her . . . feelings for the human beings closest to her," he explained. "If she lives the rest of her life with all her human feelings frozen towards everybody, it will be a tragic thing for her herself.'' His remedy was consultation with a psychoanalyst, but that Rosalind refused.[59]

Under Dr. Payne's guidance, Toynbee had by this time arrived at a psychological explanation of Rosalind's conduct, which he set forth to Gilbert Murray in his next letter.

> In her talk with me a fortnight ago, she expressly associated her marriage to me with being drawn back into the family life from which she had tried to back away by going off to London a few months before I asked her to marry me. In her mind, her present cutting of ties with me, with Philip and his family, and even to some extent with Bun [Lawrence], is a successful repetition of that break-away in 1911 which, as she sees it, was frustrated by her marriage and her having children. . . .
>
> There is a problem for every mother when her children have grown up and have reached a stage at which being mothered harms them instead of helping them. A mother does then need to find a new activity, and Rosalind has found one in Catholicism, which can occupy all her energies and gifts. But she can have this without cutting off relations with her family: she can enter into a new relation with her children, as grown-up people on a more or less equal footing with herself, and this is entirely compatible with having a new spiritual and intellectual life of her own.[60]

He could not understand why Rosalind would not agree with this view.

A year later, another seven hour talk had exactly the same upshot.

> Rosalind's terms were the same as before . . . a separate life of her own. . . . I told Rosalind, again and finally, that I would not try it on her terms. I have to admit that I cannot give her what she wants, any more than she can give me what I want. This inability to help her is grievous, because it is plain that she does need more than her religious life and activities can give her. She would like to have a husband now, she very naturally dreads the divorce, and she has a fear of loneliness in old age. . . . She looked very tired and pathetic, and I shall never forget the look on her face at the moment I left. Being unable to help someone whom one so much wants to help as I want to help Rosalind is very bitter. But one has to recognise one's own limitations.[61]

Obviously, Rosalind hid her attachment to Richard Kehoe, who, on abandoning the Dominican habit, assumed the name Richard Stafford and lived with her on a farm in Cumberland from 1956 until her death in 1967. Their relation before 1956 remains well hidden, though at some point the two of them became sexually active partners, as Richard, in his cups, blurted out to Philip soon after Rosalind's death.[62] Toynbee had no inkling of such a possibility and Rosalind and Richard may have lived through some severe emotional storms before ending up as they did. Relevant records have been carefully suppressed, and speculation is useless.[63]

Veronica accepted her future husband's enduring attachment to Rosalind with equanimity, and may well have shared his hesitancy about their projected marriage. She was forty-eight years of age and well set in her ways at the time she had tentatively agreed to marry him in December 1942. Not long afterward, Toynbee told Columba, "It occurred to Veronica that it might be our duty to part for a time, and to work on different jobs in different places, in order to make sure that the door was being held open for Rosalind."[64] But Toynbee thought this a scruple that went beyond "what conscience requires" and surviving letters say no more about it. Veronica's own retrospective account simply says: "This period of waiting was an extremely difficult period for Arnold. I gave him as much companionship . . . as was compatible with propriety."[65] His sense of propriety was quite as lively as hers. For example, he waited until a month before their marriage to hug her in public.[66]

Yet in private, slowly, their intimacy ripened. "Darling" he wrote in 1943,

> You were very dear to me last night, not that you ever are not! But somehow you were particularly sweet and kind. God Bless you,
>
> > Your loving
> > Arnold[67]

By 1945 he could write: "I haven't been cared about like that since I was on those terms with my Mother."[68] And a few weeks before their marriage he told her:

> My dear, your lovingness and staunchness and disinterestedness when it was superhuman to put right before self unreservedly as you did—these were the guardian angels that, under God, have seen me through. God Bless you Sweetkin. (And I can see you laugh as you read the little name.)
>
> > Your loving
> > Arnold[69]

Veronica's unwavering admiration[70] and affection were indeed principal supports for Toynbee throughout the tortured years of 1942–1946. But other things also contributed to a gradual relaxation of what he called his "state of arrested breakdown."[71] First of all, his lifelong habit of escaping personal difficulties by burying himself in desk work continued to serve him well. In June 1943 Toynbee moved back to London to head a new Foreign Office Research Department that merged personnel coming from the Foreign Research and Press Service, set up at Balliol in 1939, with experts from within the Foreign Office itself. This reorganization made Toynbee and those under him insiders instead of outsiders. And the busy officials of the Foreign Office soon discovered that they needed expert advice and background information from the new Research Department about a myriad of questions.

As the tide of war turned against Hitler, definition of acceptable peace terms became urgent. Future relations with the two giant powers that were emerging as victors from the war—the United States and the Soviet Union—were especially critical for Great Britain's future. Lesser, local questions arose in all parts of the world, and the men and women of talent and learning that Toynbee had gathered around him became responsible for providing the framers of British policy with all the accurate and timely information they needed for intelligent decisions. And, needless to say, those who provided information did much to shape policies by the way they framed questions and posed alternatives to the men in charge.

Toynbee did not commit himself to specializing on any particular region, as he had done during World War I. Instead, he consorted regularly with high officials, up to and including the Foreign Secretary, and edited others' reports, as well as attending to the routines of administration. He was furiously busy, but welcomed the work, deeming it important because it directly affected the prospect for a just and durable peace.

As Allied victory became certain, he grew more and more optimistic. "I have no doubt at all," he wrote to Gilbert Murray,

> that the world is going to be united politically, and war abolished—at any rate for several centuries—within the next twenty-five years. The only question is how? Cooperatively, by the League system? If we can get it done like that, we shall be able to save for the future a good deal of the liberty and variety of our world, and then civilisation may make almost unbelievable advances. That is why it has been worth fighting this war to win another chance of stabilising a League, after having thrown away our first chance. It is worth putting our last ounce of effort into this, because there is obviously a real chance of its coming off this time. On the other hand, one has to admit that history is against us. While I can think of about two dozen worlds, in conditions like ours now, having been politically united by force, I can't think of a single case of the cooperative method having done the trick. Hence I agree that we may move on to a final round between the United States and the Soviet Union. If it comes, I expect the Americans will be the aggressors . . . but that the Russians will win, because they have a more serious purpose in life and command a far larger mass of manpower, territory and resources.
>
> In England we can do something to avert that. It is infinitely worth doing.[72]

Even late in 1946, when frictions among the victors were becoming increasingly intransigent, his mood had not changed very much. "I don't take a tragic view about the prospects," he wrote. Though, he admitted, the threat of a third world war was obvious, "I also see other possible ways—for instance, a difficult but not disastrous live and let live stage. I don't believe we are doomed to meet the crowning disaster." He opposed a federated Europe. "If it were to be genuinely independent of the three Great Powers, that could only be by its becoming Hitler's Europe, run by Germany. Otherwise it must be either a Russian or an Anglo-American satellite, and that would merely make relations between the Great Powers more dangerous." He went on to argue that a worldwide United Nations was the only framework that could keep the peace, and if that failed, Europe should "stay out of the mêlée." "We are still Hellados Hellas, and the world can't yet afford to see us snuffed out."[73]

To have a real part in shaping British policy in such a time allowed Toynbee to retrieve his former confidence in the purpose and significance of his professional work—a confidence that had wavered and all but collapsed between 1936 and 1940 or 1941. This, obviously, created a powerful counterweight to his bouts of despair over losing Rosalind. By working hard he could almost blot out distractions and, not incidentally, keep Veronica beside him. He acted accordingly, and instead of allowing personal problems to interfere with his work, as ordinary mortals might have done, his remarkable lifelong discipline and habit allowed him instead to excel in handling the flood of papers that came across his desk.

In May 1945 a grateful government offered to make him Knight Commander of the Most Distinguished Order of St. Michael and St. George "in recognition of your services of state," Toynbee declined the honor on the surprising ground that such a knighthood "would be a complication in carrying on my present work," affecting his relation with "other nationalities, including Americans."[74]

In addition to his successful labors in the Foreign Office, and to Veronica's unwavering support, Toynbee was also sustained by marked improvement in his relations with Philip. The alliance Stalin concluded with Hitler in August 1939 cured Philip of his communism, and his marriage to Anne Powell three months later did something to curb his rebellious bohemianism. She had sufficient private income to allow Philip to pursue his literary ambitions without having to earn a living by regular work. Instead, he wrote novels which achieved a limited success. His father reacted to one of them as follows: "You have a great power of expression and know how to make words do what you want of them, and I should say that this power is developing fast. Passing from style . . . to contents, I shall say that the book has one weakness in common with Mummy's early novels. . . . Your young men, like her young women, belong to too small and special a world to make it easy for the admiring reader to see through the surface of them to the archetypes underneath. . . . I shouldn't be surprised if, as you go on, the experience of being 'thoroughly domesticated' . . . may not . . . give you access to the rather big and simple underlying themes. . . . You have a gift, and, if all goes well, you will do great things with it, I believe."[75]

Soon after finishing his second novel, Philip secured an army commission and, after holding various staff positions, ended the war in a junior capacity within the Economic Section of 21st Army Group Headquarters. More significant from Toynbee's point of view was the birth of his first grandchild, Josephine, in 1943, and of her sister, Polly, in 1946. Philip was, indeed, becoming domesticated, and as far as the rift that opened between his parents was concerned, he lined up entirely on his father's side.

As for Lawrence, he was far more firmly attached to his mother than to his father, but at first was too wrapped up in his own concerns to take much notice of what had happened to his parents. By the end of 1944, however, Rosalind found herself distanced from Lawrence, too, as one of Philip's letters to Toynbee attests: "Your news of Bun makes me very angry, I'm afraid. His faults were to have been shell shocked and to want to marry Jean Asquith! How can Mummy repel him on such grounds? . . . I'm afraid I think her a very wicked woman indeed."[76] In 1945 Lawrence did marry in spite of his mother's objections, and after being discharged from the army began a career at Oxford as a painter and teacher of art. His wife, a granddaughter of Britain's World War I Prime Minister, had earned an MD and managed to combine her medical practice with domestic duties even after their first child, Rosalind, was born in 1946. This gave Toynbee another grandchild to delight in.[77]

Altogether, Toynbee could feel pleased that his two surviving sons had come through the war unscathed. Both had married well and seemed satisfactorily launched on their careers. Under these circumstances, the frictions that had marred his relations with them before the war dissipated, and Toynbee could begin to cultivate a real, though rather distant, benevolence toward his grown up sons and his infant granddaughters.

In August 1945 Toynbee spent his summer holiday with his Frankland cousins, who were farmers. "I am feeling much better," he told Veronica. "Part of what has done me good is the re-discovery—so to speak—of this part of my own family, after having lived so much in Rosalind and her family."[78]

But, as it turned out, divorce from Rosalind did not require him to give up Rosalind's family. Rosalind, after all, in *The Good Pagan's Failure,* had repudiated her father before she repudiated Toynbee; and though gaps in the Gilbert Murray Papers prevent reconstruction of how the Murrays viewed the split between their daughter and her husband, by the time the divorce occurred, both Gilbert Murray and Lady Mary seem to have concluded that the main fault lay with their daughter.[79] At any rate, just three days before his second marriage, Toynbee told Murray: "I have always felt towards you and Lady Mary as though you were my parents as well as Rosalind's, and if I had lost your affection, besides Rosalind's, I should have been unable to bear so much loss and should certainly have crumpled up."[80]

Retaining connections with Gilbert Murray was doubly important for Toynbee inasmuch as his own father had died early in the war, on 24 January 1941. Toynbee attended the funeral, but, outwardly at least, he had severed all emotional ties to his father years before, and Harry's death may have been more a relief than a blow to him. Yet his father's demise made Gilbert Murray's role as surrogate father more important; it was a genuine relief, therefore, when Toynbee found that friendly relationships with the Murrays would survive his divorce from Rosalind.

Despite all these factors conducing to his psychological stabilization, Toynbee experienced one last bout of intense distress in the spring of 1946. The setting was Paris, where memories of Rosalind's collaboration with him at the Peace Conference of 1919 became especially poignant because he was participating in a second, this time largely abortive, diplomatic effort at making peace. In December 1945 the Allies had patched up their quarrels sufficiently to convene for formal peace-making in Paris in the spring of 1946. Toynbee was among those appointed to the British delegation, and he viewed his new responsibility as an appropriate climax to his official career. Accordingly, in April 1946 he arrived again in Paris to advise the new Labour government's Foreign Secretary, Ernest Bevin, about "which points of conflict we should stand our ground on, and which we should make concessions on."[81]

His official duties brought him into exalted company. "I enjoyed my dinner with the Bevins last night," he wrote to Veronica. "He is full of life and vigour and horse sense—obviously enjoying coping with difficult jobs. I asked him which was the tougher: launching a trades union or doing this. He said doing this, without hesitation."[82] Bevin showed his horse sense in replying as he did, for as Toynbee wrote on 8 May: "The Conference reached its crisis this morning. . . . You may see me again in a day or two, or—almost equally probably—the Conference may turn a summersault and begin to turn out results."[83] In fact, the conference adjourned on 16 May, reassembled on 15 June, adjourned again on 12 July, and held a final session from 29 July to 15 October 1946. Toynbee attended only through mid-July.

His natural disappointment at the inability of the Allied Powers to agree on satisfactory peace treaties was intensified both by his isolation from Veronica and by painful recollections of what had happened in Paris before. In 1919 he had been in a junior position, of course, but had suffered intense pangs all the same when Lloyd George rejected his advice. That Bevin took his advice in 1946 was no great consolation in light of the conference's overall failure, which meant that Toynbee was not going to be able to crown his professional career by contributing to a just and durable peace. That was depressing enough, but what really hurt was the fact that in 1919 Rosalind had been with him, looking forward eagerly to their life together in peacetime. "I have had quite a nasty turn," he wrote to Veronica, "partly the uncanny memories of the first peace conference, with Rosalind and the Hammonds there . . . and Allan Leeper; and partly the isolation."[84] Two weeks later, he confessed: "At moments I have been half demented though I don't believe I have shown it. . . . I don't believe I am really at all likely to crack up, though perhaps I courted more danger than I reckoned on in coming here. But then it is one of those opportunities one can't turn down."[85] By July, when he had to return to Paris, he was looking ahead. "I do get fits of very acute depression. The remedy, I am sure, is to do new things such as getting into the house, starting on the nonsense book and the survey again, going to America."[86]

When the Peace Conference recessed for a second time in mid-July, Toynbee resigned from the Foreign Office to resume his role as Director of Studies at Chatham House. The council of Chatham House was eager to start up again, and Toynbee was no less eager to return to his peacetime career. "I am profoundly glad," he told Veronica, "that I didn't become a permanent civil servant either after the last war or after this. The work to have is work in which one can achieve something by one's own efforts."[87] By far the greatest thing he had achieved by his own effort, of course, was the still incomplete *A Study of History,* and Toynbee was resolved to let nothing stand in the way of finishing that monumental task.

Immediately prior to the outbreak of war in 1939 he had, for safekeeping, sent all his notes and preliminary outlines to New York, where the Council on Foreign Relations stored the papers for him. As the time for his release from the Foreign Office neared, he asked for their return, and while he was in Paris, the first packages began to arrive. When he got back to London, he wasted no time in opening them up, and used his vacation with the Franklands to get back to work on the *Study*. "I have now digested the notes for Parts VI–VIII of the Nonsense Book," he told Veronica, "and have fitted in the jottings of the war years."[88] Ere long, he began active composition. "When I was in Westmoreland," he confided to Gilbert Murray, "I wrote four pages of the next volume of the book, and found the stuff was still there, waiting to come out."[89]

But by now the big change in his life was finally at hand, for on 13 August 1946 the divorce from Rosalind became final, and on 28 September 1946 Toynbee married Veronica Marjorie Boulter at the Registry Office, Kensington. It was a small, almost furtive, affair, legally witnessed by John S. Boulter, one of Veronica's relatives, and by Toynbee's old school friend David Davies, who had also witnessed his marriage to Rosalind.

Toynbee and Veronica had wanted a rather different sort of marriage ceremony. As Toynbee explained to another of his old schoolmates, Bishop E.R. Morgan: "We both want to have a religious marriage. We have both been brought up in the Church of England—Veronica is a clergyman's daughter— and though neither of us would pass the traditional tests of orthodoxy, we find ourselves members of the Church of England and want our marriage to be consecrated by it."[90] But the Church of England deplored divorce, and Bishop Morgan decided that he could not preside over a second wedding, even for an old school friend. Instead he invited them to attend church subsequent to a civil marriage. Accordingly, from the Registry Office in Kensington the newlyweds traveled to Winchester, where Bishop Morgan conducted a service on their behalf in the school chapel.

Some months later, Toynbee received a brisk notice that read as follows:

> Archbishop's House, Westminister, London
>
> Dear Mr. Toynbee:
>
> This is to inform you that it has been established by judicial process that the marriage contracted by you with Rosalind Murray was canonically null and void.
>
> Joseph Gerants, officialis.[91]

Thus, soon after he had begun his new life with Veronica, Rosalind and Toynbee were finally and formally sundered. At first the newlyweds had difficulty finding a place to live, since London housing was in very short supply. Toynbee therefore simply moved in with Veronica. Informing his American friend Darbishire of his divorce and remarriage, he remarked: "These ebbs and flows in life are strange. During the past few years, I have been in sight of the madness that overtook my father and the suicide that overtook Tony. Now I am on a even keel again.

"I must now catch my train: I have become a commuter for the present, but we believe we have secured a house in London. Meanwhile we are having a honeymoon in Veronica's house here, cleaning out the cupboards, blacking the shoes and doing all the little chores that it is a happiness to do together."[92]

In May 1947, eight months after their marriage, the Toynbees moved into a small house at 45 Pembroke Square, Kensington. The move involved sorting and filing old papers, among them his letters from Rosalind. Toynbee, who felt "almost as if I had died and been born again,"[93] found the experience harrowing, as is clear from the following document.

Sunday, 18 May 1947

I am winding up the last chapter of my life—a long one—and opening a new chapter. Veronica and I have just moved into this house, and I am sorting out papers that have been in store since 1939.

This bundle contains all the letters I ever had from Rosalind, from the day she accepted me after all—the 31st May, 1913—until our breach in November 1942. . . .

I do not want these letters to be destroyed at my death. I want them to be kept and handed down to one of our grandchildren . . . far enough removed from Rosalind and me not to feel the grief of what has happened to us, and yet near enough to us to feel compassion for us.

My love for Rosalind always remained what it was from the first, but this does not mean that I am less the cause than Rosalind is, of our marriage breaking down. Till the challenge of 1942–1946 gave me the choice of growing up or perishing, I remained a child: not an equal partner for her. I do not know—and no one ever will know—whether Rosalind could have had a happy lifelong marriage with someone else, or whether there was some latent strain in her that would have broken out at this stage of her life in much the same way, whoever had been her husband. I only know that she has been unfortunate, like so many members of her family, and that I failed to save her from this misfortune.

May God pity and forgive us both, as he has, I believe, shown mercy on Tony.

God Bless you Rosalind. You came into my life, and went out of it again, like some fairy creature from another world. I know how much you gave me; perhaps I did give something to you—more than you thought when you were looking back over our lives in 1942. I love you—and I am very sorry—towards God and towards you—that I failed to be the husband that you felt you wanted.

Arnold Toynbee[94]

On that sad note ended what was by far the most painful episode of his life. Slowly his feelings healed, for his life with Veronica was destined to bring surprises and satisfactions of its own.

IX

Fame and Fortune
1946–1955

Toynbee's return to Chatham House in 1946 was swiftly followed by the publication in, 1947, of D.C. Somervell's abridgement of the first six volumes of *A Study of History*. It at once became a best seller, especially in the United States, and accompanying publicity made Toynbee into a public figure for the first time. He took his new fame and fortune very much in stride, and, if anything, intensified the pace of his writing, since he refused to let multitudinous new distractions—lectures, articles, interviews—interfere with his self-imposed tasks: completion of the *Study* and catching up on the annual *Surveys of International Affairs*, which had been interrupted by the war.

His new wife, Veronica, was as eager as he to get on with these enterprises, assisting him, as always, by editing, indexing, supplying references, and filling in details. Unlike Rosalind, she had absolutely no wish for an independent life of her own, social or otherwise. Her life *was* her work. Housekeeping bored her. She was indifferent to food and dress and shrank from entertaining. Indeed, Veronica often winced at social encounters arising from her husband's fame, but came into her own in the privacy of the study. She, like him, had learned from youth to overcome her natural shyness by excelling at desk work. Her influence therefore hurried Toynbee along the path he himself wished to follow toward a speedy completion of the *Study* and the surveys in spite of all the distractions his new fame brought.

As busyness intensified, the scars of the war years healed little by little. Toynbee's new fame and fortune did not exactly fill the gap left by his loss of Rosalind, yet admiring throngs at his lectures, swarming reporters, and editors clamoring for the products of his pen, together with Veronica's steady emotional and practical support, made his life more and more tolerable. The admiration he had once hoped to wring from a reluctant and superior Rosalind now came instead from the public, especially the American public. Toynbee was aware that his new admirers were neither reluctant nor superior and often distorted his message to fit their own preconceptions. Yet he became addicted to the limelight, and gave interviews to almost all who asked, cooperating obediently with public relations experts at Oxford University Press in New York.

No other historian has ever enjoyed the status Toynbee achieved in the United States after 1947, for he suddenly became a professional wise man, whose pronouncements on current affairs, on the historical past, and on religious and metaphysical questions were all accorded serious attention by a broad spectrum of earnest souls seeking guidance in a tumultuous postwar world. Yet despite all the attention he attracted, and the hoopla of some of his journalistic exposure, he remained the eager, learned, awkward person he had always been.

His bearing undoubtedly helped to spice his reputation. American publicists were quite unaccustomed to Toynbee's manner, shaped in Edwardian England, which combined outward modesty with tacit expectation of deference. Seeming opposites, these traits mixed, like oil and vinegar, to present the journalists who flocked around him with a distinctive personality. Toynbee was refreshingly different from the actors, politicians, and other public figures with whom they usually dealt. He found the American media congenial too, less in themselves than for the good they could do by disseminating at least some of his wisdom and insight among an American public which was destined, Toynbee believed, to lead the world into the next century, however inadequately prepared it might be for that role. This matchup lasted for about a decade, though by the mid-1950s signs of disenchantment on both sides had begun to show.

Until 1955, when Toynbee retired from Chatham House, having successfully seen the final four volumes of *A Study of History* through the press in 1954, as well as preparing eleven volumes of the wartime survey for publication, his reputation in America continued to flourish, despite a mounting chorus of dissent coming mainly from British and European critics. His principal impact on American and world opinion dates from these years. For a brief while, Toynbee's voice contributed modestly but significantly to a redirection of United States' policy in foreign affairs, helping the United States to take over the role that Great Britain had played in the world before World War I. This explains much of his popularity, for the message Toynbee conveyed to his American audiences between 1947 and 1955 was almost always flattering and reassuring. He kept his doubts about the viability of democracy in a "Time of Troubles" to himself, and when he did express views that were repugnant to prevailing American opinions his words were, at first, not taken seriously, but treated instead as merely a quirk on the part of a newsworthy personality.

Of course, when Toynbee left government service, in August 1946, he had no idea of what lay ahead for him in the United States. His immediate problem was to strike a bargain with Chatham House that would allow him to complete the *Study*. He toyed briefly with a scheme to combine a fellowship at All Souls' College, Oxford[1] with diminished duties at Chatham House. The plan was to get someone else to write annual surveys for the postwar years, beginning in 1947, so that he could put his main effort into completing the *Study,* while supervising for Chatham House the work of a team of writers who would han-

Toynbee at the height of his fame
Oxford University Press photo

dle the wartime survey.[2] Such a program called for funds that were not immediately available, and no firm offer of a fellowship at All Souls' College ever
materialized. Instead, as a sort of stopgap, Toynbee accepted an invitation to
give six lectures at Bryn Mawr College (near Philadelphia) on "pieces of the
nonsense book" between 10 February and 13 March 1947.[3] Additional lectures
in Toronto (8–9 April) and other visits in Ontario and Quebec kept him across
the Atlantic until 24 April. As a result he was on the ground and readily available to journalists when Oxford University Press published the abridgement of
the first six volumes of the *Study of History* in New York at the end of March.

This visit to America also permitted him to initiate negotiations with the
Rockefeller Foundation for a grant that would allow him to combine his duties
at Chatham House with completion of the *Study* within a period of five years.
"You may be sure," he wrote to Katherine McBride, President of Bryn Mawr,
in 1953, when the last volumes of the *Study* had in fact been completed but
had yet to appear, "I have never forgotten that day in 1947 when you came to
see me about finding more time to finish my book and when you followed that
up by putting the point to the Rockefeller Foundation. These proofs of four
volumes are all due to that."[4]

The grant was due, also, to Toynbee's own address, for he sketched his
desiderata so persuasively to Joseph Willetts of the foundation at a lunch in
New York on 24 March 1947 that Willetts wrote the next day to Ivison Macadam, then chief administrative officer at Chatham House, inviting him to apply
for a grant that would allow Toynbee to come to the Institute for Advanced
Study at Princeton for three to four months each year to work uninterruptedly
at his great book. As a sweetener, Willetts also offered the prospect of monies
to hire extra help for the survey. "I hope you don't think we are trying to run
Chatham House's affairs," he concluded, or seeking "to steal Toynbee away
from the Royal Institute!"[5]

Toynbee urged Macadam to accept Willetts' offer, arguing a professional
need for annual visits to the United States. "This country is so rapidly becoming the centre of the world that it would be difficult to keep in touch with
reality without keeping in close personal contact with leaders of thought in the
U.S., and getting a first hand impression of what is going on here. Then again,
owing to changed personal circumstance, I no longer have the retreat in the
country in England which I had between the wars, and which was one of the
things that made it possible for me to write the Study as well as the Survey."[6]
Princeton, in short, was to become a substitute for Ganthorpe. Commuting
across the ocean would supersede the holidays in Yorkshire which he had devoted to writing the *Study* before the war.

Macadam readily agreed to apply for funds, and in due course the Rockefeller Foundation approved an annual grant of £2500 to Chatham House for five
years beginning 1 July 1947, to help with the wartime survey volumes, and
undertook to pay Toynbee $4500 for each visit to the United States, starting in
1948.[7]

While this arrangement was still pending, Toynbee was invited to become
Regius Professor of History at Cambridge. This was a distinguished post, ren-

dered rather more august than the corresponding professorship at Oxford by the succession of famous historians who had occupied it from the time of Lord Acton (d. 1902) and J.B. Bury (d. 1927) to G.C.N. Clark, who was about to retire when the government invited Toynbee to become his successor. The Cambridge histories—Ancient, Medieval, and Modern—that Acton had inaugurated made the Regius Professorship at Cambridge especially influential, since those stout volumes, embodying the collective scholarship of the English-speaking world, had done a lot to define the scope and content of English (and American) academic history. Had Toynbee wished, he might have used the position of Regius Professor at Cambridge to rally the resources of the scholarly world to compile a new and genuinely ecumenical history of the world.[8] But editing the work of others took second place in his ambition to finishing what he had himself begun, and the tasks of teaching and academic administration attached to the professorship repelled him. Nonetheless, he was tempted. Such an appointment would have given him an utterly unimpeachable academic pulpit from which to propagate the enlarged view of history he had himself attained. He might even have institutionalized comparative civilizational history at Cambridge, giving it an academic respectability that, in the event, was denied him by his critics.

Gilbert Murray advised him to accept; his sister Jocelyn, by now a distinguished don at Cambridge in her own right, disagreed, saying: "The more I thought about it since our talk the more I feel that you wouldn't really be happy in academic life, after being out of it for all these years. You would miss the freedom to do what is really your work . . . which the USA offers will give you."[9]

Until 18 June 1947 Toynbee was on tenterhooks, for he was genuinely attracted to Cambridge, though he clearly preferred the Rockefeller grant which had been so exactly tailored to his own wishes and ambitions. The difficulty was that he did not know for sure that it would come through. Even after he did get word that the grant had been made, he wrote to G.C.N. Clark: "I have never had a more difficult decision to make about anything in my professional life,"[10] going on to explain that if he had accepted the Regius Professorship he would have had only seven years before retiring—too short a time to make much impact on Cambridge, and at the sacrifice of his book, which he wished, above all, to complete.

In the short run, Toynbee was surely wise to opt for Princeton and Chatham House—and the speedy completion of his great work. Had he been able to create a Cambridge school of world historians, however, his influence might have been greater in the long run. A secure academic identity and prestigious position, keeping him in England most of the time, would also have lessened the tinsel with which popular journalists garnished his reputation in America, thereby exciting much academic jealousy and increasing his vulnerability to the criticism of far less famous, yet ferocious, historians. Still, since Toynbee was, temperamentally, very much of a loner, he probably judged rightly in claiming that he would be unable to make much impact on the study of history at Cambridge in a mere seven years. And at the time he made his decision he could not possibly foresee either the heights of his American popularity or the reaction against him that eventually ensued.

After the Rockefeller money started to flow, Chatham House had to renegotiate Toynbee's status with the University of London. Eventually, in December 1948, an arrangement was agreed upon, whereby Toynbee was reappointed as Stevenson Research Professor of International Relations, with the understanding that he had no duties at the university. It was a term appointment, with retirement scheduled for 30 September 1954, when he would be sixty-five. After that, the university authorities made clear, they would expect teaching and other academic service from whoever might succeed to the Stevenson Chair.

Even before his continued tenure of the Stevenson Chair at the University of London had been agreed upon, the Rockefeller Foundation's grant to Chatham House assured Toynbee of a satisfactory, if strenuous, pattern for his postwar work. He was excused from any responsibility for producing annual *Surveys of International Affairs* for the postwar years. That task was assigned instead to a much younger man, Peter Calvocoressi.[11] As Director of Studies at Chatham House, Toynbee confined his attention to the task of planning and editing a series of volumes that would deal with the wartime years, from 1939 through 1946. He quickly abandoned the format of annual volumes, preferring to treat the war as a whole, and assigning various writers to deal with particular countries or special themes such as Lend-Lease. Gathering a staff for such a task was difficult and took time.

The initial terminus for completing the wartime survey volumes was 1952, but delays multiplied because various authors failed to achieve their deadlines. In fact, the last volumes appeared in 1958, three years after Toynbee had retired. These delays puzzled and distressed him, for he expected others to write as easily and meet deadlines as readily as he had always done. At the same time, slippages in producing the survey volumes reflected the relatively little attention he gave to this side of his work after 1946. He chose writers haphazardly, often relying on friends and relations, and could not bring himself to dismiss those who wrote slowly or without much insight. Many of the wartime survey volumes therefore became a disappointing hodgepodge, without the sparkle and panache of Toynbee's prewar volumes.[12]

He gave priority, always, to his personal work on the final volumes of *A Study of History*. Most of the year he spent in London, devoting mornings at home to his writing. He then habitually arrived about noon at Chatham House, where he and Veronica lunched with other members of the staff in a basement canteen before he got down to attending to his editorial and supervisory duties and answering his mail. In the evening he returned home, and after a light supper started to take notes on whatever reading he was doing for the *Study*. While at Princeton, the whole day was devoted to writing, save for the distraction of correspondence, his dealings with the press, and public lectures on other campuses, which he limited to one per year until after the *Study* had been completed.

Rationing public appearances became necessary because of the excitement that attended the publication of the abridgement of the first six volumes of *A*

Study of History in the United States. The extraordinary success of this book and the impact it had on Toynbee's subsequent life require considerable explanation. Three separate strands came together to shape the course of events: (1) D.C. Somervell's initiative in making a condensation of the first six volumes for purposes of his own; (2) the enthusiasm Henry Luce and key figures on *Time* magazine developed for Toynbee's vision of the world, beginning in 1942; and (3) the need the American public felt for guidance in a puzzling postwar world where Russia, a wartime friend and ally, suddenly turned into a rival and potential enemy, while Germany and Japan, the nation's wartime enemies, became, if not friends at least dependents requiring American protection. Traditional U.S. attitudes toward Britain and China, too, were in urgent need of redefinition. Those Americans who embraced Toynbee's wisdom and accepted his guidance were, in effect, opting for a special relation between their country and Great Britain.

First, then, D.C. Somervell. A history teacher at Tonbridge School, he was personally unknown to Toynbee when, in September 1943, he wrote to say that he had just completed an abridgement of the six volumes that would run "in a rough estimate, to about 550–600 pages." [13] Toynbee was not pleased. "I certainly was surprised," he replied, "I think the single volume ought . . . to boil down, not only the first six volumes, but the whole book. . . . Anyone who thinks of buying a one volume epitome will expect it, I fancy, to cover the whole work." His dismay arose, also, from the fact that he had long planned to make an epitome himself, but only after working out each part of the *Study* according to plan. "A single volume embracing the whole thing will also give me an opportunity of reunifying the argument of the book. It was a unity when I wrote my (fairly full) notes covering the whole plan, but that was in 1927–28, and since then my point of view has shifted." Nevertheless, Toynbee asked to see a sample of Somervell's work and even thanked him for "taking the book so seriously and for spending so much time and trouble on it." [14]

Somervell immediately forwarded the first part of his text, and suggested that the appearance of his abridgement simultaneously with the remaining volumes of the *Study* might help readers by giving them a shortcut to understanding the work as a whole. [15] Toynbee was immersed in official work for the Foreign Officer in 1943 and put off a thorough inspection of Somervell's work for many months. "I am abashed," he eventually wrote "to find that more than a year has passed since we last corresponded. I did read about half the first batch soon after you sent it, but I then got completely submerged again." What he saw, nevertheless, pleased him so that he now proposed to edit Somervell's text with an eye to its eventual publication. [16] By the beginning of December 1944 Toynbee reported: "I have worked over your text in detail, but most of the changes . . . are miniature." [17] Somervell responded: "I am sorry my rather slap dash style should have given your pencil so much work. My excuse would be that when writing I was not thinking seriously about publication, but just amusing myself." [18]

In fact, Somervell's condensation of Toynbee's six volumes was very skillful. He relied on Toynbee's own words almost completely, adding only a little of his own to piece selected passages together. His art lay in what he omitted. By leaving out most of Toynbee's illustrations and asides—sometimes very lengthy—the argument of the book became clearer than it had been in the original. Toynbee's vision of a cyclical pattern of rise and fall of civilizations, embracing all the world, together with a clear endorsement of the truth of Christianity[19] and a tentative reaffirmation of the idea of progress,[20] made a combination eminently acceptable to American readers. Those eager to see the United States take the lead in defending peace and forestalling revolution throughout the world could wholeheartedly agree with Toynbee when, after discussing the "Rhythm of Disintegration," he declared: "We may and must pray that a reprieve which God has granted to our society once will not be refused if we ask for it again in a humble spirit and with a contrite heart."[21]

By October 1944 Humphrey Milford of Oxford University Press expressed himself "in favour of the publication of Somervell's volume, provided of course that Dr. Toynbee approves it," and suggested that royalties might be divided 2½% to Chatham House, 5% to Toynbee, and 5% to Somervell.[22] It took another year, however, before Toynbee finished the editing, despite the fact that, as he wrote to Somervell, there were "only three places in the whole thousand sheets where I have suggested really substantial changes in your draft."[23] Proofreading and indexing took additional time, so that the 600 page volume went to press only on 23 May 1946, to be published in London at the end of the year. New York publication waited until the last week of March 1947, because customs regulations made it economical to ship pages typeset in England across the ocean to be printed and bound in the United States. An initial American edition of 13,000 was quickly exhausted; in April, a second, larger printing suffered the same fate within a week of its arrival in bookstores, and subsequent printings did not begin to catch up with demand until September 1947, by which time over 100,000 copies had been sold. A spokesman for Oxford University Press confessed, not without satisfaction, that during its first three months the "book was out of stock more days than it was in stock."[24]

The American branch of Oxford University Press rose to the occasion, spending the enormous sum of $30,000 for advertising the book by the end of 1947, and arranging for extracts and essays about it to appear in such diverse places as *The Atlantic Monthly, Life, New York Times Book Review, Yale Review, Newsweek,* and *Time*. It was a publisher's dream and surprised all concerned, not least Toynbee himself, who, however, retained some residual doubts about the whole project, fearing that the abridgement, by eliminating so much of the detailed "empirical basis" of the work, reduced his history to a mere schematic pattern of rise and fall of civilizations.[25]

"I am quite bowled over by the seriousness with which the Americans have taken my work," he wrote to Gilbert Murray. "What astonishes me is the number of people of all kinds who have either read my book or anyway know quite a lot about what is in it. Evidently, they find it helps them to orient themselves."[26] He assumed, at first, that his fame would be transitory. "The Oxford Press stoked up my publicity to white heat by the American publication date of the Abridgement, but I can now look forward to lapsing into being a dowager lion." [27]

Toynbee was not to become a dowager lion in the United States until sometime in the 1960s, largely because of the way his book and ideas were presented to the American public in the pages of *Time* magazine. We have already seen how Toynbee met and impressed Henry Luce, the publisher of *Time, Life,* and *Fortune* magazines, in October 1942. Luce was a good deal more than a spectacularly successful self-made businessman and shaper of American opinion. He sought to carry on his parents' missionary tradition by shifting attention from China where they had labored, making the United States (and the world) his mission field instead.

He believed, according to a sympathetic biographer, that "man could collaborate with God in his own evolution towards higher stages of life on earth, towards an approximation of the City of Man to the City of God." [28] He also believed that the United States had been providentially chosen to lead humankind toward those higher stages of life—as long, that is, as Americans understood their destiny and collaborated properly with Providence to assure that result. Luce's role was to make sure that his fellow countrymen did understand their destiny aright. That, indeed, was what his magazines were for—not to make money, though their spectacular success made him rich when still a young man.

Luce's views both attracted him to and repelled him from Toynbee's way of thinking. As Luce explained at a dinner for his editors on 14 November 1952. "Ten years ago, I was initiated into a little cult of those who had discovered Arnold Toynbee. I liked his way of looking at things—so utterly different from all thinking then current. But then, five or six years later, when Toynbee became famous, I knew that I disagreed with him on one point, the critical point for now. Toynbee regarded America as simply a peripheral part of European civilization. I regarded America as a special dispensation—under Providence— and I said so. My spiritual pastors shake their heads about this view of mine. They say it tends to idolatry—to idolatry of a nation. I knew well the dangers of that sin. But I say we must have courage to face objective facts under Providence." Then, after expatiating on what he meant by journalistic objectivity, Luce continued: "We assert by faith one proposition: that life does have a meaning. We also acknowledge that the full meaning of human life touches the mysterious, even the mystical, because the Truth, the full Truth, about the human adventure, will forever elude our finite intelligences, however clever." [29]

These sentences show clearly enough what attracted Luce to Toynbee in the first place, and what later separated the two men. And because *Time, Life,* and *Fortune* enjoyed wide circulations among America's business and professional élites, the rise and fall of Toynbee's reputation in the United States hinged, very much, on what Henry Luce thought of him and how his magazines chose to present Toynbee's writings and other pronouncements to their readers.

The further fact that Luce's personal reactions to Toynbee's ideas moved in harmony with seismic shifts in the United States' outlook on world affairs enor-

mously magnified the effect the Luce publications exercised on American opinion. Of course, Luce was himself influential in bringing about those shifts. As
early as 17 February 1941, when he personally signed an editorial in *Life* entitled "The American Century," Luce argued that the United States was destined to serve as leader and exemplar for the rest of the world in the second
half of the twentieth century.

This editorial was part of a long debate between "isolationists" and "inter-
ventionists" that dominated American public life from 1936 to 1941. It pitted
a Northeastern establishment, to which Luce belonged, against a more conservative vision of the nation's destiny, whose spokesmen came mostly from
the Middle West. Isolationists felt that the United States should adhere to its
nineteenth century policy of avoiding foreign entanglements, and ought to influence others only by example. Leaders of the Northeastern establishment,
based especially in New York, saw that Britain was no longer able to maintain
its nineteenth century role as mistress of the seas and guarantor of an international economic order and wished to take over that role themselves. If German
and Japanese bids for world power were to be defeated, this required more
active intervention in Europe than the isolationists were prepared to accept.

In December 1941 Pearl Harbor ended the debate without definitely commit-
ting the United States to any sort of permanent engagement overseas. When
fighting ended in 1945, nearly all Americans at first expected their armed forces
to come home and demobilize, as had happened in 1919–1920. International
problems were to be handled collectively by the new United Nations. If that
worked as its supporters hoped, then the United States could expect to enjoy
the satisfactions of old-fashioned isolation together with the benefits of intervention in world affairs, and all without any special effort. But by the end of
1946, when peace treaties with Germany and Japan were not forthcoming from
the meetings in Paris, this pipe dream had begun to dissolve. Accordingly, the
isolationist–interventionist debate had to be resumed, and with far greater urgency than before.

By 1947 Foreign Minister Molotov's obstructive tactics at the Paris Peace
Conference convinced nearly all Americans that the Soviet Union was the principal obstacle to a just and speedy postwar settlement. Moreover, Soviet intran-
sigence was made more ominous by the emergence of new threats to peace and
order in the form of revolutionary communist movements. Wherever Americans looked in Europe and Asia, they saw local communists making bids for
power, sometimes as puppets of the Soviet Union, sometimes on their own—at
least ostensibly.

In such a situation, what was the United States to do? Here was an agonizing
choice for a great many Americans, and it all came dramatically to a head on
12 March 1947 when President Truman asked Congress to approve grants of
aid to the governments of Greece and Turkey so that they could better with-
stand internal and external communist challenges. Five days later, on 17 March,
the editors of *Time* set out to explain what was at stake and make the destiny
of their country clear. They did so by publishing a portrait of Toynbee on the
cover of the magazine that showed him looking fixedly ahead, against a back-

ground of climbing figures, scaling and falling from almost perpendicular cliffs. Looking inside the magazine one discovered that the climbing figures symbolized civilizations, but for those too hurried to search inside, the editors summarized Toynbee's message with a caption under his portrait that read: "Our civilization is not inexorably doomed." That, in a nutshell, was the message projected by the 1,500,000 copies of *Time* that circulated in March 1947.

It was just what the American public wanted to hear. Response was extraordinary, for Toynbee's vision of the past, as digested in the pages of *Time,* seemed to answer questions about international affairs that had been distressing and confusing the more thoughtful members of the American public since the end of the war. Six weeks afterward, a "Letter from the Publisher" (in the issue for 28 April 1947) declared:

> Neither *Time* nor, so far as I am aware, any other of our magazines has ever before published a story quite like the Toynbee article. . . . We made reprints available to anyone who wanted them, and offered them to others whom we thought would be interested. To date, the response has been overwhelming. It has come from professors of history, philosophy and anthropology, from deans of American colleges and universities, heads of public and private schools. . . . The governors of seven states have been heard from, so have businessmen, Congressmen and just plain citizens, radio broadcasters, journalists. . . . In particular, the clergy have been strongly represented.
>
> This reaction has been of unusual interest to *Time*'s editors. For them, the story of Historian Toynbee and his work in progress was an unusual challenge and opportunity. . . . Periodically he was suggested as a cover subject by one or another of *Time*'s editors who had profited from reading him, but the moment never seemed quite right. It seemed eminently right when the Oxford University Press announced that his six-volume work would be made available to laymen in abridged form—for Toynbee's fresh viewpoint on history was more than ever applicable to the difficult problems of today.
>
> Last fall the story was turned over to *Time*'s Special Projects department. . . . A researcher set out to read all of the 3,488 pages of *A Study of History,* as did the writer assigned to the story. Both of them saw Arnold Toynbee when he arrived here this year to lecture. The writer's problem was how to make this monumental material and complicated thought briefly and clearly communicable. Of the many expressions of approval *Time* has received one, perhaps, illustrates best the kind of reaction *Time*'s editors hoped for in doing the Toynbee story: "In these days when Americans are called upon to make decisions of direct consequence to the whole world, they should understand something of the nature and course of civilization. *Time* and Toynbee have helped to fill that need."

No doubt, the self-gratulatory tone of this report should be discounted, but hard facts support the view that this article did affect American opinion in remarkable degree. *Time* received no fewer than 14,000 requests for reprints of the article, and the no less extraordinary sales of Somervell's abridgement of *A Study of History* likewise attested the impact of *Time*'s cover story. The New York branch of Oxford University Press sold 129,471 copies in the first year and 214,544 hard cover copies in toto.[30] Thanks to *Time* and to Henry Luce, and literally overnight, Toynbee thus became famous in the United States. Then, more slowly and guardedly, his fame percolated from America throughout most of the world.

His life was never again the same, but before considering the personal and career repercussions that stemmed from the article, a more careful examination of how *Time* chose to make Toynbee's "complicated thought briefly and clearly communicable" is called for. Whittaker Chambers, one of the magazine's nine senior editors, was the writer assigned to the Toynbee cover story.[31] Chambers was a repentant ex-Communist and was soon to attain national prominence by serving as principal witness in the trial and conviction of Alger Hiss, a former State Department officer, for espionage, in 1949–1950. He was a man who took ideas seriously, whatever the tortured intricacy of his personality. Having repudiated Marxism in 1938, he was looking for a substitute world view in 1946–1947 and took to Toynbee like a duck to water. By the time he had read his way through all six volumes, Chambers was prepared to argue that Toynbee had supplanted Marx as the best available guide to the meaning of history and the destiny of humankind. That, at least, is what he set out to show in the *Time* article.

His article began by referring to Britain's withdrawal from Greece and the Empire as "a crisis in Western civilization itself. . . . The U.S. must, in Britain's place, consciously become what she had been in reluctant fact, since the beginning of World War II: the champion of the remnants of Christian civilization against the forces that threatened it." That required an understanding of civilization and of history. "But most Americans had no more idea that there is a problem of history than that there is a problem of evil." Fortunately, "The one man in the world best equipped to tell them was in the U.S. last week. Professor Arnold Joseph Toynbee, cultural legate from a Britain in crisis to a U.S. at the crossroads, was delivering six lectures . . . to the young women of Bryn Mawr College."

Chambers went on to call Toynbee's book "the most provocative work of historical theory written in England since Karl Marx's *Capital*" and declared that he had surpassed Spengler, having "found history Ptolemaic and left it Copernican." The bulk of the article consisted of a sketch of Toynbee's life followed by a summary of the argument of the *Study* that emphasized the image of climbers Toynbee had used to describe the way human societies respond, or fail to respond, to challenges facing them. (The rock climbers on the cover were a visual rendering of this metaphor.)

In conclusion, Chambers wrote: "Our civilization, says Toynbee, is in its time of troubles (he dates them from the wars of the Reformation), perhaps towards the end of them. He finds bleak comfort in the thought that as yet no universal state has been imposed despite Napoleon's attempt, and two attempts by the Germans. But from the vast design and complex achievement of *A Study of History* one hopeful meaning stands out: not materialist but psychic factors are the decisive forces of history. The action takes place within the amphitheater of the world and the flux of time; the real drama unfolds within the mind of man. It is determined by his response to the challenges of life; and since his capacity for response is infinitely varied, no civilization, including our own, is inexorably doomed.''

Thus *Time* presented Toynbee's vision of history as a call to action—American action, accepting the challenge of defending a civilization that was not Western, or as Toynbee often said "post-Christian,'' but specifically Christian. (The article, discussing Toynbee's account of would-be saviors appearing amidst civilizational breakdowns, asserted: "Only one transfiguring Savior has ever appeared in human history: Christ—the highest symbol of man's triumph through ordeal and death.'')

Clearly, Luce's ideas as well as Toynbee's went into the preparation of the story, and *Time*'s writers and editors significantly distorted Toynbee's ideas to fit the immediate situation of March 1947. And, of course, it was exactly that distortion that induced so many Americans to seek answers to uncertainties about themselves and the world in Toynbee's pages. This sort of slippage between author's intent and reader's reaction is normal in written discourse, as anyone who has ever read reviews of his own work will know. The way Toynbee was launched on American public consciousness in March 1947 was not a plot, but an unusually dramatic example of the normal, selective sort of response readers and reviewers always indulge in. Chambers simply set out to make Toynbee's ideas "communicable"—and he succeeded, beyond anyone's expectations.

Even before the extraordinary impact of *Time*'s cover story had become evident, Henry Luce asked Toynbee to write an article for his other mass circulation magazine, *Life*. The assigned subject was the future of Western civilization; but when Toynbee submitted his thoughts on the matter, Luce decided not to publish it after all. This, in all probability, was the point at which Luce discovered how his vision at the destiny of the United States differed from Toynbee's. At any rate, the explanation for the rejection relayed to Toynbee via the Oxford University Press publicity director was that "Mr. Luce felt your article dealt more with the future of mankind, rather than of Western civilization."[32] Instead, *Life* editors prepared a series of questions for Toynbee to answer, hoping thereby to pin him down to more specific prophecy about the future of the United States than Toynbee was, in fact, willing to make. Nonetheless, the magazine paid him the princely sum of $1500 for an article eventuating from what amounted to a tug of war between Luce and Toynbee over how to envision America's destiny.[33]

For a long time, the discrepancy between Luce's missionary brand of American nationalism and Toynbee's emphatic rejection of what he called "tribe worship" was muted by the fact that until the mid-1950s all the major initiatives undertaken by the United States in Europe and Asia met with Toynbee's approval. To be sure, he disliked the purge of communists and "fellow travelers" that got into high gear in Washington after October 1949, when the Chinese communists came to power. One of his friends, Owen Lattimore, was among the victims, and Toynbee publicly, but discreetly, deplored his dismissal from Johns Hopkins University.[34] But the landmarks—the Marshall Plan, 1948–1952, which did so much to restore West European prosperity; the NATO military alliance, which committed the United States to the defense of Western Europe after 1949; and the Korean War, 1950–1953, fought in the name of the United Nations to repel aggression by the communist government of North Korea—all commanded Toynbee's sometimes reluctant support.

"I suppose it is the first phase of a coming American world empire," he wrote to Gilbert Murray in May 1952. "Her hand will be a great deal lighter than Russia's, Germany's or Japan's, and I suppose these are the alternatives. If we do get an American empire instead, we shall be lucky." [35] As relations between America and Russia hardened, he even contemplated the possibility of World War III with some equanimity. "After all it need not take the 170 years it took the Romans for the Power which has already become the destined bringer of peace and unity to the world to realise what its destiny is. If we could move now at one jump from the morrow of Zama to the morrow of Actium, telescoping the two decisive battles into one, we might be able to get unity and peace at a sufficiently low cost to make it possible for our world then to take a great step forward." [36] In other moods, he was more optimistic. "I believe," he wrote to a Canadian correspondent, "that the Western World is going to crystallize around the US into something like a single community—which might, in my view, start us off on a new and very promising chapter in our history. . . . If things go well, I hope to see Democracy and Industrialism, between them, make it possible for the benefits of Civilisation, hitherto confined to small minorities, to extend to much wider circles, including, sooner or later, the depressed peasants of Asia and Africa." [37]

Yet Toynbee had real reservations about the Americans. As a good classicist and cultivated Englishman, he distrusted democracy. The red scare that spread across the United States in 1949–1950 struck him as "a bit terrifying; the unforgivable sin is Communism today, but it might suddenly be something else— who knows what?" [38] "An oligarchy with a sense of enlightened self interest is probably the best government attainable, but I fear we can't get back to that," he wrote to Gilbert Murray. "When it is a question of governing the mass of the human race, probably a benevolent dictatorship is the least bad form of government that is practical politics." [39] His great quarrel with democracy was that popular government led to all-out war. "Western states (including the US) become more chauvinistic as and when their governments become more democratic," [40] he wrote to an American admirer.

In public, Toynbee contented himself with making speeches and giving interviews that endorsed American policies in Europe and Asia in fairly general terms. He was heartened by the way the Marshall Plan provoked the beginnings of what became the European Economic Community. "Isn't Europe moving fast towards unity now?" he wrote to Gilbert Murray in 1952. "It looks as if the good that was in the evil the Germans did is now going to prevail—but that seems almost too happy to be true." [41] But even a united Europe seemed tiny to him in comparison to those beyond its pale. "We shall, I imagine, have to go through a hundred years' siege of the West by Russia, Asia and Africa," he wrote to the same confidant, "but we may hope that, if the western countries unite, we shall be able to hold the walls and at last bring the siege to a good, undramatic, inconclusive ending. After all, the West's 400 year siege of the rest of the world has ended inconclusively."[42]

Unlike most Americans, he never thought that the most dangerous challenge came from Russia. At a Book and Author's luncheon in New York in 1949 he compared the Russians to "a catfish in a barrel of herring," for, just as fear of the catfish[43] could make herring swim vigorously and keep them healthy, so rivalry with Russia might provoke Europe's unification and hasten justice for the working class. More important still, it might persuade the West to do something to satisfy the demands of the peasant majority of humankind in Asia and Africa.[44] As he explained in 1952: "Western imperialism, not Russian Communism, is Enemy No. 1 today for the majority of the human race, and the West hasn't woken up to this."[45] Two years later he was a little less alarmed: "When the painful process of liquidating West European imperialism has been completed, the Asians' and Africans' determination to have equality with the rest of us—which is the biggest force, I believe, in the world today—will, I should imagine, turn against Russia."[46]

Power politics, he thought, was ultimately subordinate to religion since only with appropriate religious commitment could the West hope to win over the angry Asian and African majority. "I think we have it in our power, *if* we have the will, to make a decisively preponderant part of the World crystallise round America," he wrote to Columba. "But I don't believe this can be done except on a religious basis: I mean, I don't think anything except Christianity + Islam + Hinduism + Buddhism can defeat Communism. The common platform of the religions is their belief (i) that Man is not God (ii) that spirit, not matter, is ultimate. These are two pretty big planks: can the religions stand together on them?"[47]

In a Voice of America broadcast, Toynbee explained in more general terms how religion and politics intertwined: "Man is challenged ultimately by God. The challenge may take the form of some challenge from other people or from something in physical nature. But at the back, it works out into being a kind of encounter between God and man. It is a very familiar religious idea. It goes back to the prophets of Israel and Judah. . . . It is the profoundest, most illuminating view I know of the way human affairs work."[48]

Toynbee's religious views and feelings were, nonetheless, in a state of flux. As Rosalind receded into the past, his gravitation toward Catholicism waned and Toynbee's valuation of the more tolerant Indian religions rose. "Our aim, no doubt, should be to combine Indian Catholicity with Palestinian Monotheism," he summed up in 1954.[49] And he reproved Columba, saying: "Christians don't effectively become humble by transferring their pride to the Church. . . . All the higher religions make the same claim to unconditional allegiance of all mankind, and declare they have absolute authority. I think they are all revelations of different aspects of the same truth, and all of them are clogged with silt and flotsam and jetsam that they have picked up on their way through the world."[50] Writing to Gilbert Murray he was more explicit: "What all the religions need is a great winnowing out of permanent truth in them from the accidental accretions. . . . This kind of truth can only be expressed in myth, and myth has to take its colours from the everyday life of the particular time and place, so that the eternal truths are always needing new suits of mythical clothes. Unfortunately this doctrine is detestable to ecclesiastical administrators.''[51] A year previously (perhaps partly in jest), he agreed with Gilbert Murray on the necessary character of a future world religion: ''It must be tolerant (i.e., un-Jewish), undogmatic (i.e., un-Greek) and unsuperstitious (i.e., un-Hindu). I should, I am sure, be burnt by all the representatives of all the traditional forms of religion if they had a free hand again for that—and, of course, by the Communists, who are superstitious as well as dogmatic and practical. With these reservations, I am pro-Asian.''[52]

Toynbee never wavered in the belief that the spiritual reality that had comforted him in moments of personal crisis, first in 1929 and again in 1939, was at the center and core of the universe. He felt, moreover, that his own encounters with that Being were part of a long tradition of God's self-revelation to specially sensitive individuals, chosen for reasons known only to God, and prepared to receive His message by some sort of special suffering. He never called himself a prophet, and was rather embarrassed when others did so. Yet he never repudiated the title either, and sometimes came close to claiming the role. In a lecture to the British Psycho-Analytical Society on 30 November 1949, for example, he declared that Poetic Truth and Scientific Truth were but aspects of an inaccessible Ultimate Truth, and since human beings needed both of the available forms of truth, we must allow ''the Unconscious to speak for herself, now and in future, as of old, through those ancient channels of Poetry and Prophecy. These traditional forms of expression have been part of Mankind's indispensable equipment for understanding and managing itself ever since Man became human.''[53]

That, assuredly, was what he did in completing the final volumes of *A Study of History,* most transparently in the extraordinary final section entitled ''The Quest for a Meaning behind the Facts of History.'' What he had ventured only hesitantly and with some embarrassment in 1930 and 1931, when he resorted to metaphor from Goethe's *Faust* to explain how the growth of civilization took place, became his preferred mode of discourse, since he had come to believe that metaphor and myth conveyed a Poetic Truth that could bring men closer to God than baldly scientific discourse could ever do. And that, being the true goal of human life, and the specific goal of historical study, had become his overarching purpose. In his own words: ''The meaning behind the facts of History, towards which the poetry in the facts is leading us is a revelation of God and a hope of communion with Him.''[54]

Insofar as Toynbee made it his purpose to reveal God to mankind at large, he did indeed become a prophet, and exposed himself to accusations of megalomania. The accusation acquired further plausibility because of all the adulation he encountered in American lecture halls and drawing rooms after 1947. As we shall see, critics soon set out to cut him down to their own size. But Toynbee's private conduct showed no sign of megalomania. Instead, he remained courteous, even hesitant, when meeting strangers, listened to what others had to say, and expressed his own opinions moderately. His religious faith had much to do with his demeanor. The loss of Rosalind had chastened him profoundly. "To break out of self-centredness seems to me, more and more, to be the whole of the Law and the Prophets," he wrote to his school friend David Davies.[55] And again: "To get rid of self centredness implies, I think, finding a true and better positive centre: Nirvana or God (two names for the same reality, I believe). One cannot replace something so positive as the self except by something still more positive. 'Trying to get rid' is the road, the reality behind the phenomena is the goal—so I guess."[56]

Aspiring to such an ideal, Toynbee therefore eschewed overt self-assertion, even assertion of the correctness of his ideas. "What I am trying to do," he wrote to the French intellectual Raymond Aron, "is to encourage people to take more interest in looking at history as a whole, as part of a unified study of human affairs. The more people who enter this field the sooner my own work will be superseded, and this is what I should regard as a sign of success."[57] And he echoed Marx, by declaring: "I am not a 'Toynbeean' in the sense of wanting to establish any kind of cut and dried orthodoxy."[58]

On public occasions, he sometimes deliberately tailored his remarks to flatter his audience. In Britain he defended the United States even while complaining that "we are at America's mercy";[59] in Sweden he argued that Britain ought to imitate the Swedes of the eighteenth century who in becoming a small power had wrung "moral victory from material defect";[60] broadcasting in Arabic he declared that Arabs could teach the world how to combine loyalty to a local community with loyalty to a world community, because of their mix of local sovereignties with attachment to the world of Islam as a whole.[61] Above all, he was polite to Americans, responding patiently to journalists' questions and refraining from overt criticism as a matter of courtesy more than of policy.

The storm of publicity that broke over Toynbee's head in March 1947 reached another peak in 1954 when the final volumes of the *Study* were published. During the six intervening years, he led a hectic life, migrating annually between London and Princeton, trying to find time to finish the *Study* amidst a host of distractions. But distractions, especially when they took the form of invitations to lecture for fat fees on various American campuses, were hard for Toynbee to turn down. He liked the money and he liked the adulation, however much he pooh-poohed its more extreme manifestations. In 1948 he got into trouble by accepting two such invitations, one at Columbia University in New York and another at Birmingham, Alabama. He figured, initially, that he could undertake these lectures without delaying his writing by presenting segments of the unpublished portion of the *Study* to his audiences. But both invitations required publication of the lectures, and Oxford University Press objected to premature publication of portions of a book that was already under contract. After some acrimony, a compromise was reached whereby Columbia University Press published 400 copies of the Bampton lectures, titled "The Prospects of Western Civilization," while Alabama was fobbed off with a promise of offprints of "The Inspiration of Historians" when that portion of the book was ready for final publication.

Toynbee's scheme for lecturing on his forthcoming book also ran afoul of the Rockefeller Foundation, whose officers objected to his collecting fees while also in receipt of payments from them for the completion of the *Study*. Joseph Willetts presented Toynbee with a choice between giving up his stipend from the foundation or giving up lecture fees. Toynbee responded contritely: "My purpose in accepting the Foundation's generous grant . . . is to get on with the writing of my book, and not to earn additional money." [62] Thenceforth, until the Rockefeller grant expired in 1954 Toynbee made it a policy to refuse fees and accept only one invitation to lecture on another campus during his annual visits to the United States. He ordinarily lectured on some recently completed segment of his book, and when accepting such invitations, carefully ruled out separate publication.

Travel was another distraction Toynbee found hard to resist. As with his lecturing in the United States, after 1947 he rationed himself to a single trip each year. Thus in 1948 he visited Greece and Turkey, lecturing almost daily at stops along the way. In 1949 he did the same in Holland, Germany, and Denmark; in 1950 he went to France and to Holland again; he visited Ireland and Spain the next year, and Switzerland and Canada in 1952. In 1953 he traveled to Rome as one of six "wise men" invited to take part in a Roundtable Discussion organized by the Council of Europe. Toynbee spoke on "The Spiritual and Cultural Problem of Europe" in the company of such figures as Robert Schuman and Alcide de Gaspari, former foreign ministers of France and Italy, who had led the first stages of Europe's postwar economic integration. Alone among those attending the Roundtable, Toynbee was granted an interview with the Pope, thanks to preliminary arrangements made by his friend Columba. He was much gratified, even though he protested: "How preposterous to live in this worldly splendour. I had just time to get back . . . in order to be entertained at lunch by Prime Minister Pella. I was next to De Gasperi, who is also one of 'the Six'. I like him and Schuman particularly." [63]

By this time, Toynbee's fame had spread throughout Europe, as attested by the fact that Akademician G. Alexandrov of the Soviet Union found it necessary to denounce him in the *Moscow Literary Gazette* on 10 February 1954 for his "toadying obsequiousness towards the United States" and for advocating the "destruction of the national sovereignty of all countries and peoples." [64]

Toynbee's travels and other public appearances, giving lectures and interviews wherever he went, resulted in a stream of articles, published in newspapers, magazines, and sometimes as independent pamphlets. Some he wrote himself; some were transcriptions of talks prepared in advance; and some were off-the-cuff answers to journalists' questions. Most dealt with themes from the *Study* or with questions of current public concern. The scale of his output was extraordinary, as the following tabulation of his published articles suggests:

1946	2	1951	8
1947	9	1952	12
1948	11	1953	19
1949	9	1954	24
1950	8	1955	30 [65]

Until August 1951 Toynbee put his main literary effort into finishing the *Study,* and the upsurge of his journalistic output after that date reflects in part the release of energies his completion of that vast task allowed, and in part his indulgence in quasi-autobiographical public remarks on the occasion of the publication of the concluding four volumes of his monumental work in 1954.[66]

In 1948, when Toynbee's fame was still fresh, Oxford University Press published a collection of some of his more significant recent lectures and articles under the title *Civilization on Trial.* All but three of the reprinted articles had been written in 1946 or 1947 and reflected Toynbee's upbeat message for Americans in those years. "What shall we do to be saved?" he asked, in an article originally published by *The New York Times.* "In politics, establish a constitutional co-operative system of world government. In economics, find working compromises . . . between free enterprise and socialism. In the life of the spirit, put the secular super-structure back onto a religious basis." Then, after wondering out loud about how these challenges might be met, he concluded: "These riddles may be hard to read, but they do tell us plainly what we need to know. They tell us that our future largely depends upon ourselves. We are not just at the mercy of an inexorable fate."[67] No wonder, therefore, that this book was also a popular success, especially in the United States. Like the abridgement, and in far shorter compass, it told Americans what they wanted to hear and felt they needed to know.

The same cannot be said of Toynbee's next important literary effort. In the interval between completing the writing of the *Study* and publication of the four stout volumes, numbered VII–X, the British Broadcasting Corporation invited him to deliver the Reith lectures for 1952. These lectures, named for the first director of the BBC, were intended to crown the broadcast year by delivering an important message to the British nation. Toynbee sought to oblige, entitling his lectures "The World's Encounter with the West."[68]

Assigned prime listening time on the Home Service, they attracted a large audience and stirred considerable indignation among patriotic Englishmen who felt that Toynbee's portrait of the British Empire, then in a state of rapid dissolution, was utterly unfair. This was because he dwelt upon the West's aggressiveness in times past and on the pain and indignation Western superiority had created in Asian and African minds, while saying nothing of the benefits, or presumed benefits, that westerners had brought to their colonial subjects. Moreover, he treated Russia as one of the victims of Western aggression, viewing the current quarrel with Russia as, at bottom, a religious competition that pitted a Marxist materialist heresy against the West's spiritual Christian heritage—a heritage that had already been foolishly rejected by a secularized West.

His lectures were thus an indictment of European (and American) smugness vis-à-vis what Kipling had called "the lesser breeds without the law." Most Englishmen found Toynbee's tone hard to take at a time when the communist threat seemed acute and when the future of the British Empire was still up for grabs, even though India, Burma, and Ceylon had become independent, and Malaya was in the throes of a communist guerrilla. The London *Times* published an editorial on the day after the final lecture, attacking the way Toynbee had treated communism as a "spiritual force"; but the most conspicuous assault came several months later when a Catholic journalist, Douglas Jerrold, published a pamphlet, *The Lie about the West,* denouncing Toynbee as a traitor to Christian civilization and a catspaw for communism. A generally favorable review of Jerrold's broadside in *The Times Literary Supplement* triggered two months of heated debate in the form of letters to the editor, which were subsequently gathered together and published separately.[69]

Toynbee stuck by his guns, denying that he was a catspaw for anybody, least of all for communism, which, he asserted, was "far too narrow and too badly warped to be likely to prove permanently satisfying to human hearts and minds."[70] Father Columba and others came to Toynbee's support; but in general, it seems fair to say that his reputation in England suffered from the way he had tried, in his own words, "to give my fellow 'Franks' a bit of a jolt" by taking a "look at the encounter between the World and the West through the eyes of the great Non-Western majority of mankind."[71] The *Daily Telegraph* summed up the debate in an editorial on 21 June 1954, attacking Toynbee for substituting vapory Buddhism and watered down Christianity for good old-fashioned patriotism.

Privately, Toynbee was both hurt and puzzled. "I soon saw," he wrote to one of his gentler critics, "that a lot of people, including many who did not start with any hostile feeling, had mistaken my meaning in my Reith lectures, and ever since I have been trying to make out how it was that I failed to make myself clear."[72] But it was too late to make amends. After so conspicuously provoking the wrath of his fellow countrymen, Toynbee's reputation in Britain suffered something of an eclipse, even though in the United States he continued to command respectful deference until the early 1960s.

Another, less conspicuous but in the long run more significant sort of criticism dated back to 1948, when Toynbee and a Dutch historian, Pieter Geyl, addressed the question: "Is there a pattern to the past?" on BBC's Third Program. Geyl questioned the empirical basis of Toynbee's *Study,* pointing out various errors of fact and doubting the value of such concepts as Challenge and Response, on the ground that Toynbee's evidence did not support the conclusions he had drawn. Toynbee apologized for obvious mistakes, but defended himself by saying that the "job of making sense of history is one of the crying needs of our day." He acknowledged "many different ways of analyzing history, each of which is true in itself and illuminating as far as it goes, just as, in dissecting an organism, you can throw light on its nature by laying bare either the skeleton or the muscles or the nerves or the circulation of the blood." For himself, "I should be well content if it turned out that I had laid bare one genuine facet of history, and, even then, I should measure my success by the speed with which my own line was put out of date by further work by other people in the same field."[73]

Geyl's tone was moderate and he always remained respectful when criticizing Toynbee. But his suggestion that the whole magnificent edifice of *A Study of History* was not based on accurate historical facts roused wide response in academic circles. Before Toynbee became famous, academic specialists had paid almost no attention to him, expecting to learn nothing about their own particular fields from a history that embraced the world. But after 1947, every self-respecting historian in the English-speaking world suddenly had to have an opinion about Toynbee. Most took the obvious path of looking in his pages for what he had to say about what they knew best. Usually they found something to criticize, for, even if there were no outright errors, Toynbee inevitably left things out, selecting only some of the facts known to the specialists in order to support whatever point he was making. Consequently, an undertow of criticism and disparagement (often fueled by simple jealousy of Toynbee's fame) began to gnaw at his reputation in American and English academic circles.[74]

Such carping was counterbalanced by awe and admiration for the breadth and depth of his learning, and when volumes VII–X of *A Study of History* finally came out in 1954, the 2688 pages of these four volumes put Toynbee's extraordinary erudition again on display in a fashion that was indeed dazzling. "We are here in the presence of one of the 'Great Books' of our century, or of any century," said one reviewer;[75] this judgment was widely shared, even by those who found much to disagree with in Toynbee's pages.

His published bibliography lists twenty-eight reviews and no fewer than thirty-two articles about these four volumes, but much else was written about the *Study* as a whole after 1954, and in many languages besides English. Thus of the eighty articles listed in his bibliography as dealing with *A Study of History* in general, thirty came out between 1954 and 1956, presumably triggered by the last batch of volumes.[76] Nearly all reviewers found something to reject or even deplore in what Toynbee had to say; but that did not prevent them from admiring the grandeur of the conception and the extraordinary learning he exhibited in many different fields.

These volumes are indeed a remarkable *tour de force,* even though signs of the haste with which Toynbee produced them abound.[77] He anticipated a mixed reception. "I am not much looking forward to publication," he wrote to Gilbert Murray, "as I do not expect this batch of volumes to be so kindly treated as the first two batches were."[78] In retrospect he found much to regret, in particular, an overwrought style.[79]

Yet in November 1954, when he arrived in New York for the American publication of his volumes, he readily lent himself to a new burst of publicity, orchestrated by Oxford University Press. Newspaper interviews, together with newsreel, radio, and television appearances, focused more on the man than on the book, for the simple reason that popular journalism could not cope with so vast and complex a text. Toynbee cooperated by providing much family and personal information. His sister Jocelyn was outraged by an article published in Henry Luce's magazine *Life.* "You've no right to use our parents and our family in this way for your own publicity," she wrote; and he replied, a little lamely, "I do not seek this kind of publicity, but I cannot avoid it, at any rate not in America."[80]

Time chipped in with a ten column review; other American popular magazines—*Reader's Digest* and *The New York Times Magazine* chief among them—also devoted special articles to Toynbee on the occasion of the publication of these volumes. The publicity director for O.U.P. in New York pointed proudly to three newspaper cartoons and no fewer than 1473 clippings about Toynbee that her office had inspired or helped to implant between November 1954 and January 1955.[81]

It was an amazing sendoff for any book, and especially for such a massive and erudite work. Yet it was, or turned out to be, a last hurrah for Toynbee's American reputation. The gap between what he had to say and what Americans wanted to hear was widening, and the brittle effort to treat Toynbee as a newsworthy personality condemned him to almost instant obsolescence, since that kind of journalism discards the old as readily as it hails the new, feeding, as it does, on the perpetual discovery of new personalities to make into public idols—or scandals, as the case might be.

There was another, more significant reason for the eventual diminution of Toynbee's American reputation. He had been touted in 1947 as "the one man in the world best equipped to tell [Americans] what a civilization is or what to do with one." [82] But Toynbee had changed his mind about civilizations and their fundamental meaning between the time he wrote his first volumes and the time he wrote the final four. Readers in search of unambiguous truth were bound to be bothered by such mutability. If he had been mistaken in 1934, obviously he might still be mistaken in 1954. The quasi-scriptural authority with which *Time* and Luce had sought to clothe Toynbee's version of history, and which many Americans had accepted more or less at face value, could not survive that sort of revisionism, particularly when the author's own admission of past error was abetted by a chorus of hostile academic critics pointing to other, often trivial, sorts of errors.

Toynbee's revision of his former views turned upon the relation between civilization and religion. As we have seen, his personal encounters in 1929 and, more powerfully, in 1939 with what he recognized as a transcendental spiritual reality, and his flirtation with Catholicism after 1937, had convinced him that the core and meaning of history turned upon a slow and painful clarification of the true relation between God and Man. His own sufferings and (very limited) encounters with God were part of the process—a very small part. After such experiences, his writing, naturally, had to take account of what he had learned, even if it required him to contradict some of what he had said and believed before.

Accordingly, when he came to "Universal Churches" in Part VII of the *Study*, Toynbee altered his initial plan. In the 1920s, Toynbee had viewed churches as "chrysalises" capable of transmitting bits and pieces of a disintegrating civilization to an "affiliated" civilization, which could only grow by responding successfully to challenges, including the challenge of escaping from its ecclesiastical chrysalis. As he wrote in volume VII, "The writer of this Study had to confess that he, too, had been satisfied for many years with this rather patronizing view of the churches' role and nature." He still thought it was "true as far as it went"; but being a partial truth, it was also profoundly misleading, since their role as chrysalises constituted only "a small and unrepresentative facet of the whole truth about universal churches." [83] The whole truth about churches and the religions they embodied was that civilizations existed to serve them, not vice versa. Civilizations were not, as he had formerly assumed, humankind's supreme work of art. Rather, the supreme end and goal of human life was a more perfect communion with God, and this was achieved, however imperfectly, by the various higher religions.

Toynbee's transvaluation of values between the time he planned his great work and the time he completed it was so far-ranging that he might have been wiser to abandon his original outline entirely in 1946, and write a new and different book to explain his revised vision of the pattern and meaning of history. This would have involved overt repudiation of his earlier volumes of course and he was not ready to do that. He still adhered to most of what was there: the notion of separate civilizations, their pattern of growth, breakdown, disintegration, and much else. Moreover, he had already done a lot of work on the concluding portions of the *Study,* so when his extensive notes and outline came back from their wartime refuge in New York he never really considered a new start. Instead, he bent every effort toward completing the work along the lines originally contemplated, making only such changes as his new point of view required.

The result, all the same, was a rather clumsy hybrid. Many passages of the last four volumes stemmed from his initial, this-worldly frame of reference, and read as though human ingenuity in solving problems—especially problems of politics and war—were the mainsprings of history. Toynbee retained keen interest in such matters and was therefore capable of writing for volume VII a learned, technical monograph on the Achaemenian empire of ancient Persia featuring geography and military technology as key factors, or of devoting most of volume VIII to a sophisticated description of how Westerners in the modern era had disrupted other civilizations in Russia, the Balkans, India, the Middle and Far East, and the Americas. Most of volumes IX and X, analyzing relations between civilizations in space and time, also leave God out in explaining what happened. Toynbee's accustomed erudition and occasional flashes of insight or suggestive generalization remained as dazzling as ever in such passages.

But these segments, fleshing out his original scheme, are intermingled with other passages in which Toynbee explored his new religious and transcendental vision of the human condition. He quite failed to eliminate a profound difference in tone between the two sorts of discourse. Moreover, the new matter disfigured the book by introducing a gaping logical flaw. That was because his revised statement of the true relation between civilizations and religion, if applied rigorously, would banish most of the matter in his older discourse to the same realm of the irrelevant and meaningless to which he had formerly condemned national history.

He half recognized the dilemma but did not act on it by condensing or simply cutting out the lengthy passages of his volumes that treated secular history as significant. In his own words: "Now, however, that our Study has carried us to a point at which the civilizations in their turn, like the parochial states of the Modern Western World at the outset of our investigation, have ceased to constitute intelligible fields of study for us and have forfeited their historical significance except insofar as they minister to the process of Religion, we find that, from this more illuminating standpoint, the species itself has lost its specific unity. In our new list of societies arranged in a serial order of ascending value, the primary and secondary civilizations appear as separate categories, differentiated from one another and located on different qualitative levels by the difference in value between their respective contributions to the achievement of bringing the higher religions to flower. As for civilizations of the third generation, they are now right out of the picture." [84] If so, one might ask, why spend nearly all of volume VIII on Western civilization's encounter with the rest of the world? It was after all the leading member of the "third generation of civilizations," whose meaninglessness might be expected to banish them from his book.

Toynbee's answer was: "My sense of the command laid upon me to write out my book is as urgent as ever it was. As I see it, the book is a kind of thinking out loud of the thoughts of secularized Western Man now that he is near the end of his secular tether." [85] In other words, his mission was both to explore secularized thought and to proclaim its corrective. Inconsistencies arising from such a mix were not directly his responsibility. Others—future generations perhaps, or some new revelation from God—would have to knit together what Toynbee left asunder.

Significant additional costs were implicit in Toynbee's revised intellectual posture. A view of history that devalued—indeed, deplored—everything that had happened in the Western world since the secularization of thought in the seventeenth century was bound to displease and disappoint Toynbee's American admirers. What room did his new vision of the meaning of history leave for the United States, whose founding fathers were bearers of the secularized vision of human affairs that Toynbee had come to deplore? Toynbee also offered cold comfort for those Christian true believers who might have agreed with his indictments of secularism. "Uniformity is not possible in Man's approach to the One True God," he wrote, "because Human Nature is stamped with the fruitful diversity that is a hallmark of God's creative work." [86] Moreover, "There has been no religion in which . . . persecution of all heterodox opinions without regard for the consequences and without shrinking from any crime has been and remained so dominant as it has been in Christianity in all its manifestations." [87] Such sentiments set him in opposition to nearly all Americans who were not already repelled by his radical devaluation of modern civilization.

The contradictions of Toynbee's final volumes and the way in which the evolution of his views cut him off from the American public took time to sink in. Many reviewers recognized Toynbee's reversal of valuation as between civilizations and religions, and his heterodoxy toward all established religions was never in doubt. But the deeper contradictions of his book remained largely hidden. Reviewers commonly reflected their own predilections by dwelling on only one aspect of Toynbee's thought. Most simply dismissed (and perhaps did not even read) what they did not already sympathize with. No one seemed to be bothered by the logical incoherence between his secular and religious approaches to history.

Toynbee was not bothered either, largely because as he worked on these final four volumes and enjoyed the tumult of his public fame, the urgency of his personal religious quest ebbed slowly away. That slow ebb did not prevent him from inserting the revised judgments arising from his anguished quest of the decade 1937 to 1946 into final volumes of the *Study*. Yet the fading of his religious anxiety also allowed him to return to his notes of 1927–1929 and to revive his old concern for secular, civilizational history with something like his old zest. In effect, he lived in both worlds at once and felt no need to reconcile his old with his new point of view.

To understand his changing state of mind we must look at how Toynbee's personal and family life changed after his fame and fortune burgeoned so suddenly after 1947. When American royalties began to roll in, he decided to invest part of his new wealth in two Westmoreland farms, purchased and managed for him by his cousin Raven Frankland. Rooms in one of the farmhouses were set aside for his use, and he spent several summer vacations there with Veronica. But it was all a rather pitiful imitation of what he had known before the war at Ganthorpe. He explained to an old school friend: "When you do things to a place—as luckily we can here, owing to the windfall of American royalties—you strike roots, and this is a very happy thing for me, after having the touch of the experience of a 'displaced person.' "[88] But the attractions of Westmoreland fell far short of Ganthorpe, and time spent there never meant very much to him.

That was partly because, as we saw in the last chapter, his marriage to Veronica in 1946 was not, to begin with, any real substitute for what he had lost when Rosalind left him. Veronica's role in his life expanded, of course, but did not alter in essentials. She remained his admiring assistant in all his literary endeavors. Veronica also took responsibility for running the household—minimally, since desk work always took precedence for her as for him—and accompanied him on his travels and on public occasions as well. These roles enlarged the emotional support Veronica had previously provided to him, and Toynbee soon became completely dependent on her quiet, everyday efficiency. But Veronica never took Rosalind's place. He never adored her. Instead, adoration ran the other way, insofar as Veronica's crisp common sense allowed her to adore anyone.

It is tempting to suggest that God took Rosalind's place in Toynbee's emotional economy. If so, it is not surprising that as the wound of Rosalind's desertion healed, the urgency of his search for God diminished. Slowly the wound did heal, but it took longer than he expected. In July 1947 Toynbee visited Ampleforth again, and initially reported to Veronica: "I am serene here— happy at regaining a connexion I value very much, and that I shall now be able to keep, when I thought it had been lost."[89] But the next day: "I was attacked a bit yesterday by the Ganthorpe wound. . . . I am very happy in my relation with the monks: I don't think Columba is any longer expecting to convert me (the others never did) and he seems content to share the things we have in common.''[90]

The "Ganthorpe wound" Toynbee experienced at Ampleforth in 1947 was triggered by a rather curt letter from Rosalind in which she warned him to stay away. "In case you might have it in mind . . . to go over either to Ganthorpe or Castle Howard, I am writing now to beg you, most earnestly, *not* to. Now that you have a new home of your own and a new wife . . . I can't feel that it is asking too much of you to be content to disappear from the picture.''[91] Such brusque banishment from a place he loved hurt him deeply. "Those four days [at Ampleforth] were happy and sorrowful,'' he wrote to Columba afterward, "but sorrowful in retrospect, happy in prospect.''[92] He sought to exorcize his hurt by composing two poems, one in Greek and one in English. The following extract from the English poem registers the intensity of his reaction:

> O God, I was a harp in Thy hands,
> And Thou hast broken my strings,
> And now Thou biddest me play for Thee new music on a stringless lyre . . .
> Help me to make Thy music on my broken strings.
> In these latter days Thou hast marked me too as Thine;
> Thou hast printed on these ageing hands and feet
> The stigmata of Thucydides and Dante.
> Make me too Thine instrument, as Thou madest them . . .
> My strings are broken, my roots are riven, I bleed.
> O Lord, I bless Thee for these Thy grievous wounds;
> For whom the Lord loveth he chasteneth,
> And scourgeth every son whom He receiveth.[93]

Toynbee thus privately was moved to compare himself with Thucydides, Dante, and Christ. The immodesty of such comparisons presumably held him back from publishing this poem as he did the Greek poem he composed on the same occasion in his book *Experiences*. It expresses his desolation in almost equally forceful terms, but exposed him less, thanks to the disguise of a learned tongue.

Thereafter, Toynbee refused subsequent invitations to Ampleforth until 1954. "I found myself too dangerously attacked by old memories,'' he explained.[94] Contacts with Rosalind were few, and remained at least mildly abrasive. In November 1947 she wrote: "I am very glad you have got so much American royalties—it is good to have dollars now,'' and then suggested that since she was hard up because of the cost of repairs at Ganthorpe, he should give Lawrence £150 and lend him £50 more to relieve his debts.[95]

In 1951 Rosalind's finances were remedied by the gift of land and other forms of wealth from her mother because the Murrays decided to minimize death duties by transferring the bulk of their property to their two surviving children—Rosalind and their youngest son, Stephen—while they were still alive. Thenceforward, Rosalind was wealthy, yet just three months before she came into her new property, when her finances were still in disarray, she reproached Toynbee for not giving her everything she had a right to in the division of their property after the divorce. This stung him to the quick, and he replied with a full and precise accountancy.

She did have the grace to apologize, saying that she now agreed he had been "perfectly fair," but went on: "As you know, I never bothered much about accounts . . . so it was a rather natural mistake to make when I found suddenly, as I did, that I had hardly any capital left to think I had been hardly treated—but I am very sorry that in this I blamed you unfairly, and I am certainly glad to know that I was wrong."[96]

Soon thereafter, Rosalind moved to one of the farms in Cumberland that she had inherited, and lived there with Richard Stafford. She thereby distanced herself from Ganthorpe, which, she explained to Toynbee, would go to Lawrence when he was ready and she no longer needed its rental income. This pleased him. "I find I am able to put my treasure with Lawrence and Jean and little Rosalind living there sooner or later," Toynbee told Veronica.[97]

After 1951, therefore, both Toynbee and Rosalind settled into new routines of life, geographically removed from one another and from Ganthorpe, which they both had loved though for quite different reasons. Little by little, Toynbee was able to forget the wife he had lost. They rarely met, though at the end of 1953, when Rosalind had a operation for goiter, Toynbee called on her in the hospital, finding her "haggard."[98] On the other hand, when she was hospitalized again in 1954, this time for appendicitis like that which had killed her sister Agnes,[99] Toynbee was out of the country and did not see her.

By then, Toynbee's wound had finally healed. The final volumes of his great book were about to appear. His life with Veronica was more and more satisfactory—busy to the breaking point of course, but full to bursting with flattering encounters. Accordingly, in August 1954 he visited Ampleforth again, taking the precaution of asking Veronica to accompany him. This time he was not

devastated, as I was last time (which, as you know, is what made me keep away so long). I have managed to put to work a good deal of painful experience by transmuting it into the last volumes of the nonsense book, in faith that one will have made some use of one's life if one succeeds in helping other people to get infinitesimally nearer to a vision of God through history. And my feeling towards Rosalind—as I believe I can truly say, when I test it, as I was doing yesterday— is love combined with contrition at not having been able to give her what she needed. . . . I am surprised at the strength of the feelings that objects in the landscape . . . call up in me. Something out of me had projected itself into them and animated them—I suppose this is the pagan worship of Nature. Perhaps it happened because human relations at Ganthorpe were never more than partly happy for Rosalind and me, so one's feelings ranged out and attached themselves to woods and hills. One must push on through them, for I am sure the road through them, as through all God's creatures, leads to God.[100]

While Toynbee was thus slowly and painfully detaching himself from Rosalind, he maintained somewhat distant but cordial relations with his two surviving sons. In particular, he achieved a new kind of bond with Philip when Philip's wife, Anne, broke away from him in February 1949. "It is very bitter and terrifying,'' Philip wrote, ''to be suddenly deprived of wife, home and children in this way—and though the fault has been largely mine (in the past) this does little to console me.''[101] Toynbee could certainly sympathize, and he was able to help as well. Philip, who had been living on his wife's income while writing novels and living a bohemian life in London literary circles, now needed a job. Toynbee therefore introduced him to David Astor, editor of the *Observer*. As a result, Philip was offered the post of Middle East correspondent for the *Observer* and gladly headed off for Cairo in March 1950.

A year later Toynbee reported: ''The news from Philip is good. He has done well for the *Observer,* and they are taking him on permanently. He has felt the pleasure of making a success of a job and of being valued by the people employing him. He has married again—at Tehran—a girl called Sally Smith in the U.S. Embassy of Tel Aviv: she was secretary to the second in command there. I like her: she is straight forward and, I should think, stalwart. I am of course heartily sorry that his marriage with Anne broke up, but, as there was no mending that, I am not sorry about Sally Smith.''[102] Sally Smith did indeed turn out to be stalwart, modulating the tempestuous swings of mood that characterized Philip throughout his life. She also raised three children, including Toynbee's only grandson, Jason (born 1953). On returning to England, Philip became book review editor for the *Observer* and wrote reviews almost every week for the rest of his life. This made him an important figure in London literary circles and allowed him to write books and poetry of his own in spare time. It was a career for which he was well suited. Toynbee was pleased.

Lawrence and his wife, Jean, also achieved satisfactory careers at Oxford. Lawrence painted and taught art and his wife practiced medicine—while also raising six daughters and somehow keeping house. They remained closer to Rosalind than to Toynbee, but he visited them from time to time and maintained amicable and relaxed relations with them.

Because Philip's divorce scattered his grandchildren among three separate households, Toynbee made a point of trying to bring all the cousins together by inviting them to attend the circus in his company. As he told Columba: ''I took a party of eleven to the Circus on Friday. The number will increase year by year, as more grandchildren reach circus-going age.''[103] Attending the circus became an annual ritual. Some of Toynbee's grandchildren enjoyed it, although others found it to be a rather strained occasion. He quizzed them about their progress in school and the like, but seldom established any real rapport; and the lunch Veronica provided beforehand was, at least sometimes, too scanty to satisfy the appetites of growing children.[104]

But if his grandchildren sometimes felt awkward in his presence, his sons did not. Philip wrote after Toynbee's death: ''He proved a delightful grandfather to my own and my brother's children. They were not his direct responsibility; they would not distract his attention from his work; they would not be a constant drain on his financial resources. This enabled him to show them the gentleness and kindness which were so deep a part of his true nature. He enjoyed few things more than taking a party of grandchildren to the zoo or the circus.''[105]

Thus Toynbee's relations with his sons and grandchildren achieved a satisfactory equilibrium by the early 1950s, while Veronica looked after him diligently, both in the everyday housekeeping sense and by assisting and accelerating his literary labors. More relaxed personal relationships therefore conspired with his public success to diminish the inner turmoil of the decade 1936–1946, and as that turmoil quieted so did the intensity of his religious quest. He continued to believe that humanity's evolving relation to God constituted the central drama of history, but he no longer spent much time worrying about his own personal relation to a spiritual reality that became less and less like the God of Christian theology and a good deal more Indian (or perhaps more Iranian) as he grew older.

In 1954 Toynbee turned sixty-five and was due to retire from his professorship at the University of London. But the wartime survey volumes were not yet complete, and he very much wanted to finish that project before surrendering his post as Director of Studies at Chatham House. The Rockefeller Foundation came to the rescue with additional grants, and Toynbee therefore did not retire until June 1955, when his duties as editor of the wartime survey volumes were substantially complete, although the last two of the eleven volumes did not actually appear until 1958.

Toynbee's approaching retirement provoked discussion within Chatham House about how to fill his place. This, in turn, occasioned reexamination and some rather sharp criticism of the way Toynbee had functioned as Director of Studies. In May 1954 the resulting controversy boiled over into the pages of the *Manchester Guardian*. Toynbee was outraged and demanded that the Chatham House Council "censure me or clear me." He angrily enumerated all he had done for Chatham House: enriching the Institute with a total of £15,175 in royalties from his books, and writing and publishing under the institute's seal thirty-four survey volumes and his *Study of History* along with two volumes of documents on International Relations.[106]

The council appointed a subcommittee to look into the whole matter, and on 21 June 1954 it reported unanimously "that Dr. Toynbee has carried out these duties with great distinction," even though his successor would have to teach at the university in addition to what Toynbee had been able to do. Not surprisingly, the council was unable to find a suitable candidate for such a Herculean job and in 1956 decided to separate the duties of the Director of Studies from the Stevenson Professorship. Yet when the respected historian Geoffrey Barraclough accepted the post in 1956 he found he could not produce an annual survey volume and also teach at the University of London. He therefore let the *Survey* fall further and further behind events, until he resigned in despair. Thus in the event, Chatham House discovered that Toynbee's performance had in fact been unique. No one could fill his shoes, and no one did. Instead, the annual surveys had to be abandoned, but not until yet another Stevenson Professor, D.C. Watt, had found the task impossible.

Despite the dust-up surrounding his retirement, Toynbee did not completely sever his connection with Chatham House. "What Veronica and I are hoping for," he wrote in 1953, "is that we may continue here on an 'emeritus' basis

X

Toynbee as a World Figure
1956–1965

Toynbee timed retirement from Chatham House to coincide with the completion of his editing of the wartime *Survey of International Affairs*. Thereafter, other literary enterprises needed only sporadic attention. An American friend and admirer, E.D. Myres, was preparing a volume of maps and gazetteer for the *Study;* Somervell was at work on a condensation of volumes VII through X; and Toynbee's Gifford lectures, published in 1956 with the title *An Historian's Approach to Religion,* were at the printer. But proofreading and editing the work of others could be handled by mail. Accordingly, Toynbee set out on a trip around the world, lasting from February 1956 until August 1957.

One purpose was to see places he had written about in the *Study* but had not visited, with the idea that firsthand experience and consultation with archaeological experts on the spot would help him to revise the great work more authoritatively. As far back as 1952 Toynbee had applied to the Rockefeller Foundation for a grant to visit Mexico, Guatemala, and Peru "before setting to work on this second edition,"[1] and the terms of the grant were later renegotiated to allow him to visit Asian sites as well.

A second and no less compelling purpose of the trip was to see as much of the contemporary scene as possible, thus enriching his familiarity with all the issues that he and Veronica had written about in the annual surveys. Toynbee intended, particularly, to look for promising religious responses to the worldwide breakdown of civilizations that he had diagnosed. A final, and unexpressed motive may well have been to distance himself from the tide of criticism that had welled up at Chatham House and pervaded the Anglo-American intellectual scene, without, as yet, doing much to dim his popular reputation in the United States.

Characteristically, Toynbee arranged to supplement his stipend from the Rockefeller Foundation by contracting to write articles for the *Observer,* the same Sunday newspaper that published Philip's book reviews.[2] In addition, he arranged for a multitude of lectures along the way, sometimes requiring publication afterward. Ten weeks spent lecturing at different Australian universities (May–August 1956) and a course of lectures at the American University of Beirut (March–August 1957) were the principal but by no means the only academic engagements of the trip.

Toynbee's decade of exposure to high-voltage American publicity had made him famous enough to attract the attention of local journalists and dignitaries wherever he went. As he explained to an old school friend: "The further I go, the better I seem to be known, and this piles up honorific ceremonies and social engagements, which are the last thing I want, as seeing Peru is a strenuous job in itself."[3] He often appeared on radio and television, and prime ministers and presidents in Latin America and southeast Asia went out of their way to give him full-scale VIP treatment. This, of course, increased and expanded his fame. The trip therefore made him more of a world figure than before, and in Japan it laid the foundation for his subsequent career as a wise man from the West.

Toynbee approached the trip with apprehension. "It was odd," he confessed, "retiring from Chatham House after thirty-two years; and I was so business like in winding up my affairs that, when at last I started off, I felt as if I were going to my death."[4] This reprise of anxiety on the eve of travel to strange lands proved better founded in 1956 than had been the case in 1911 and 1929, for Toynbee's physical strength was not what it had been. Strenuous trips to archaeological sites of the Peruvian coastlands brought on a heart problem that compelled him to abandon plans for going to Bolivia. Yet after a week's rest, he took the risk of ascending the altoplano to visit the Inca heartland and see Cuzco, Machu Picchu, and Lake Titicaca.[5]

Veronica had stayed behind to finish up editorial work on the last survey volumes, but she joined her husband, according to plan, by taking passage on a ship bound for New Zealand via the Panama Canal. After visiting Jamaica, Colombia, Ecuador, and Peru he boarded the ship she was on just before it passed through the Canal. Thereafter Veronica exercised a restraining hand on his perennial enthusiasm for exploring new sites and landscapes. "I get so fired up," he confessed, "with a wish to see the whole of the infinite Earth that I overtax myself unless I have a loving hand to hold me back gently. I am still a child."[6]

Veronica did not enjoy travel as he did and came down with jaundice in Baghdad. From this time on, nagging health problems never completely disappeared. They both suffered from high blood pressure and accumulated a variety of other ailments. The only important episode for Toynbee in the first decade of his retirement was a prostate operation he underwent in 1957. Veronica's appendix had to be removed in 1959 and she was slow to recuperate.

Partly for reasons of health, the trip around the world was not a great success. The goal of preparing for a second, corrected edition of the *Study of History* eventually went by the boards, since Toynbee found that thorough revision of so vast a work was too big a job, even for him. Instead he wrote another volume, numbered XII, and entitled *Reconsiderations*,[7] in which he responded to criticisms and showed how thirty years of archaeological work and other scholarly discoveries had affected the substance of his earlier volumes. But he settled down to this task only in 1958, since distractions of travel and absence of library resources made serious scholarly work impossible in course of the tour itself.

Another depressant for Toynbee was the Suez crisis of 1956, when Britain, France, and Israel conspired to attack Egypt in retaliation for that country's expropriation of the stockholders of the company that had built and operated the Suez Canal since 1869. The attack on Egypt was, however, called off almost as soon as it began because of American and Russian opposition. Toynbee, in Japan at the time, reacted sharply. He wrote back: "Well, my gloomiest forebodings . . . have been far surpassed. Half my friends and about a third of my children's friends, have given their lives to prevent aggression, and now our own country has been turned into an aggressor by Eden."[8] Gilbert Murray, however, backed the attack, and the last letter Toynbee ever sent him regretted their difference, saying: "It is strange how, having the same fundamental principles, we can see the same events in such different lights."[9] Lady Mary, Gilbert Murray's wife, had died in September 1956 when Toynbee was in Indonesia; news of Gilbert Murray's death, in May 1957, reached Toynbee in Tehran. Toynbee naturally regretted the loss of the man who had become "a second father"[10] to him, and the fact that their correspondence had ended with sour disagreement over Suez presumably made the loss more poignant.

Toynbee had another grievance against Eden's government which may have accentuated his reaction to the Suez crisis. In June 1956, while he was in New Zealand, he was invited to become a Companion of Honour. This was a dignity bestowed by the British Crown at the suggestion of the government. An official letter reached him when he was having breakfast in Auckland, with "a sausage half way to my mouth." The messenger who brought it from the High Commissioner of New Zealand wanted a Yes or No on the spot. Toynbee was chagrined to discover that he felt peeved at not being offered a higher honor, but, swallowing both his sausage and his pride, accepted what had been offered to him all the same. Still, it was a humiliating and irritating experience. "I had fancied that I didn't care a straw about such things," he confided to his old school friend David Davies, "and then found, when the letter came, that I had quite a twinge at being classed second instead of first. . . . It is quite common to be rated lower at home than abroad, and if one did turn out to have been underrated by the authorities, that would be a boomerang, in the long run, for them. Well, it is humiliating to have felt about it as I did, and it is a relief to confess it to you."[11]

Toynbee's outward bearing usually obscured the pride he took in his learning. But he could afford to be frank with Davies, and on this occasion shed the mask of modesty he had learned to wear as a youth at Winchester and Balliol, where the ideal and presumption of "effortless superiority" required self-deprecation to make superiority seem effortless. Toynbee's egotism had been bathed for a decade in endless compliments and much hyperbole coming

Veronica and Arnold Toynbee arriving in New York by sea
Oxford University Press photo

Father Columba and Toynbee
Lawrence Toynbee photo

from journalists and ordinary citizens. To be sure, academic and professional criticism had reached something of a crescendo after the publication of the final four volumes of the *Study* in 1954, and while he was in Beirut, in June 1957, the cruelest and wittiest attack of all came from the pen of Hugh Trevor-Roper, newly appointed Regius Professor of History at Oxford. This took the form of an article in the magazine *Encounter,* entitled "Arnold Toynbee's Millennium."[12] Trevor-Roper mocked Toynbee by treating him as an upstart prophet, and invented an Anno Toynbaei (A.T. for short) for the religious era Toynbee was supposed to have inaugurated.

Being a skillful caricature, the article struck home. Father Columba thought it "scandalous and blasphemous" and refused to read it.[13] The editor, Stephen Spender, sent Toynbee an advance copy of the article and explained: "We decided to publish this, though after a good deal of hesitation, because we thought that despite its virulent manner, it is a serious article."[14] He went on to ask for a reply, but Toynbee declined, and years afterward simply noted "On the article as a whole, no comment."[15] On a still later and rather happier occasion, at a public celebration of his seventy-fifth birthday, he jocosely referred to Trevor-Roper and himself as Pope and anti-Pope, declaring that when "saying something ex-cathedra, we are, jointly and doubly, infallible."[16]

It seems clear that this article, coming as it did from a man whose post at Oxford gave him a claim to primacy among English historians, cut Toynbee to the quick. Together with a supercilious review of his final batch of volumes from the pen of a young anthropologist named Philip Bagby that appeared in the *Times Literary Supplement,* Trevor-Roper's attack turned professional opinion among the younger generation of historians in the English-speaking world decisively against Toynbee's global and prophetic vision. Mockery was more effective than criticism, since it denied the intellectual significance of what it mocked. Toynbee's reputation among historians has yet to recover from Trevor-Roper's dismissive wit. Indeed, a principal purpose of this book is to try to establish a better balance between the popular adulation on the one hand and the professional hostility on the other that closed in on Toynbee after the mid-1950s, obscuring his real accomplishment and the long-range importance of his work for the scholarly study of the past.

Another depressiong experience for Toynbee during his world tour was to discover that Asian religious leaders were not notably more ecumenical in outlook than Christians and Jews of the Western world. "I have been busy here," he wrote from Burma, "meeting Southern Buddhist monks, to try to find out whether their older and more serious form of Buddhism can fill the great spiritual blank in modern life any better than Northern Buddhism does in Japan. The results are rather discouraging."[17]

Yet, even though Japanese Buddhists disappointed Toynbee, there were people in Japan who reacted very positively to what he had to say. This was not fully evident at the time. But fertile seeds were sown during his two month sojourn in Japan (October–November 1956) when, to all outward appearances, Toynbee merely did what was usual for him to do: visiting Buddhist monks, meeting professors of history, lecturing at universities, and appearing on national television. He was based at the (Rockefeller-financed) International House in Tokyo, making trips as far as the northern island of Hokkaido, Kyoto, and elsewhere.

It is impossible for an outsider, unfamiliar with nuances of Japanese thought and feeling, to understand exactly why Toynbee provoked the response he did. Moreover, his elevation to the level of a Western guru did not occur until the 1970s. But it seems clear that his later fame among the Japanese was triggered by his visit in 1956. One key may lie in Japan's readiness to accept his utterly familiar denunciation of nationalism—a theme he had preached ever since the 1920s. At any rate, it is a fact that when he appeared on national television (with attendant translator) he declared: "I believe the Japanese people are going to make a significant contribution to our common effort to overcome our traditional nationalism." [18]

It was flattering, of course, to be told that Europe had something to learn from Japan. Moreover, Toynbee's denunciation of nationalism as a false religion probably carried far more weight in Japan than was the case in Europe. After all, under American tutelage, the Japanese had ostentatiously repudiated the overtly religious form of nationalism that had prevailed among them during the prewar era. Under Japan's postwar constitution, the Emperor expressly disclaimed the divinity attributed to him by the prewar Shinto religion. Just what if anything of Shinto faiths and rituals could claim continuing respect among the people of Japan remained profoundly problematic. Yet there was no obvious substitute at hand.

In postwar Japan, the only available alternative vision of history and human affairs was revolutionary Marxism, which had lurked underground before the war and emerged to a kind of academic acceptability after 1945. Perhaps Marxism could satisfy rebellious youths and threadbare professors, but it accorded ill with Japan's burgeoning commercial and industrial success in a capitalist world market. Marxism also belonged to the Russians, whereas Japan from 1945 had found itself firmly and for the most part willingly on the American side of the Cold War. In this situation, a man whose vision of history was neither Marxist nor nationalist may indeed have seemed specially worth listening to, all the more because his American following guaranteed, so to speak, Toynbee's intellectual importance and social respectability.

At any rate, International House in Tokyo arranged for the translation and publication of nine lectures Toynbee gave while visiting Japan in 1956. They addressed miscellaneous themes: "Can We Learn Lessons from History?," "Will Democracy Survive in the Atomic Age?," "The Spiritual Challenge of the Ideological War," and the like. This little book, whose title may be translated as *The Lessons of History,* met with some success, selling 8500 copies in the first few weeks. Efforts to translate Toynbee's big book were already under way, but *The Lessons of History* appears to be the first significant publication of anything from his pen in Japanese. In addition, he was introduced to a far wider public on New Year's Day, 1957 when a Tokyo newspaper with a circulation of some 8 million published an article he had written on current affairs, treating his remarks as a sort of keynote for the coming year.

Thenceforth, Toynbee's name was familiar to a considerable number of Japanese. A significant company of intellectuals, who regretted the moral vacuum arising from repudiation of the nationalist Shinto faith and who felt dissatisfied with Marxism, turned to Toynbee for answers to their discontents, and did so in sufficient numbers to establish a transmission belt along which his ideas began to penetrate the country. But this took time, and a full translation of *A Study of History* did not appear until 1968–1972.

Nowhere else in his long itinerary, from Peru to New Zealand, then Australia, Indonesia, Vietnam, Japan, Hong Kong, Siam, Burma, India, and the Middle East, did Toynbee leave such lasting traces. All the same, his private distresses and disappointments along the way constituted an odd counterpoint to the public triumphs accorded to him almost everywhere. Simultaneously, as his *Observer* articles made clear, he took real delight in visiting ancient monuments and looking over the ground where the civilizations he knew so well had flourished. So the trip was not a failure, even though he had to give up his original plan of using it to prepare for a second edition of the *Study*. By August 1957, when he got back home, he was more of a world figure than ever before, and, being officially retired, was also jobless. As such, he was free to choose a new style and pattern for his life.

Toynbee set himself three goals for his retirement: to continue scholarly activity, to get rich, and to see even more of the world than he already had seen. The two latter goals could be achieved by lecturing overseas, and particularly in America. His regimen for the next decade therefore involved yearly visits to the United States, where he launched himself upon a furious round of lecturing, aimed, quite crudely, at collecting as much money as possible. He supplemented this kind of travel by accepting invitations to visit other lands and places, the more exotic the better.

Yet, between times, and even while visiting the United States, he continued scholarly work. Struggling to find time for writing was nothing new to him of course. He had written the first six volumes of the *Study* in vacation time after all; and if one thinks of his lecturing tours as equivalent to his former duties at Chatham House, one can say that his regimen was in fact unchanged. This was how he viewed it himself. Contemplating retirement, and the diminished income that would ensue, he explained to Gilbert Murray that he would henceforth have "to earn our bread and butter by visits to American universities." [19]

He might better have said that he intended to earn cake and caviar in America, for in fact Toynbee was well off. Chatham House pensions were certainly meager, but Toynbee could count on substantial book royalties and he had already accumulated a considerable capital sum from past earnings. Yet the more money he had, the more he worried about it, and he never felt safe against the risk of penury in old age. This, too, confirmed to deeply ingrained habit, for Toynbee had been haunted by money fears throughout his life. His parents' desperate effort to maintain middle class status on inadequate income had bitten into him as a child, inspiring his intense effort to win a scholarship to Winchester. The whole shape of his life descended from that youthful response to his parents' lack of sufficient money to maintain their inherited standing in society, and an obsessive fear of running out of funds never left him. It had accorded ill with Rosalind's insouciance in money matters, which often irritated and occasionally humiliated him in his middle years. As we saw, he had responded, just as he did to his parents' difficulty, by working harder, hoping to compensate for her extravagances by his journalistic and other miscellaneous earnings.

Veronica, however, was almost as chary of spending money as he was. Accordingly, after 1947, when his income from royalties and lecture fees swelled enormously, Toynbee had no further cause for concern. Despite stiff British income taxes, he accumulated capital rapidly and put some of his newly disposable wealth into land by buying additional farms. But a lifetime of anxiety over money could not be relieved by mere facts. Toynbee could always imagine running short of funds. In particular, he was haunted by fear that tax inspectors might find him at fault, and assess enormous penalties.

He filled out his own tax returns, and undoubtedly took advantage of every device he knew to minimize his liability. But he had one particular reason for uneasiness. This arose from the fact that in 1954, in anticipation of another peak in royalties from the publication of volumes VII through X of the *Study*, he had signed a contract with Oxford University Press whereby the Press paid him a fixed annual stipend (which could be increased at discretion) while holding back surplus royalties in a special standing account.[20] His future books were reserved for Oxford University Press and all past and future royalties were to come under this umbrella contract.

Obviously, the arrangement was mutually advantageous. The Press assured itself of a convenient sum of working capital as well as a flow of future manuscripts, while Toynbee escaped liability to income tax surcharges in years when his royalties peaked. But was it legal? Tax experts both in Britain and the United States said yes, but Toynbee felt there was something underhanded about the deal, and dreaded being found out by some crusading tax inspector. No accumulation of capital could still that fear. The more money he made, the more he worried about the vast penalties he might have to pay if found out.

His obsessive concern about money made him oddly, sometimes crudely, eager to get more. Lecture fees were the biggest bonanza available to him, for innumerable American universities were willing and eager to pay large fees to attract a lecturer as famous as Toynbee had become. Other organizations, the oil company Aramco, for instance, were sometimes also willing to pay extravagantly for his services.[21] As long as Toynbee's visits to the United States had been funded by the Rockefeller Foundation he was prohibited from accepting fees for lecturing, but after 1954 the grant expired and this restriction was removed. Accordingly, in the autumn of 1955, before starting on his world tour, he paid a short visit to the United States with the object of making as much money as he could by lecturing.

He was spectacularly successful. In six weeks he delivered a total of forty-three lectures and talks before an enormous variety of audiences, appearing on radio and television, before women's clubs and the boys at Groton School, and at a roster of prestigious American universities, including Brandeis, Chicago, and Columbia. The high point of the tour came in Minneapolis, where Toynbee addressed an overflow audience of some 10,000 people gathered into the two largest halls of the University of Minnesota on "The New Opportunity for Historians." His fee was $2500, and people came through heavy snow from hundreds of miles away to hear what for most of them must have been a disappointing discourse on the possibilities of world history in the twentieth century.[22]

His foray in 1955 showed Toynbee how lucrative the American lecture circuit could be, and when he got back from his world tour he was eager to exploit its possibilities. He hit upon the device of accepting a teaching appointment at one or another American university for a term or part of a term, with the purpose of using it as a base for lecturing elsewhere, wherever suitably attractive fees were available. Accordingly, in 1957–1958 he accepted an invitation from Rice University in Houston to spend five weeks on campus and deliver two to three lectures each week. But the trip had to be delayed because of his prostate operation, and when he showed up on 12 December 1957 the nearness of Christmas holidays and Toynbee's other lecturing commitments combined to reduce his formal appearances at Rice to a mere six lectures. Needless to say, this disappointed his hosts.

From Rice Toynbee went to Washington and Lee University in Lexington, Virginia, where his friend and admirer E.D. Myres was chairman of the Department of Philosophy. There he gave fifteen lectures, reflecting upon what he had seen during his recent travels round the world. But he simultaneously mounted a long series of "side shows," giving at least twenty-five lectures off campus before leaving the United States early in June. In subsequent years Toynbee visited the University of Pennsylvania (1961), Grinnell College in Iowa (1962–1963), and the University of Denver and New College, Florida (1964–1965). Each time he used his host institutions as a base from which to travel far and wide in pursuit of additional lecture fees. Insofar as his object was to make money, he certainly succeeded. At the end of his stay at Grinnell, his Iowa bank balance stood at $28,684.72—an accurate measure, presumably, of what he had accumulated during the preceding three and a half months.[23]

But despite financial success, Toynbee was wearing out his welcome, and on two distinct levels. First of all, in his eagerness to make money he risked offending his hosts. Frequent absences from the university or college that had invited him to teach for a term meant that instead of getting the benefit of Toynbee's whole time and energy his hosts only saw him occasionally. Even those who invited him for a single lecture sometimes felt short changed as well, if only because he insisted that Veronica accompany him and that her way should be paid as well as his own, even though American academic practice dictated otherwise. His aim was to make sure that all his expenditures were covered by others during his visits to the United States. He came very near to achieving that goal, but infringed American academic custom in doing so. A consequence was that host institutions did not ask him back, and as the years passed, his unending pursuit of lecture fees took him to campuses of lessening prestige.

Intellectually, Toynbee also wore out his welcome by repeating himself. Many of the lectures he gave went over familiar ground, adding little or nothing to what he had already said in print. He operated with about half a dozen stock lectures. The roster was never fixed, and was altered somewhat from year to year. But the two books that engaged his mind between 1957 and 1965— *Reconsiderations* and *Hannibal's Legacy*—did not lend themselves to popular lecturing. His repertoire of lectures therefore did not reflect any serious new thought.

It is true that Toynbee accepted a series of lecture appointments requiring subsequent publication, and for these occasions he duly composed fresh lectures. But these were chores, undertaken for pay and executed in haste. The short books that resulted added little to his repute. Moreover, they touched on themes that offended nearly all Americans, for Toynbee cast doubt on the viability of democracy in an atomic age,[24] deplored Christianity's arrogance and intolerance,[25] reproached Americans for their affluence and for betraying the principles of the American revolution,[26] endorsed the righteousness of revolution in Latin America, including Castro's revolution in Cuba,[27] and reaffirmed his belief that a world empire was the only way to stave off destructive war.[28] Flashes of his old brilliance occasionally enlivened these published lectures, most notably an analysis of Latin American urbanization and peasant unrest in the lectures he gave in Puerto Rico. In that case he was exploring a subject he had not touched before and showed undiminished power of processing new information and synthesizing it in haste, just as he had done year after year when writing the annual surveys. But most of the time he merely embroidered old themes and engaged in increasingly harsh criticism of the foreign policy of the United States government.

Toynbee's mounting criticism of America's role in the world was partly triggered by public events, like the effort in 1961 to overthrow Castro's government in Cuba by contriving the Bay of Pigs invasion. In addition, he was disappointed by the character of the students he met, beginning in 1957–1958 with his visit to Rice, when ill health (he was not fully recovered from his prostate operation when he arrived) and the self-satisfaction of oil-rich Houston made for a sour encounter. "I must confess," he confided to an old friend, "that since coming to America this time I have been shaken in my assumption that this is the American century, and I find myself feeling that Communism's chance of winning the missionary war—which is the war we really have to reckon with—is better than I had thought. . . . I have been shaken (i) by the levity of the American reaction to Sputnik; (ii) by the iron grip of the rich on American life: it is done through 'democratic' machinery by pressure groups, but is nonetheless oppressive for that."[29] Americans, he felt, were spoiled by their easy wealth, thinking themselves "automatically privileged," and entitled to lead a "rather pampered and intellectually indolent" life.[30]

With each passing year, his disenchantment with the American way became more emphathic. In 1961, when he took part in a nationally broadcast symposium on world strategy with such dignitaries as Henry Kissinger, Adlai Stevenson, and Admiral Lewis Strauss, he declared: "The U.S. has become the champion of vested interests. As a result she has lost the leadership of the revolution that was launched by her one hundred and eighty years ago. But the American Revolution is continuing, and will continue, with or without America." At stake, according to Toynbee, was "a share in the benefits of civilization for the depressed two-thirds or three-quarters of the human race. Will the U.S. resume the lead? Or will affluence prevent it? This, I believe, is the crucial decision the American people have to make today."[31]

By 1965 he was even more blunt. In an interview for the magazine *Playboy,* which was, however, not published until 1967, he said, among other things, that communism had more to offer Latin America than anything the United States could provide, since "Latin America really needs a drastic revolution . . . putting down the selfish ruling minority." He expressed impatience at "the diversion of energy and interest to the space race" and asserted that the United States was "the leader of a world wide counter-revolutionary movement in defense of vested interests." Moreover, "Madison Avenue [i.e., American advertizing and the values it propagated] is more of a danger to the West than Communism."[32] In May 1965 he summed things up in an essay for the *Observer* entitled "America Needs an Agonizing Reappraisal," declaring that the United States "is up against the determination of the non-Western majority of mankind to complete its self-liberation from Western domination. . . . America, without realizing what she has been doing, has made herself the heir of British, French, Dutch and Japanese colonialism."[33] The basic idea of his jeremiads was far from new, for he had proclaimed essentially the same message in 1918–1922. Then, however, he had viewed the British Empire as the principal target of the anger of the majority of mankind; now it was the United States.

Privately he endorsed unilateral nuclear disarmament (at least for Britain), even at the risk of a communist victory. "Is the Communist 'scourge of God' going to descend on us?" he asked Davies rhetorically. "If it just descended on us without the human race being destroyed, that would be quite a relief: the door would then be open again. And I would back our granddaughter's granddaughters to outlast Communism."[34] Some years later he told another correspondent that the United States had no right to the Panama Canal and was wrong to intervene in Vietnam.[35]

Few in America responded to Toynbee's reproaches in the 1950s, although some sectarians welcomed his assaults on the establishment. World consciousness of the sort Toynbee advocated was rare; and the few Americans who concerned themselves with problems facing the peasant majority of humankind thought they knew the answer: all that was necessary was to transplant American-style democracy and free enterprise to other lands. Confidence in American know-how and the effectiveness of foreign aid was at a peak after the undeniable success of the Marshall Plan in Europe (1948–1952) and the defeat of communist guerrilla war in Greece. In Asia, the United States, with help from other United Nations forces, had turned back an organized communist attack in Korea (1950–1953), and British successes against communist bands in Malaya seemed to show that the Greek formula for winning guerrilla war could be transferred to Asian ground as well, while Japan's economic success showed that industrial skills, too, could be transplanted readily enough. Americans were simply scornful of French difficulties in Vietnam until after they got seriously involved there themselves in the 1960s. To be sure, Castro's victory in Cuba was a nasty setback, but elsewhere in Latin America the United States seemed to be holding its own. In such a climate of opinion, Toynbee's warnings fell on deaf ears.

Thus, a decade after popular adulation in the late 1940s had created a popular image of Toynbee as a wise man come from afar, his role in American consciousness twisted itself around to make him over into a kind of court jester to the republic. As such, he remained a public figure and continued to command considerable attention among journalists and broadcasters, who systematically played up his more extreme statements because they were attention-catchers. Besides, a jester might still be wise, even if he dispensed a bitter wisdom, verging on madness. But he might also be a fool. Men of affairs preferred to treat Toynbee as a fool, and no longer listened respectfully.

The diminution of Toynbee's stature in American eyes became irremediable after 1961, when he aroused widespread Jewish antagonism by equating the moral conduct of Israel with that of Nazi Germany. Controversy blew up suddenly on 26 January 1961 when a student newspaper at McGill University in Montreal, Canada, reported that in the course of an informal question and answer meeting with students he had insisted that "the Jewish treatment of Arabs in 1947 is comparable to the Nazi murder of six million Jews." [36] News services picked this up and within a few hours headlines blossomed throughout the English-speaking world. Within hours, the Israeli ambassador to Canada invited Toynbee to debate, and he agreed. They met on 31 January at Hillel House on the McGill campus, where they firmly but politely explored their differences. Toynbee had to admit that Arabs had committed atrocities too, and that Nazi atrocities were on a far bigger scale than what had happened in Palestine; but he reaffirmed his view that some of the "massacres by Israeli armed forces in Palestine do compare . . . in moral quality with what the Germans did," having been planned in advance, carried through deliberately, and aimed at eliminating an ethnic group from a particular piece of territory. He referred specifically to deeds of extremists like the Stern Gang and Irgun, and admitted "I do not know how far the Haganah [the official armed force of the nascent Israeli state] was also implicated." [37]

Toynbee had obviously struck a raw nerve. Jewish dissatisfaction with the way he had treated their history in medieval and modern times was already at the boil. As far back as 1934 he had asserted that Jewry was a fossil, surviving from an otherwise extinct Syriac civilization. Such an unflattering label rankled. His longstanding opposition to Zionism had also aroused counterattack by figures as prominent as the future Israeli Minister for Foreign Affairs Abba Eban. [38]

When Toynbee sought to paint what he thought might be a flattering picture of the future role of the Jewish diaspora, he simply provoked more anger. In

1960 he had published an article in which he hailed the Jewish diaspora as the wave of the future, since "ubiquitous but non-monopolistic religious associations will, I believe, be the standard type of community in our Atomic Age."[39] Then, after the clash in Montreal had exasperated Jewish opinion against him, he incautiously returned to this theme when addressing an anti-Zionist organization called the American Council for Judaism in mid-May 1961. Toynbee argued that for the past twenty-five hundred years the Jewish communities of the world had been poised awkwardly between remaining closed ethnic communities and becoming open religious bodies. He then challenged his hearers to go the whole way and, incidentally, repudiate Zionism which, he declared, expressed an identical point of view with anti-Semitism since both assumed Jewish segregation. This occasioned a full-page story in *Time,* which quoted him as saying: "If the abandonment of ethnic reservations were complete and genuine on both sides, the traditional caste barrier between Jews and non-Jews would be likely to be broken down by more and more frequent intermarriage," whereupon "the great spiritual treasures of Judaism" could become "one of the common spiritual possessions of the human race."[40]

After that, the fat was really in the fire. Toynbee was roundly denounced by Jewish leaders for tampering with a faith not his own. The most judicious expostulation came from a Christian writer on Judaism, James Parkes, who explained that Toynbee's fundamental error was to think that the prophets, not the Talmud, constituted the core of Judaism, because he assumed that religion turned on a relationship between God and individual human beings, whereas Judaism was based on a collective connection to God. He concluded by saying that Toynbee "allows his great reputation to be exploited by partisans, whether Arab or anti-Zionist Jew. . . . Oh the pity of it, the pity of it. . . . The whole weight of his learning has been thrown on the side of division, of misrepresentation, and of falsehood."[41]

In the dust-up provoked by Toynbee's remarks in Montreal, a well-known Christian theologian from Union Seminary in New York, Reinhold Niebuhr, told journalists that "in spite of his best efforts at objectivity, Toynbee has a deep-set prejudice against the Jews."[42] Another respected Christian scholar and expert on Palestinian archaeology, W.F. Albright, had written in 1955: "There can be no doubt in my mind that he [Toynbee] is definitely hostile to Jewish culture and unfair to recent developments."[43]

Such accusations haunted Toynbee thereafter, and he was never able to disentangle himself from lingering controversy about his attitudes toward Jews and Jewish culture. As he often pointed out, he was no anti-Semite in the obvious sense, counting such figures as Lord Samuel and Lewis Namier among his friends.[44] Moreover, he was rather embarrassed to be reminded that in 1918–1920 he had backed the Zionist program. He explained to the researcher who had uncovered his role in the post-World War I settlement in Palestine: "Your well considered and well documented opinion that the things I wrote during World War I actually influenced what happened surprises me. . . . However, on second thoughts I am inclined to think that you may be more right about this than is credible to me. I think so because of the levity, the shortness of view, and the naivete of the people who were then in power. I shared this naivete of course. . . . No one then realized the strength and dynamism of the Zionist movement."[45]

While it is wrong to call him anti-Semitic, it remains true that Toynbee was profoundly unsympathetic to Judaism and to Jews. As a child he learned to associate the poor immigrant Jews of London with sharp commercial practice, and his early Christian indoctrination branded the Jews as stubbornly proud for refusing to accept the truth of the Gospel. His formal education simply left Judaism out after the birth of Christ. Only late in life, and largely as a result of the controversy that started after 1955, did he acquire a decent familiarity with the history of medieval and modern Judaism. In all these respects he was completely representative of his time and social milieu. But it remains true that such a background did not prepare him to sympathize with Jews or Judiasm; and when Jews were able to set up the state of Israel by force in 1947–1948, Toynbee viewed their victory as an unjustified act of aggression and another manifestation of the wickedness of nationalism and collective self-worship. Moreover, he felt that the victorious Israelis were repudiating the universalism of prophetic morality, thereby betraying Judaism's crowning achievement. "The tragedy of recent Jewish history," he wrote, "is that, instead of learning through suffering, the Jews should have done to others, the Arabs, what had been done to them by others, the Nazis."[46] Yet, as he said in private, by deploring atrocities against Palestinian Arabs "I was paying the Jews the highest compliment I know. I was expecting of them a higher standard."[47]

Toynbee adhered stubbornly to that sort of reverse discrimination all the rest of his life. "The right of the Jewish state to exist," he wrote in 1971, "is as I see it, conditioned on their being able to establish a state of their own without dispossessing, or otherwise injuring, other peoples."[48] He had not asked the same of the Turks in 1922; and when Arab–Israeli war broke out anew in 1973 his partisanship even led him to write to a Syrian general: "I send you my heartfelt wishes for an Arab victory."[49] He consistently refused to visit Israel, because, he said, he would not be "able to have a fair opportunity of seeing both sides."[50] Yet in 1971 he wrote to the *Times* of London protesting that "the appearance of the city [of Jerusalem] has been irreparably damaged" by clearances the Israelis were making in front of the Wailing Wall.[51] In this and other cases, Toynbee seems to have been quite won over to the Arab point of view.

It is hard to doubt that he developed a sense of personal grievance against the array of Jewish apologists for Israel who dominated American opinion about Middle Eastern affairs; and on occasion he allowed his exasperation to show in the form of petty counterattacks of the kind just quoted. But he was careful not to record his feelings in any fashion that could be used to justify tagging him "anti-Semitic." In private he warned an anti-Zionist Jew: "If the American public were suddenly to realise that Zionism, and American support for it, has dealt a major blow to America's international interests by throwing the Middle East into the arms of Russia or China or both, the American people . . . might suddenly turn against not only the American Jewish Zionists, but also against all Americans of Jewish religion.''[52] That was as far as he felt it safe to go in criticizing the commitment of American Jews to Israel.[53]

Toynbee's quarrel with Zionists attracted favorable attention throughout the Moslem world; it was especially welcome to Arabs, among whom he became something of a hero. Consequently, as his American welcome wore thin, his popularity increased in Moslem countries, both because of his criticism of Israel and because of his criticism of the West for not doing more to atone for past injustices toward Third World countries. This was not yet evident in 1957, when he visited Saudi Arabia under the auspices of the Aramco Oil Company, since argument over his views about Judaism were then still confined to specialized journals and a limited circle of indignant Jews. Three years later, however, he was treated royally when he lectured for a month at the University of Peshawar in Pakistan and spent additional weeks of travel inspecting historic landmarks on either side of Hindu Kush as a guest of the governments of Pakistan and Afghanistan. Stopovers in India before and after his stay in Pakistan gave Prime Minister Jawaharlal Nehru a chance to compete with General Mohammed Ayyub Khan, President of Pakistan, in extending official honors to Toynbee.

Early in his tour he wrote back' "I am still living in comic grandeur. I am not yet used to having a sentry with fixed bayonet at my gate.''[54] Yet he soon came to enjoy the privileges showered upon him. Toward the end of his travels, when crossing the border between Afghanistan and Pakistan, he was met by an honor guard, and protocol required him to dismount from his Land Rover and review the detachment. This extraordinary role for a private citizen must have provoked memories of his childhood play with toy soldiers. Toynbee was delighted. "I have never enjoyed a journey so much, not even walking about in Greece in 1911–1912," he confessed.[55] Well he might, for the Afghan government sent special work parties to repair roads he wished to travel over, and the King put a helicopter at his disposal for trips that were too long by road. Toynbee's health had improved by this time, and he showed remarkable endurance for a man of his age. Once, in India, when he missed an air connection, he endured 135 miles by jeep in hot sun in order to appear on time for a lecture engagement. He was rather proud of this feat, telling Davies the next day that he was "feeling much less like a boiled vegetable after a long night's sleep.''[56] Veronica, recuperating from an appendectomy, had stayed behind, so her restraining influence was lacking.

What delighted Toynbee, of course, was not merely the deference paid to him. That was pleasant but inessential. What aroused such intense satisfaction was his opportunity to explore the historic borderlands between the Middle East and India with the help of every facility at the command of the Pakistani and Afghan governments. His imagination fed on scenes he witnessed. In Multan, Pakistan, for example, he reflected on how the British, who had conquered the place in 1846, had there for the first time intersected the path of Alexander, who had "rashly led the storming party and was knocked on the head. His Macedonians only just managed to save his life." But the contemporary scene was never far from Toynbee's mind. In the same letter, he wrote of Multan: "Its charm lies in the swarming crowds of people and animals. Across the yard there are four buffalo cows who give me milk. . . . And there are strings of camels with bead necklaces, buffaloes with necklaces and garlands round their humps, magnificent sheep with their heads dyed saffron and their backs dyed blue, and of course all manner of goats and donkeys. I am still a child; I enjoy all this hugely."[57] His combination of historical awareness and sensitivity to the current scene was nicely captured in what is by far the best of his travel books: *Between Oxus and Jumna: A Journey to India, Pakistan and Afghanistan,* published in 1961.

As far as contemporary affairs were concerned, he drew two lessons from this trip that confirmed views he had already expressed repeatedly. First, "In each of these countries, government and people alike are intent, today, on giving the mass of the people some share in those benefits of civilization that, till now, have been the monopoly of a privileged minority."[58] Second, he commended the authoritarian military government of Pakistan for its "basic democracy" and General Ayyub Khan in particular for holding out "a hand of friendship to India" in hope of reaching "an agreed settlement of their dispute over Kashmir."[59] British and American styles of parliamentary government, he suggested, could not work in lands lacking the necessary social underpinnings, whereas benevolent military dictatorship might, by allowing local management of local affairs to flourish, actually represent the needs and wishes of the peasant mass better than the Congress Party with its electoral machinery did in India.[60]

His tolerance of authoritarian military regimes in countries with a "smaller supply of honest and competent people than parliamentary democracy requires"[61] smoothed Toynbee's way in those Arab lands ruled by upstart military regimes of one sort or another. Before he visited Egypt in December 1961 the British Ambassador in Cairo assured him that "the people here regard you at present as a kind of St. George to the Zionist dragon."[62] This was Toynbee's first visit to the land of the Nile, and as he traveled upriver he visited all the standard archaeological sites, stopping along the way at various universities to deliver lectures on the ancient and medieval history of the Middle East. In addition, he did not dodge the delicate task of speaking on the Palestinian problem as a guest of the municipality of Cairo. His lectures were subsequently published in Arabic (1961) and a selection of them also appeared in English as *The Importance of the Arab World* (1962).

This visit was followed by a trip to Morocco in 1962, where Toynbee lectured to the King and a select audience in the royal palace, and by a far more extended tour of the northern parts of Africa in 1964, which took him to Nigeria, Sudan, Ethiopia, Egypt, and Libya. Once again, he lectured on university campuses wherever he went, visited historic sites whenever possible, and interviewed such figures as Gamal Abdel Nasser, President of Egypt. His impressions, written initially for the *Observer* were later gathered into another little book, entitled *Between Niger and Nile* (1965). For the most part, Toynbee contented himself with geographical and picturesque description in these essays. But he also tried to make sense of all the diversity he encountered by emphasizing political aspirations that ran throughout the region. His conclusion was not optimistic, for Toynbee admitted that striving for Arab and/or Moslem unity on the one hand was sure to conflict with the pan-African ideal of negritude on the other. Nor could he muster much hope that Christian Ethiopia might remain above the battle and play a mediating role, though he endorsed that pipe dream on the last page of this little book.

The European continent constituted a third theater in which Toynbee's reputation reached new heights after his retirement. In 1959 he traveled to Rome to address a meeting of the Food and Agriculture Organization of the United Nations on "Population and Food Supply." He came out so strongly for birth control that the international civil servant who had invited him to speak, an Indian named B.R. Sen, felt it necessary to tone down the prepared text to minimize offense to the Vatican.[63] The precarious balance between food and population became a continuing concern for Toynbee thereafter, and along with the human costs of urbanization constituted a growing point for his thought about contemporary affairs.

But it was not these new thoughts that impressed Europeans. Rather it was Toynbee's cyclical view of the rise and fall of civilizations as well as deeper theoretical issues underlying *A Study of History* that attracted most of their attention. A sort of horizon point for Toynbee's reputation in high intellectual circles on the continent came in July 1958, when a group of carefully selected professors and journalists assembled at a retreat in Normandy to discuss his ideas. Toynbee was invited to attend and spent two weeks defending his book against various kinds of misunderstandings. Raymond Aron, a distinguished French man of letters, presided while designated rapporteurs from Germany, Poland, France, and the United States took turns delivering papers about aspects of Toynbee's thought. Discussion followed each paper, and Toynbee offered a formal rebuttal to some of the papers as well.

To judge from the published record of the conference, there was little real meeting of minds. Toynbee agreed, more or less, with what an American, Owen Lattimore, had to say. They were old friends after all. But French and Polish spokesmen posed abstract philosophical questions, often with a Marxist coloration. Toynbee relied upon a different mode of discourse—his term "spiritual" particularly bothered the French—and the two parties were ill prepared to understand one another. Only the Germans resonated to aspects of Toynbee's thought, but in Toynbee's own view they quite misunderstood him. Professor O.F. Anderle distinguished himself by rather dogmatic assertions about how the comparative study of civilizations should be pursued—ideas which he attributed to Toynbee, but which Toynbee failed to recognize as his own. "The result was unfortunate," Toynbee explained afterward. "Dr. Anderle was aggressive, and this put people's backs up."[64]

Subsequently, Professor Anderle devoted his energies to organizing an International Society for Comparative Study of Civilizations and invited Toynbee to become its president. When Toynbee declined, a Harvard professor of sociology, Pitrim Sorokin, was persuaded to assume that dignity while Anderle became secretary general. Toynbee attended a meeting of the new organization at Salzburg in October 1961 which addressed "Die Problematik der Hochculturen." But he felt little sympathy for the way Anderle ran things and dissociated himself from the enterprise thereafter.[65]

Difficulties in attuning Toynbee's thinking to the high intellectual traditions of continental Europe were to be expected and did not affect his access to German mass media. He often appeared on one or another of the regional radio or television networks of West Germany, speaking usually on current affairs. In addition, he visited Germany five times in the decade 1955–1965 to address popular and university audiences. Toynbee's education had attuned him to German culture. He preferred Goethe to Shakespeare, for example, and his command of spoken German was excellent. He was, therefore, very much at home in Germany.

The Germans responded by regarding him in about the same way that the English did. Both nations, as he grew older, treated him as a kind of cultural monument. It was a comfortable niche, allowing the public to admire him while discounting his views on public policy and current events. Oddity was expected from artists, and as Toynbee was assimilated to their company the estrangement from his countrymen occasioned by his Reith lectures faded away. In addition, his growing alienation from the United States strengthened Toynbee's standing in Europe. Meanwhile, for reasons of their own, the Japanese were beginning to idolize him from afar. As a world figure, Toynbee therefore led quite different lives in different lands. The world's cultural diversity made anything else impossible.

His negative response to Professor Anderle's effort to organize the study of world history contrasted with Toynbee's enthusiastic response to another intellectually ambitious enterprise known as the Institute of Ekistics. (Ekistics, a new coinage, signified the study of patterns of human settlement and resulting social interactions.) The Institute of Ekistics was the creation of a Greek architect and city planner named Constantine Doxiadis, who used part of the wealth he had accumulated by building and designing new cities in Pakistan and elsewhere to organize an American-style "think tank" where, he hoped, solutions to social problems arising from runaway urban growth could be explored by an international array of the best minds he could attract to the task.

In childhood, Doxiadis had arrived in Greece as a refugee and had personal experience of the disruption of life inherent in moving from a village or small town into the maelstrom of a disorderly city. He was convinced that replicas of village community life had to be created within sprawling metropolitan centers to permit satisfactory human relations to arise within the urban environment. This in fact is what had happened spontaneously in makeshift suburbs around Athens after 1922, where refugees from different parts of Anatolia had clustered, like with like. Doxiadis remembered this experience when he won a contract from Ayyub Khan's government to build public housing for the swarming poor of Karachi, Pakistan. He therefore created simple, cheap houses, grouped around mosque and marketplace, in what amounted to a simulacrum of rural village structures.

During his state visit to Pakistan in 1960, Toynbee inspected and was favorably impressed by the new housing at Karachi and learned something about its designer. As a result, in 1962, when Doxiadis invited him to join a cruise in the Aegean and take part in what he called the Delos Symposium, Toynbee was eager to do so. Other famous people joined the cruise, meeting each day to discuss the problems of cities and other world issues. Few of the invited guests had given much thought to urban problems beforehand, so Doxiadis' ideas tended to prevail. This was evident at the close of the cruise when ten "wise men," including Toynbee, subscribed to the following declaration. "At the base of all human settlement lie the neighborhoods and villages. These must be identified, resurrected or created."

Doxiadis also believed that urbanization was destined to increase, creating vast megalopolitan clusters, in which a majority of humankind would have to struggle toward a new way of life. He even spoke of a future Cosmopolis in which all human beings would participate. Toynbee's longstanding commitment to world government and his exposure to the slums surrounding Latin American and other Third World cities sensitized him to such projections of the future. Accordingly, when Doxiadis invited him to take active part in the Institute of Ekistics by looking at urban patterns of the past and asking how human beings had built physical frameworks for living, he was swiftly won over to this new sort of historical inquiry. He became consultant for one of Doxiadis' projects at the Institute of Ekistics, aimed at creating a complete inventory of information about the layouts of ancient Greek city states and in 1963 he undertook the editorship of a book entitled *Cities of Destiny,* using some of Doxiadis' ideas as his guiding light.[66] Toynbee returned for subsequent Delos Symposia in 1966, 1970, and 1972, and he maintained high enthusiasm for Doxiadis' ideas and accomplishments throughout the rest of his life, praising him as "the first intellectual pioneer who has succeeded in reuniting human studies into an intelligible whole."[67] Doxiadis reciprocated, admiring Toynbee's historical knowledge and exploiting his name in an effort to establish the new discipline of ekistics.

In spite of all the time and energy that Toynbee devoted to lecturing, traveling, and commenting on current affairs after 1955, this remained essentially a sideline for him—a way to make money, to enjoy the fruits of his fame, and to toy with new ideas. Nonetheless, it was not what mattered most. What did matter was, first, to settle accounts with all those who had attacked *A Study of History.* This meant accepting his critics' corrections when they were right and rebutting their arguments when they were wrong. His second aim was to write a book of exemplary scholarship so as to discountenance all those nitpicking experts who had impugned his command of factual detail. Only so, he felt, could his reputation as a scholar and historian be thoroughly vindicated; only so could he disperse the cloud of professional criticism that *A Study of History* had provoked in the post-World War II decade. The theme to use in exhibiting his scholarly prowess and exactitude was predetermined for Toynbee by the fact that in 1919 he had turned away from exploring the consequences of the Hannibalic wars in order to undertake what became *A Study of History.* Toynbee always liked to finish what he had begun. This was a chance to do so and simultaneously to redeem his scholarly reputation. Given such intense motivation, it is not surprising to find that Toynbee threw himself into the twin scholarly tasks of his retirement with enormous energy, although, as we shall see, the intellectual task of remodeling his original synthesis defeated him.

He began the first part of his dual task in 1958 when visiting Washington and Lee University. "I have started to read my book of nonsense," he confided to Davies. "I haven't read it before except in proof. A trunkful of criticism is on its way to Virginia by sea, and when I have read these too, I hope to start having my second thoughts. I am trying to read it as if I were Geyl or Trevor-Roper. It is an odd and rather amusing job."[68] At first he projected a new edition of the entire work, changing passages as seemed necessary; and his papers contain a few notes that aimed at this goal. But when Toynbee began to peruse the "trunkful of criticism," and realized how extensively he would have to rewrite the original text if he were to satisfy his critics and himself, he gave up the idea of a second edition and instead decided to write a book about what changes he would and would not make if he were to do the whole thing over. He devoted less than two years to this task and by September 1959 had a completed manuscript that was published—all 740 pages of it—in 1961 as *A Study of History*, volume XII, *Reconsiderations*.

The first third of *Reconsiderations* is devoted to theoretical questions, philosophical and structural; the book then turns to particular topics and ends with fifty-five pages entitled *Ad Hominem*, in which Toynbee tried to view his great work as a product of time and place and the personal experiences of his life. No summary is really possible, since Toynbee arranged the text around objections various critics had made to his book and those critics had, of course, converged upon him from many different directions and with entirely disparate reproaches. Sometimes Toynbee contented himself with citing contradictory criticisms; sometimes he agreed that the way he had argued was at fault. Almost never did he find it possible to assert that his original formulation was in need of no alteration.

With respect to the most important distortion inherent in his original outlook, he agreed that he had underestimated the role of diffusion among human societies and accepted the reality of what the American anthropologist A.L. Kroeber had felicitously dubbed "stimulus diffusion." Toynbee concluded, rather lamely: "I still need to reconsider the whole question in the light of new knowledge and new ideas that the last thirty years have brought with them." With respect to another especially vulnerable proposition of his original work he decided: "We must conclude that the conscious continuity of the Syriac Civilization did not survive the fall of the Achaemenian Empire—always excepting the still unbroken tradition of the Jews and the Samaritans." This makes it hard to view the Moslem califate as a Syriac universal state, but Toynbee did not expressly retract that implausible view. He did retract his whole notion of "arrested civilizations" as simply erroneous and repeatedly sought to meet his critics halfway, returning soft answers to barbed attacks and manifesting a detached, irenic disposition throughout.[69]

This was a genuine part of his personality, just as his ecumenical religiosity was. Both shine through when he redefined civilization "as an endeavour to create a state of society in which the whole of mankind will be able to live together in harmony, as members of a single all-inclusive family. This is, I believe, the goal at which all civilizations so far known have been aiming unconsciously, if not consciously." Belief in God and faith in progress lent a remarkably rosy color to the conclusion, where Toynbee wrote: "Though the goal of mankind's continuous and increasing endeavours is hidden below our horizon, we know, nevertheless, what it is. . . . Its intellectual goal is to see the Universe as it is in the sight of God, instead of seeing it with the distorted vision of one of God's self-centred creatures. Human nature's moral goal is to make the self's will coincide with God's will, instead of pursuing self regarding purposes of its own."[70]

Necessary steps along this path were renunciation of war and acceptance of an obligation for mutual help. Toynbee saw progress in both directions, for "the World's most powerful nations and governments have shown an uncustomary self-restraint on some critical occasions. They have given priority to their sense of responsibility for avoiding a world war." And "advance towards greater social justice through an increase in human kindness has been taking place in two fields simultaneously: as between different classes in a single country and also between different countries. . . . The relatively rich minority of the human race has now recognized that it has an obligation to make material sacrifices in order to assist the relatively poor majority."[71]

Such professions of faith sound old-fashioned. In a sense, Toynbee did return to many of the assumptions and commitments of his youth in his old age. But there were differences. Progress was not material progress, and Anglican Christianity was not the truth. But then, neither was *A Study of History* the truth. By the time Toynbee had agreed with some points made by his critics, met them halfway on others, and left questions unresolved in still other instances, little was left of the original, and no new vision of human history as a whole emerged from his *Reconsiderations*. In effect, he backed away from his original mission when he gave up the plan for a second edition of the whole work. Faith now sufficed to tell him where the human adventure was headed; but the intelligible pattern of the past he had striven so mightily to create lay fractured in his hands, and he could not put it back together. Instead he invited others to take up the task and work out a more accurate vision of humanity's history than he had been able to do in the 1920s. Nonetheless, he insisted, as strongly as ever, that a view of the whole was necessary to give meaning to more detailed historical research.

By 1961, when *Reconsiderations* appeared, professional debate over what Toynbee had achieved (or perpetrated) in *A Study of History* had pretty well exhausted itself. Toynbee's general tone invited reconciliation, except for those who found his pages on "History and Prospects of the Jews" objectionable,[72] and in fact there was little further debate. Historians were not impressed one way or the other. Universal history was thoroughly out of fashion in academic circles, and the ridicule Trevor-Roper had poured on Toynbee's religious preoccupations could not be counteracted by Toynbee's readiness to reconsider points large and small as demonstrated so freely in this final volume. More and more, Toynbee simply sank from sight among academics while his public and popular reputation melted away more slowly. One of his most persistent critics, Pieter Geyl, summed up his personal reaction to *Reconsiderations* as follows: "Toynbee appears to be more accessible to reason that I had expected. . . . He is doing his best to be a historian, but first and foremost he is still a prophet."[73]

That sort of patronizing condescension spurred Toynbee to press ahead with the second phase of his self-imposed program of scholarly work. Accordingly he began writing *Hannibal's Legacy* the day after he completed *Reconsiderations*. He had first explored the impact of the Punic Wars in lectures at Oxford in 1913–1914 and had preserved notes for the book he had then projected. In returning to classical scholarship in 1959 he was therefore picking up threads he had dropped forty-five years before. It was, for him, a joyous return and he worked on the book with an amazing energy in every spare moment of his time until October 1963, when the massive manuscript was at last ready for the printer. "I feel apologetic about the length," he wrote to his publisher. "This is partly due to my wanting to guard myself against the kind of criticism that I have had from the specialists in the past. At the cost of length, I believe I have managed to take account of pretty well everything of importance that has been published about this piece of history within the past fifty years, so that critics will not be able to say 'he hasn't mastered the work that the specialists have done.' . . . The general interest of the book is, as you say, the question: Is war an instrument of policy that can produce any results except destructive ones?"[74]

In the preceding decade, Oxford University Press had found the flow of manuscripts from Toynbee's pen excessive. As the publisher explained to him in 1962: "Sales of your new full-scale books are diminishing. . . . Too frequent publication and the publication of all your lectures tends to detract [sic!] public attention from your more important writings and thus to reduce sales."[75] Two books per publishing season was judged to be the maximum that could be handled comfortably,[76] and the Press actually held back some of Toynbee's books to conform to that schedule.

The massive manuscript of *Hannibal's Legacy* dismayed the experts, who foresaw a very limited market for such a work. They therefore asked a scholar familiar with Roman history to recommend cuts. Toynbee was (more or less) grateful. Acknowledging the critique, he confessed: "I always find it particularly difficult to see for myself what cuts to make in my own work."[77] Yet he did not accept everything his critic had suggested, insisting, for instance, on retaining a long and rather muddled annex about the origins of the Etruscans that had been marked for excision.

In due course *Hannibal's Legacy* was published in two stout volumes, totaling 1395 pages. It is a monumental work, ranging across most of Roman republican history, with penetrating side glances at Carthage and the Hellenistic monarchies with which Rome contested control of the Mediterranean. Toynbee had indeed mastered all the significant scholarly writing on the subject that had come out in his adult lifetime, piling it on top of a thorough mastery of the relevant classical texts that he had achieved in his youth.

Specialists were impressed. The anonymous reviewer of the *Times Literary Supplement* called it "a great synthesis, solidly based on the most scrupulous scholarship. Those historians who thought that Professor Toynbee's bold venture into universal history had ruined him as a serious historian have been utterly refuted."[78] The worst anyone could say was to criticize it for excessive length. "There is, in fact, new insight in this book, but the proportion seems far from justifying its huge bulk, much of which consists of epitomizing the researches of others."[79] That was, indeed, the case, for in many passages Toynbee adopted the practice he had followed in his *Reconsiderations*, summarizing what others had said about a given question and then adding comments and judgments of his own or, sometimes, simply leaving the issue undecided in a kind of limbo. The book is therefore a prolix, accurate mirror of the state of historical scholarship, interesting only to specialists.

Toynbee had hoped for a wider audience. He conceived the book as a parable for the times, showing how an enormous victory that raised the Roman Commonwealth to sovereignty throughout the coastlands of the Mediterranean had nevertheless been a defeat inasmuch as it broke up that Commonwealth, producing social catastrophe and civil war followed by a peace of exhaustion under imperial dictatorship. This story, he felt, had a special meaning for the second half of the twentieth century, when Cold War between the United States and Russia was threatening to turn into a final showdown. Even without atomic weaponry, war *à l'outrance* between Rome and Carthage, Toynbee believed, had inflicted such moral and social wounds that winners became losers, compelled to surrender all they most prized, at least as far as public affairs were concerned.

But the message Toynbee hoped to communicate was in practice drowned out by the book's scholarly apparatus. Men of affairs and public-spirited citizens were unwilling to wade through the pages of specialists' debate to discover what was, in fact, an already accepted proposition among ancient historians. *Hannibal's Legacy* therefore was only partly successful. It did help to establish or reestablish Toynbee's reputation among classicists as a great scholar, but it completely failed to reach the general public. Scholarly writing almost never does, but Toynbee had been led to expect otherwise by the extraordinary response to *A Study of History*. To be sure, most of that response had been triggered by Somervell's condensation, and Toynbee hoped, initially, that a condensation of *Hannibal's Legacy* might bring his parable to public attention. But the experts at Oxford University Press were not enthusiastic and the project was eventually dropped.

One must admire the magnitude and copiousness of *Reconsiderations* and *Hannibal's Legacy* and wonder at the unwearied energy Toynbee put into writing them. But both fell far short of his initial hope and intention. In an intellectual sense the first decade of his retirement was, on balance, a time of failure. In *Reconsiderations,* he himself shattered the wholeness of the vision of human history set forth in the first three volumes of *A Study of History* without being able to put anything comparable in its place, while *Hannibal's Legacy* muffled occasional flashes of his old brilliance in a heavy fog of mere erudition.

Toynbee's family life in these years may also be described as a partial success. Relations with his sons were cordial, though he continued to have far closer connection with Philip than with Lawrence, who remained Rosalind's favorite. Grandchildren—eleven in all—he saw occasionally and entertained with ritualized trips to the circus or pantomime at Christmastime. But Toynbee's real life was with Veronica; her faithful love, care, and admiration, mimicking the love, care, and admiration his mother had lavished on him in boyhood, provided a firm support for his daily routine. He recognized the debt he owed her, and reciprocated with affectionate admiration. But Veronica was not Rosalind. The high drama of his middle years was missing from his second marriage. In that sense it remained no more than second best.

Friction with Rosalind could never be entirely avoided. Sparks flew over two issues: how to describe Gilbert Murray's attitude toward religion and how to apportion property between Philip and Lawrence in their wills. The first was trivial, though symptomatic. Rosalind had asked a Roman Catholic priest to come to her father's deathbed where he administered extreme unction. Rosalind therefore believed that in his last moments of life Gilbert Murray had been reconciled to the church into which he had been born. Others thought she deceived herself. When a Catholic chaplain at Oxford boasted of Gilbert Murray's deathbed conversion, considerable public controversy ensued, which Rosalind found intensely repugnant. Toynbee, of course, stood completely aside from this tempest in a teapot, being overseas at the time. The question of what to say about Gilbert Murray's religion arose for him only when he agreed to write a preface to a book of memoirs that appeared as Gilbert Murray, *An Unfinished Autobiography* (1960). Rosalind insisted that he change his original wording, and after several exchanges he wrote in exasperation: "I am not going to discuss my draft further with anyone."[80]

Far more important was the question of wills and how possessions ought to be divided between Philip and Lawrence. Rosalind wished to assign most of her landed property to Lawrence, reserving something for Richard Stafford, with whom she was living, and leaving remnants to Philip. Philip's prospective heritage was substantial, but less than equal, and he deeply resented his mother's partiality. Toynbee sympathized with Philip, and in 1959 revised his will, bequeathing two-thirds of his estate to Philip and one-third to Lawrence in order to compensate for the inequity Rosalind had created.

But the prospect of inheriting property someday was not the same as possession, so when Rosalind gave Ganthorpe to Lawrence in 1959 and he left Oxford to live there in 1965, Philip's jealousy flared afresh. "What is it that has upset Philip now?" Rosalind inquired of Toynbee in 1963. "I feel that he has a sort of morbid obsession about being dispossessed of some (imagined) fortune. Do you remember his fantasies as a boy that there was a sum of money put by for him (he didn't know how or by whom) but it was there for him when he grew up. . . . If I am right about this, no further adjustments or alterations will satisfy him. He is bound to go on finding reasons for his always feeling dispossessed. (Is he still the 'dispossessed' Baby?)''[81] Philip's complaints may, indeed, have reflected the traumatic emotional wound inflicted on him when he was sent off to school just when Lawrence, as a newborn infant, usurped nearly all of his mother's love and attention. But protests merely annoyed Rosalind and confirmed her preference for Lawrence.

Toynbee occupied the sidelines, regretting the controversy but unable to do much to mollify Philip's indignation. He did assign him an outright gift of £3000 in 1965 when Lawrence moved into Ganthorpe. This gave Philip tangible possession of part of what otherwise remained only a distant prospect of inheriting wealth after his parents' deaths. But Ganthorpe was valued at £5075 according to Toynbee's calculation of 1959,[82] so his gift did not really equalize the heritage. In any case, mere money was not the same as the physical reality of Ganthorpe, with its proximity to the splendors of Castle Howard that had so colored Philip's childhood and youth.

Philip had some consolation inasmuch as his career achieved greater prominence than Lawrence's. His book reviews for the *Observer* made him an influential figure on the London literary scene. On the occasion of another, subsequent quarrel over money he told his father: "As a professional man, it would be false modesty on my part not to say that I have been at the top of my little tree for quite a long time."[83] But book reviewing was not enough to satisfy his ambition. "I have the strongest possible feeling that I am in the world in order to get something done," he once explained, "and I flatter myself that it is something important."[84] For a while, he took the path of political protest, playing a prominent role in Lord Russell's Ban the Bomb agitation. And, as we saw earlier, some of Philip's horror at the prospect of atomic annihilation rubbed off on his father, with whom he had many political conversations. Toynbee's criticism of the United States for its atomic policies, imposing an unfair risk of "annihilation without representation" upon the rest of the world, reflected Philip's influence, but the agitation had no tangible effect.

Philip preferred literary fame anyway, aiming to write a "nonsense book" of his own that might match or exceed his father's achievement. "I've taken an extraordinary, and perhaps quite dotty decision," he confided, "which is to write my book in verse. So far I've translated about thirty pages. . . . And the odd thing is that passages that looked stilted and extravagant in prose look perfectly acceptable, at the least, when there are capitals at the beginning of each line."[85] The first fruit of his ambition appeared in print in October 1961. But the mock heroic verse of *Pantaloon, or the Valediction* did not fit public taste, and Philip was soon compelled to confess: "I shall have to plug in the knowledge that probably I shall never have in my lifetime the kind of general hurrah which I'd hoped for."[86] Two subsequent installments of his poem were also published—*Two Brothers: The Fifth Day of the Valediction of Pantaloon* (1965) and *Views from a Lake: Seventh Day in the Valediction of Pantaloon* (1968)—but they met with an even less favorable response, and Philip was unable to find a publisher for the remaining segments of the poem. Clearly, the effort to equal or surpass his father's literary achievement by writing a great poem fell far short of his hopes. Toynbee sympathized, both with Philip's ambition and with the hard handling *Pantaloon* received from reviewers. He, too, had suffered from critics, so Philip's much harsher experience had the effect of strengthening the bond between them.

Another link between Toynbee and his son was the fact that after abandoning politics Philip became interested in religion. In 1962 he conceived the idea of producing an instant book by recording conversations between himself and his father on religious and related subjects. He actually secured a contract with a publisher for such a text before he approached Toynbee, who thought, apparently, that his conversation with Philip would be one of several similar conversations to be made into an anthology. When, however, it developed that Philip was planning to publish an edited version of their conversation as a separate work, Toynbee protested vigorously, on the ground that his longstanding contract with Oxford University Press and his immunity from income tax on accumulated royalties would be endangered. Though he told Philip: "This is going to embarrass you and, I am afraid, annoy you too,"[87] Philip returned a soft answer and when it appeared that Oxford University Press did not want to have anything to do with the dialogue, and would not regard its publication by Nicholson and Weidenfeld as an infringement of Toynbee's contract with them, the book duly appeared as Arnold and Philip Toynbee, *Comparing Notes: A Dialogue across a Generation* (1963).

As might be expected from the way the book came into being, Philip's questions gave Toynbee a chance to repeat what he had said elsewhere on matters of religion, morality, and public affairs. Differences of view between father and son were minimal. The book is therefore a monument to Philip's growing admiration for, and sympathy with, his father. As a result, the quarrel over its publication was as evanescent as it was sharp.

Lawrence, too, had public opportunity to come to terms with his father, for he was commissioned by Chatham House to paint Toynbee's portrait at the time of his retirement. The resulting likeness reflects Lawrence's greater detachment, showing, as it does, an elderly, angular, and anxious figure, rather than the oracle Philip consulted in 1962. Toynbee certainly liked the way Philip had begun to defer to his wisdom and knowledge, but he was well content with Lawrence's cooler attachment. Philip's temperamental oscillations were, after all, very hard to handle. Lawrence remained, as always, a quite satisfactory son from his father's point of view.

As Toynbee's seventy-fifth birthday approached, Oxford University Press conceived the idea of organizing a formal luncheon in his honor, inviting some of his friends and some of his leading critics as well. Toynbee prepared a brief essay, "Janus at 75," which the Press published for the occasion,[88] and made a graceful speech, in which he acknowledged his debt to Humphrey Milford, who had first agreed to publish his great work, and to David Somervell, who had made the abridgement. Somervell could not attend, having just suffered a stroke, but he wrote, saying: "I have derived much more fame and fortune from missing out parts of your book than from writing any of my own." Toynbee commented: "This is witty, though not true. What is true is that David Somervell did me an invaluable service in making cuts in my work I wouldn't have made myself." [89] At ease among friends, Toynbee also made jesting references to Geyl's "fast trigger" as a critic, and to Trevor-Roper's ridicule. Altogether, it was a warm and cheerful occasion, matched by a leader in *The Times* that appeared on his seventy-fifth birthday. The writer praised Toynbee's effort to see things whole, faulting contemporary academics for abandoning his ecumenical aspiration and criticizing new nations for not heeding Toynbee's strictures against narrow nationalism. [90]

Thus as old age closed in around him Toynbee became a prophet not without honor, even in his own country. Though his stature in America was diminishing, he had become a world figure. No other historian, and few intellectuals of any stripe, have even approached such a standing; and, as we shall see in the next chapter, his reputation in Japan was just beginning to bloom in 1965 as he entered upon the last decade of his life.

XI

The Closing Decade
of a Busy Life
1966–1975

Not long after *Hannibal's Legacy* had come out, Toynbee confessed to his friend Father Columba: "I am in a low state, which I hope to struggle out of. At bottom, no doubt, it is the difficulty of coming to terms with old age. I don't dread death . . . but I do dread staying alive without being still able to use my capacities at full stretch. At the moment, with one book in the press and two others finished and being typed, I have no agenda—for the first time since 1913! I find this disconcerting. . . . Well, no doubt I shall come up again, but at the moment I am a bit up against it, and public affairs don't help—they are pretty glaring evidence of the reality of Original Sin."[1]

Morale soon revived. As he had done since childhood, Toynbee took refuge in work, exorcising dread of an incapacitated old age by adhering rigorously to old habits. As he explained to his friend: "How have I completed my main agenda? Because I have schooled myself to write every day, whether I am in the mood or not, and because, each morning, since about the age of sixteen, I have started, bent forward to run the hundred yards when the pistol goes off, i.e., at 7:00 A.M. (a late hour by your standards.)"[2] Since he wrote rapidly and seldom revised, his output of books and articles remained extraordinary. Oxford University Press adhered to a policy of stockpiling manuscripts so as to publish not more than four of his books each year. But his contract with Oxford University Press permitted him to edit and contribute to books for other publishers, and he did so in old age, most notably for Thames and Hudson.

The extraordinary literary output of his last years did not depend entirely on Toynbee's continued adherence to old habits. New technology also came into play, for the advent of portable tape recorders allowed him to generate instant books simply by answering questions others put to him. His son Philip had resorted to this device in 1962 and an interviewer for Radio Free Europe did likewise in 1974;[3] but it was Toynbee's Japanese admirers who exploited the potentiality of this form of publication to the full, arranging interviews lasting as long as six or seven days. With suitable editing, Toynbee's replies to questions they framed became articles and books that were fine-tuned to contemporary Japanese sensibilities by the way they had come into existence. The first of these extended interviews appeared initially as ninety-seven daily installments in a mass circulation newspaper. They kindled an extraordinary response that ensured Toynbee's fame and influence in Japan after their publication in 1970.

Until 26 March 1969, when Toynbee suffered a coronary thrombosis that laid him low for several weeks, he maintained the rhythm of life he had established after retirement in 1956. This meant annual visits to the United States to earn money by lecturing, and travel to other parts of the world when suitably enticing invitations came along. But his heart attack in 1969 required him to cancel a scheduled trip to New York for the celebration of his eightieth birthday. Thereafter, he traveled less and less and gave up nearly all lecturing. Despite efforts to conserve energy, he found it difficult to sustain his accustomed routine of writing. Nonetheless, he persisted, so that two books were awaiting publication along with one he had edited when, on 3 August 1974, at age eighty-five, his longstanding problem of high blood pressure resulted in what he had most feared—an incapacitating stroke. These last books came out only after his death, which occurred more than a year later, on 22 October 1975.

Longevity and continued literary productivity brought Toynbee fresh honors in the last decade of his life. But family tensions surfaced anew following Rosalind's death in 1967, when her inequitable disposition of property and Toynbee's decision to move into a cottage that Lawrence proposed he should build at Ganthorpe, provoked Philip to make financial demands that Toynbee was unwilling to accede to. Reconciliation of a sort occurred shortly before Toynbee's stroke made him unable to communicate with those around him. But old passions, like molten lava, had surfaced once again to disturb his last days, making his exit from the world less benign than he would have wished or, in fact, deserved.

Only three of the nineteen books he published, dictated, or edited in the years after 1966 involved new historical reading and research, and because he had completed his own agenda, they were all stimulated by outsiders. Two reflected Toynbee's recent involvement with Constantine Doxiadis and the Institute of Ekistics, while the third (and more important) book derived from his undergraduate days when he had been asked to try his hand at editing Byzantine texts.

Cities on the Move (1970) and *An Ekistical Study of the Hellenic City State* (1971) constituted Toynbee's contribution to ekistics. They combined Toynbee's lifelong interest in human geography with Doxiadis' ideas about human reactions to changing urban design. The freehand sketches of ancient fortifications and citadels with which Toynbee adorned letters to his mother during his walking tour of Italy and Greece in 1911–1912 were prototypes of the professional drawings of town layouts that constituted the backbone of *An Ekistical Study of the Hellenic City State*. But whereas in 1911–1912 he was sensitive mainly to considerations of military defense, in 1971 he was looking for social, psychological, and political significance in the way ancient Greek towns were built. This book illustrated Toynbee's interest in and mastery of detail, especially with respect to the ancient Mediterranean world. His other master trait—the drive to find an intelligible order amid the confused clangor of history and geography—found expression in *Cities on the Move,* in which he explored the human and physical reality of five kinds of cities: city states, capital cities, holy cities, mechanized cities, and the "Coming World City," which, he declared, would be "an accomplished fact by the closing decades of the twentieth century."[4]

Toynbee had already used almost the same classification of cities for a book he had edited, *Cities of Destiny* (1967). This was the first of three similar volumes, each lavishly illustrated, that he presided over for the publisher Thames and Hudson. The idea for each book came from the publisher, but the execution was left largely to Toynbee; and in fact he sometimes put very considerable effort into revising chapters contributed by writers whom he, or the publisher, had recruited.

His editing reached an acme of intensity when he rewrote about a quarter of a chapter on Jewish–Roman relationships that had been prepared by an Israeli scholar for *The Crucible of Christianity* (1969). Toynbee's emphatic intervention frightened the Thames and Hudson staffer in charge of the book, who feared the charge of anti-Semitism,[5] but in the end the author capitulated under the weight of Toynbee's erudition and indignation.[6] Later, when Toynbee took on another editing job for Thames and Hudson and produced a third coffee-table book, entitled *Half the World: China and Japan* (1973), he played a comparatively passive role, since he knew less about the subject, and the after-effects of his heart attack in 1969 had weakened him permanently.

These publishing ventures all paled before the really substantial scholarly work of his old age, *Constantine Porphyrogenitus and His World,* published in 1973. This work was well along at the time of his coronary, and after recovering from that attack it seems likely that he abandoned part of his initial plan for the book. At any rate, Toynbee's pages on Byzantine civilization are uncharacteristically brief, and he skips over ecclesiastical affairs entirely. He explained in the preface: "I have gone into the most detail in discussing interests of Constantine that happen to be mine, for instance, the East Roman army corps districts, the Slav settlements south of the Danube . . . , and the Empire's foreign relations."[7]

The effort to complete this book strained Toynbee to the limit. "The crowd in the British Museum reading room is now so great," he explained to Philip, "that one has to rush early in the morning for a seat—no seat number, no book—so what used to be a peaceful occupation has become strenuous and agitating—especially when the mini-cab failed to turn up. I have stood it all right, but I am glad it is over. It was quite a test."[8]

Cutting things short, as he seems to have done, meant a lopsided book, despite its 768 pages. All the same, much of what Toynbee had to say was new and commanded expert admiration. Reviewers were generally favorable, and one American Byzantinist praised his "wonderful book" for the "interesting and novel things in every chapter."[9] For a man of his age, it was indeed a remarkable achievement. The book was inspired initially in 1911 by a tutor at Oriel who had suggested to the promising undergraduate that he might want to edit the emperor's writings. It was sustained in his old age by the fact that Toynbee was able to identify himself with a man born to the purple who was nonetheless, in Toynbee's own words, "a natural-born scholar," whose reign turned out to be a "bed of thorns," and who was often pushed to action by a strongminded wife.[10] Toynbee freely confessed: "I lose my heart to the Porphyrogenitus," though he called him "incurably muddle headed" when it came to the way he organized his writings.[11]

Three of Toynbee's other books were efforts to tidy up and reformulate work he had done earlier. *Some Problems of Greek History* (1969) falls into this category, comprising disconnected, highly technical inquiries into questions that had interested him since his undergraduate days, with the incongruous addition of two sprightly essays in hypothetical history that imagined what might have happened if Philip of Macedon had not been assassinated, or if his son Alexander had lived to old age. Toynbee had a serious aim in composing these essays. As he explained to his editor, these *jeux d'ésprit,* constituting part IV of the book, "actually deal with a much bigger and more serious question than Parts I–III. These are about details of historical fact; Part IV is about the nature of human affairs: determinism vs. unpredictability. I have been accused, mistakenly, of being a determinist, and Part IV is a vindication of unpredictability and, incidentally, of myself."[12]

Toynbee also tried to justify himself on a more massive scale by undertaking a new condensation of the ten volumes of *A Study of History.* As we know, Toynbee had originally intended to prepare a summary himself, only to be forestalled by D.C. Somervell's skillful surgery on the original text. Toynbee returned to his initial ambition at the suggestion of Thames and Hudson, whose formula for successful coffee-table books subordinated text to lavish illustration. The idea was that an illustrated version of Toynbee's great work would sell anew, and Thames and Hudson offered to select the illustrations and arrange for an editor to assist Toynbee with condensation of the text. Matters were successfully cleared with Oxford University Press[13] and the project got under way in April 1970 when Jane Caplan, a young Ph.D. candidate from Oxford, was entrusted with the editorial task.

Toynbee wrote some new passages, especially in the first part of the book, but Jane Caplan wrote rather more and did a great deal of updating of footnote references and the like on her own. She worked very fast and Toynbee, weakened by his coronary attack, approved what she submitted to him with little alteration. By August 1972 the work was completed. Bidding adieu, Miss Caplan wrote: "It has been a profound privilege for me to have been associated with you in this project; I found it immensely exciting, and I owe to you two of the most absorbing and satisfying years of work I could hope to experience."[14] She was a very good editor, but neither she nor Toynbee was genuinely capable of condensing the vast bulk of *A Study of History* while also

accommodating all the things he had learned since the original volumes came
out, though that was what the foreword claimed for the book. Toynbee had
already failed to rebuild the structure of his great work, producing an amor-
phous twelfth volume of *Reconsiderations* instead. In his extreme old age, neither he nor the youthful Jane Caplan was able to put Humpty Dumpty together
again. The volume advertized as *A Study of History,* a new edition, revised
and abridged by the author and Jane Caplan (1972), must therefore be judged
an intellectual failure, despite the handsome illustrations that occupy almost
half of the book's 576 pages.[15]

Toynbee devoted greater effort to a different way of reordering the subject
matter of his *Study.* Having discovered the futility of trying to condense or
rewrite the original, in October 1972 he set out instead to compose a narrative
history of the world. Less than two years later, when a debilitating stroke suddenly halted his writing, this manuscript was far enough along to permit its
posthumous publication as *Mankind and Mother Earth* (1976).

The feat of writing a world history of 641 pages in so short a time would be
amazing for a person half Toynbee's age, and is doubly so for a man in his
eighties suffering from chronic ill health. He drew on a lifetime's reading and
reflection, of course, and must have written almost entirely from memory. His
aim of putting all the human past together into an intelligible whole was utterly
characteristic of the man. Nonetheless, Toynbee once again failed to shape his
vast knowledge into a viable synthesis. Passages of political and military narrative, often highly condensed so as to become a mere jumble of proper names,
disfigure a great many pages of *Mankind and Mother Earth.* In his old age
Toynbee's memory tended to revert to thought patterns of his early childhood,
when wars and battles, kings and emperors paraded through his imagination.
He juxtaposed this skein of events—"a tale told by an idiot," as he says
himself[16]—with accounts of religion and, every so often, interjected remarks
about human impact on the biosphere. These disparate themes simply do not
cohere, and like Jane Caplan's new edition of *The Study of History,* Toynbee's
dying attempt to make the world make sense historically must be judged a
failure. The effort was heroic nonetheless. Every so often something of his old
fire broke forth in the form of a shrewd observation of unexpected comparison.
On the other hand, simplistic moral explanations of human conduct seem almost like hostile caricatures of his former self.[17]

Whether Toynbee realized the failure of these two final efforts to rescue his
vision of the past from the shambles in which he had left it with the publication
of *Reconsiderations* in 1963 is impossible to tell. In his old age, he embarked
on reminiscence,[18] and, with Veronica's encouragement, tended to treat himself as a public monument. Philip also admired his father extravagantly when
he was not quarreling with him. Many thousands of others, particularly Japanese disciples, were even more extravagant in their praises. Under these circumstances, Toynbee's modest demeanor could not disguise a diminishing capacity for culling dead horses from the stable of his favorite cliches.

In this he was perfectly normal. Self-criticism and fresh thought commonly
dry up with age; and what is remarkable about Toynbee is not that he repeated

himself glibly and often—as he did—but that he kept on picking up new ideas, without, however, fitting them into the older framework of his thought. His enthusiasm for ekistics was the most obvious demonstration of Toynbee's continued openness of mind; to this he added in the last decade of his life an awareness of ecology (thanks to Philip's conversion to that cause), an expanded consciousness of the value of pictorial art as a historical source (owing, no doubt, to his collaboration with Thames and Hudson), and a nodding acquaintance with the thought of Teilhard de Chardin.

But the most important new conclusion he reached in his last decade of life concerned the nature of the spiritual reality that he had encountered in 1929 and again in 1939, and whose supreme significance for human life he never afterward doubted. Recall that when Rosalind left him, Toynbee struggled with the problem of how God, being good, could permit such an evil to occur, and it was on this issue that he finally rejected Catholic theology. "Does submitting to God's will mean acquiescing in something that cannot be God's will? This baffles me," he wrote.[19] By the 1970s he had worked out an answer, arguing that spiritual reality "is human-like, not in being a personality like a human being, but in being non-omnipotent, like a human being. . . . I believe it is trying, as we try, to make things better. Christian theologians would probably classify me as being a Marcionite."[20] To Father Columba he wrote: "My conception of the inconceivable ultimate spiritual reality is no doubt more Indian (i.e., less in terms of a personal God) than Judaic. . . . I believe that every living creature is a temporary splinter of ultimate reality, and is re-united with this at death. 'Reunited' covers several superficially different beliefs: e.g. personal immortality, personal annihilation, and nirvana. I am content with 're-united.' "[21]

Awareness of his own diminishing energy and of the approach of death may have stimulated Toynbee to make explicit his notion of a struggling spiritual reality seeking to do good in a wicked world. The fact that in 1966 an enterprising publisher asked him to address attitudes and ideas about death may also have helped to crystallize his ideas. At first, Toynbee wished to write the book himself, but when difficulties arose over his contract with Oxford, he agreed to serve as editor and principal contributor instead, writing only about a third of *Man's Concern about Death* (1968). A follow-up book, arranged by the same editor, appeared after Toynbee's own death, with the title *Life after Death* (1976). In the lead essay of this book, as well as in the earlier volume, Toynbee suggested the possibility of some sort of immortality, in the form of reunion of each separated individual human spirit with the transcendent reality lurking within or behind the material universe.

It seems worth suggesting that in his old age Toynbee reverted to the antithesis between "Life" and "Mechanism" that he had borrowed from the philosophy of Henri Bergson in his youth. Many of his remarks in these two books fit the Bergsonian antithesis, but Toynbee was apparently unaware of the echo, and instead preferred to use Buddhist terms—karma, nirvana, and annihilation of desire. Nevertheless, the struggling spiritual reality of his latest writings seems close kin to Bergson's "Life," struggling valiantly against dead "Mechanism,'' forever and ever, yet capable, given sufficient time, of establishing an ascending curve of creative evolution, to borrow the title of Bergson's most famous book.

In terms of Toynbee's career, the important consequence of his revised religious outlook was the way that it appealed to large numbers of Japanese. For it was Toynbee's articulation of moral rules for everyday behavior based on his personal religious faith that made him famous and influential in Japan. To be sure, there were those who responded to the more secular side of his historical vision. As early as 1954 a leading Tokyo businessman, Yasuzaemon Matsunaga, decided that *A Study of History* ought to be translated into Japanese. He explained his motives to a journalist as follows: "Liberalism is my principle . . . I reject chauvinism. That is the reason I have made up my mind to have the work of Professor Toynbee translated into the language of our people." [22]

While this monumental enterprise was still coming to fruition (in the form of no less than twenty-four volumes published between 1968 and 1972), a Japanese version of Somervell's abridgement appeared commercially in 1966. This volume, in effect, entered into competition with a translation of Marx's *Kapital* which was made at about the same time. An enthusiastic journalist, reviewing Toynbee's book, described his excitement at discovering that Western civilization was not the "sole civilization mankind has ever had." He went on: "I am literally captivated by this book." Toynbee, he declared, had cured a "mental crisis" arising from his youthful addiction to Marxism. *A Study of History,* he concluded, "was my salvation." [23]

Toynbee's rising reputation in Japan was much enhanced by a third visit he made to that country in November and December 1967. His principal host on this occasion was Kyoto Industrial University, where he lectured on "The Coming World City" and "Mankind's Future." After thus introducing ekistics to Japan, he embarked on a lecture tour of other leading cities. The climax of his visit came on 6 December when he lectured in the Imperial Palace, Tokyo, before the Emperor, the Prime Minister, the Minister of Education, and other high dignitaries.

On shipboard, returning to England, Toynbee wrote a lengthy article summarizing his impressions of Japan on the basis of the three visits he had made there in 1929, 1956, and 1967. But the newspaper that had hitherto served as Toynbee's principal vehicle in Japan, *Asahi Shimbun,* turned it down, and a rival newspaper, *Mainichi Shimbun,* published it instead. *Mainichi Shimbun* subsequently took over the role in Japan that Henry Luce's magazines had played twenty-five years earlier in the United States, becoming the main channel through which his fame was spread to the Japanese public.

This shift of patronage within Japan was matched by a shift of emphasis. Toynbee, in effect, abandoned the merely negative role of serving as an antidote to Marx and took on a more positive role as a revered religious and moral teacher. His article "Impressions of Japan" helped to set the new tone, for Toynbee put special emphasis on the Japanese socioreligious scene. "In Shinto," he wrote, "I have met with a living religion of the same kind as the extinct pre-Christian Greek and Roman religions. Greek and Roman religion is congenial to me, and this may lead me to over-estimate the virtues of Shinto as well as of Buddhism.'' He suggested that Japan suffered from a profound ambivalence in its relations both with China and with the West, because feelings of inferiority mingled incongruously with a sense of national superiority. "In Japan, as elsewhere," he observed, "urbanization seems bound to disrupt family life." As for industrial management, it "may be more authoritarian, but it is also more paternalistic and more human; and if I am right about this, it is another reason for expecting that in Japan 'labour troubles' are not going to be extreme." But what he saw in Japan, above all else, was the contrast "between the technological virtuosity of present day Japanese life and its spiritual vacuity."[24]

After Toynbee's 1967 visit, his following in Japan seems to have bifurcated. On the one hand, in April 1968 those interested in his historical ideas and in counteracting Marxism founded a Toynbee Society, drawing membership initially through the Letters to the Editor page of *Asahi Shimbun*.[25] A self-made industrial tycoon, Konosuke Matsushita, was a prime mover in this enterprise, providing funds for meetings and publications in the early stages. Later, the Toynbee Society qualified for a state subsidy, and it continued to flourish after Toynbee's death. Lectures, assemblies, and seminars under its sponsorship sought to "study and disseminate Toynbee's ideas," according to one of its directors. In addition, the society published three different periodicals, *Modern Age and Toynbee* (circulation 2000), *Toynbee Studies* (circulation 300), and *Toynbee and I* (circulation 600).[26] A well-informed outsider described the society as follows: "It is a place where men of practice, intellectually oriented businessmen and bureaucrats meet with philosophers and miscellaneous scholars ranging from liberal to conservative, but not radical leftist. . . . It plays a role in the Japanese intellectual establishment, albeit old-fashioned."[27]

As long as Toynbee lived, the society asked him for annual messages of "advice to Japan." Toynbee regularly complied, writing back in 1971, for example: "My advice to Japan today is that she should continue firmly to follow the pacific and constructive policy that she has pursued since the end of the Second World War."[28] Two years later the society asked that he record his message so they could hear his voice at their annual meeting. Toynbee again obliged by telling them not to rearm and to try to demechanize and deindustrialize.[29]

Clearly, the Toynbee Society remained an elitist organization with limited reach into Japanese society. As such, it was less influential than Toynbee's other following, which was seeking specifically religious and moral guidance. This aspect of Toynbee's career in Japan broke like a tidal wave in the fall of 1970 when for ninety-seven days *Mainichi Shimbun* published his answers to all manner of questions about how to live well that were put to him by Kei Wakaizumi. Wakaizumi was professor of international affairs at Kyoto Industrial University and a specialist in Japanese–American relations. The two met in 1967. Toynbee, an elder statesman in the field of international affairs and a generation older than Wakaizumi, made a great impression on the Japanese professor, who was deeply distressed both by his personal disillusionment over

United States policy in Vietnam (a view he shared with Toynbee) and by the storm of campus disturbances that were then spreading throughout much of the world—a storm with which Toynbee often expressed sympathy.

Wakaizumi asked himself: "Why does the young generation manifest grave misgivings about existing value systems? Why has one part of that generation chosen to escape from reality, while another has gone the way of radical opposition?"[30] Toynbee, he found, could answer these questions, and even supplied hope of reconciliation with the rebellious students. This was because, according to Professor Wakaizumi, he not only offered everyone a shining moral example, but had also thought through all the moral and religious uncertainties that bothered both Wakaizumi and his students. In his own words: "To put it frankly, Professor Toynbee comes close to an utmost perfection of character and consistent achievement that few men have attained. The thing that impressed me most was that despite advancing age, he was still driven by an incredible desire for knowledge and a spirit of inquiry. Moreover, I was struck by his insight into the future of humanity, from the expansive point of view of world history, as well as his unlimited affection for young people. . . . From then, on whenever I went to Europe, I invariably called on Professor Toynbee, not only to discuss my area of specialization—international politics—but also to deal with perhaps the broadest question that one human being who lives in the rapidly changing present can pose: How should man live? It was this eternal theme that turned into our main topic of exchange."[31]

Having profited personally, Wakaizumi decided to share his illumination with others. After consulting some of his students, he formulated a series of questions to put to Toynbee. He sent them on ahead, so that Toynbee could prepare his thoughts before Wakaizumi arrived in person with his tape recorder. *Mainichi Shimbun* financed the whole undertaking, offering Toynbee a fee of £3500 for his services, and specifying that each question and answer should total about 2000 words to make a suitable newspaper installment.[32] The two conversed for seven days, three hours a day. Then translators and editors, working from the tapes, duly produced daily installments that ran in the *Mainichi Shimbun* from 24 August until 9 December 1970.

The articles roused an unusually strong response, and as the series was drawing to a close, Wakaizumi proposed that Toynbee answer a few supplementary questions sent in by readers. Toynbee agreed, with the proviso that he would only respond to a dozen new questions.[33] The articles were then gathered into a book, which sold "extremely well" according to Professor Wakaizumi's report.[34] Subsequently, a part of what Toynbee had dictated was rearranged and published in English under the title *Surviving the Future* (1971). In addition, excerpts from the English text (but arranged differently from the chapters of the book, and with some grammatical simplification) were published in Japan as a series of pamphlets for use by students of English.

The questions addressed in *Surviving the Future* center on moral issues, but as Toynbee was always eager to make clear, moral choices ultimately depended on religious commitments. "Religious beliefs are answers to questions that cannot be answered in scientific terms," he explained. "However, I believe that the higher historic religions all convey permanent truths, and all give counsel, advice and precepts for action that are permanently valid. They all counsel us to try to overcome self-centredness and to surrender ourselves to love, and they point out practical ways of acting on this counsel. . . . We now need to disengage these permanent truths and precepts from the temporary forms that are the traditional expressions of them. We need to re-express them in forms of their own, forms which will no doubt become out of date and in their turn will have to be re-expressed again by our successors.''[35]

Toynbee resorted to unabashedly mystical language in describing the soul's relation to the "supra-personal spiritual presence behind the universe." "I believe," he said, "that personal human individuality is acquired at the price of being separated from this supra-personal reality. I feel that this price is high, and I am therefore glad that it has to be paid for a limited period only.''[36] Political themes also came up, provoking Toynbee to deplore race feeling and reaffirm the eventual necessity of a world government. He also discoursed on the meaning of history, human nature, the shape of the future, and things in general.

Throughout their dialogue, Professor Wakaizumi played to perfection the role of deferential disciple, while Toynbee accepted the role of accredited sage with gracious equanimity. Given the disorientation that the Japanese people had suffered from their defeat in World War II, and the subsequent social and economic transformation of the country, it is not entirely surprising that Toynbee's opinions struck a responsive chord. Here, at least, was a man who addressed abiding uncertainties with unusual assurance and an affable seriousness. His advanced age and vast learning gave additional weight to his remarks. In effect he became a new bodhisattva, and Kei Wakaizumi hailed him as such. Many of *Mainichi Shimbun*'s readers presumably reacted in similar fashion, fitting Toynbee's words and person into their Buddhist heritage. Only so, I think, can the warmth of the Japanese response to Toynbee's artfully encapsulated wisdom be accounted for.

Professor Wakaizumi, however, offered a different explanation soon after Toynbee's death: "For us Japanese he was a great man who came to understand Japanese culture and religion. . . . It was his non-Europe–centered stance, with heavy emphasis on the future potential of East Asia, that made such a great appeal to Japanese scholars as well as the thinking public.''[37]

Soon after *Mainichi Shimbun* established his reputation as a living sage and oracle, Toynbee attracted the attention of another powerful group, known as Soka Gakkai. The name may be translated as "Value-Creating Association." Soka Gakkai was the most successful of the new religious bodies that sprang up in Japan after 1945, claiming, in 1987, 8 million adherents in Japan and 1.3 million in 115 different countries abroad.[38] The organization, established in 1930, was affiliated with a sect of Buddhist monks who trace their version of the faith to a holy man of the thirteenth century named Nichiren Daishonin. Nichiren declared that Buddhahood was open to anyone who disciplined himself and developed his Buddha nature by chanting excerpts from the Lotus sutra, morning and evening. Since the chant was a conspicuous ritual act, fol-

lowers were compelled to mark themselves off from others. As a result, Nichiren's followers turned into a rather aggressive sect, claiming that all other forms
of religion were false. It flourished mainly among peasants and provoked (or
at least was associated with) a number of rural revolts. As a result the sect was
officially suppressed in the seventeenth century but survived underground, and
was again allowed legal expression in the twentieth century.

Nonetheless, a revolutionary taint remained, and in 1943, when the leader
of Soka Gakkai openly attacked Japan's official Shinto faith, he was jailed,
whereupon the organization almost disintegrated. Revival came after 1945 and
got into high gear when Daisaku Ikeda (b. 1928) became the third president of
Soka Gakkai in 1960. In postwar Japan, Soka Gakkai appealed especially to
migrants from the Japanese countryside who found life in the big cities especially difficult. A tight-knit, semimilitary style of organization and recruitment,
together with daily chanting and massed public ritual acts on special occasions,
gave otherwise lost souls a new kind of primary community. Success was spectacular, and when Soka Gakkai launched a new political party in 1964 it swiftly
became the third largest in the country. Nevertheless, Soka Gakkai suffered
from the disdain of the wealthier, more educated, and more fully urbanized
elements of Japanese society. Ikeda made great efforts to overcome the handi-
cap, founding a university in 1971, for example, and going to great lengths to
arrange and then publicize his meetings with world figures like Chou En-lai of
China and Henry Kissinger of the United States.

Wakaizumi's dialogue with Toynbee gave Ikeda the idea that he, too, might
add a cubit to his stature by engaging the Western Sage in discourse about
ultimate questions and things in general. Accordingly, he approached Professor
Wakaizumi for an introduction, and in due course arranged for a series of
interviews with Toynbee, which took place in May 1972.[39] In return for what
turned out to be six days of recorded interviews, Ikeda offered Toynbee a fee
of £500 a year for seven years; and in addition gave Veronica a "gift" of
£3500, making a total of £7000.[40] In anticipation, he wrote: "Our meetings
will become the greatest historical event in my life."[41] Subsequently, he published an edited version of their conversation in Japanese and arranged for
translations into English, French, German, Chinese, Korean, and Portuguese.
References to and excerpts from Ikeda's dialogue with Toynbee abound in publicity materials Soka Gakkai has issued since the original publication. It seems
clear that Ikeda was pleased by the way Toynbee received him as an equal.

From Toynbee's point of view, Soka Gakkai was exactly what his vision of
the historical moment expected, for it was a new church, arising on the fringes
of the "post-Christian" world, appealing principally to an internal proletariat,
and deriving part of its legitimacy from an ancient and persecuted faith. Comparisons with early Christian history fairly leap to mind, and in a preface he
wrote for the English translation of one of Ikeda's books Toynbee explicitly
compared the world mission of Soka Gakkai with the Christian church on the
eve of its coming to power in the Roman Empire.[42] When an Englishman living
in Japan reproached him for his association with Ikeda,[43] Toynbee defended
himself, writing: "I agree with Soka Gakkai on religion as the most important

thing in human life, and on opposition to militarism and war.''[44] To another remonstrance he replied: "Mr. Ikeda's personality is strong and dynamic and such characters are often controversial. My own feeling for Mr. Ikeda is one of great respect and sympathy.''[45]

Convergence of East and West was, indeed, what Toynbee and Ikeda sought and thought they had found in their dialogue. In a preface, written in the third person, Toynbee emphasized and tried to explain this circumstance. "They agree that a human being ought to be perpetually striving to overcome his innate propensity to try to exploit the rest of the universe and that he ought to be trying, instead, to put himself at the service of the universe so unreservedly that his ego will become identical with an ultimate reality, which for a Buddhist is the Buddha state. They agree in believing that this ultimate reality is not a humanlike divine personality.'' He explained these and other agreements as reflecting the "birth of a common worldwide civilization that has originated in a technological framework of Western origin but is now being enriched spiritually by contributions from all the historic regional civilizations.''[46]

Toynbee said little or nothing to Ikeda that he had not said before, though he occasionally went out of his way to flatter his interlocutor. For example: "The Buddhist analysis of the dynamics of life, as you explain them, is more detailed and subtle than any modern Western analysis that I know of.''[47] Toynbee also had praise for Japan. "The Western course is heading for disaster,'' he said. "I believe that the Japanese people can lead mankind into a safer and happier path.''[48] Once in a while a flash of his old inventiveness can be glimpsed, as when he declared: "We ought to aim not at gross national product but at gross national welfare.''[49] But, at least in the short run, the value of this book for Ikeda and his movement was far greater than anything it did for Toynbee's reputation.

It seems probable that the Toynbee–Ikeda dialogue will dwindle into insignificance in time to come, for the sect's basic ritual was defined centuries ago, and Ikeda has published a book, *The Human Revolution,* setting forth his own authoritative statement of doctrine. Still, Ikeda has had to adjust some aspects of the sect's Buddhist inheritance, especially in the overseas branches of the organization. His dialogue with Toynbee is the longest and most serious text in which East and West—that is, Ikeda and a famous representative of the mission field that Ikeda sees before him—have agreed with each other. In the unlikely event that Soka Gakkai lives up to its leader's hopes and realizes Toynbee's expectations by flourishing in the Western world, this dialogue might, like the letters of St. Paul, achieve the status of sacred scripture and thus become by far the most important of all of Toynbee's works. On the other hand, if the sect's future is one of decay and disintegration, it will remain, with *Mankind and Mother Earth,* a posthumous monument both to Toynbee's weakening powers and to his unwearying aspiration for articulating an all-embracing truth.[50]

Books were, of course, only part of Toynbee's literary output in his last years. Scores of articles, interviews, radio and television talks, and book reviews continued to flow from his pen. He also maintained a vigorous personal correspondence, answering almost every letter sent to him, even those from unknown correspondents, until his stroke in 1974 made further writing impossible. His unflagging industry was indeed extraordinary, all the more so because, until the final weeks of his eightieth year, when he suffered his coronary, he maintained the hectic travel and lecture schedule he had established after his retirement from Chatham House in 1956. In 1966, for example, he visited Ethiopia at Easter, went to Greece in midsummer to attend another of the Delos cruises and seminars organized by Constantine Doxiadis, and then embarked on a whirlwind lecture tour of Brazil, Uruguay, Argentina, and Chile between August and October.

Yet his old zest for seeing new places was diminishing. "I have been told by doctors to take things a bit easier," he wrote to Columba, "as I have a slight irregularity . . . of the heart. I don't want to leave Veronica solitary, and I find that, after all, I have a number of bits of work to do. . . . Apart from these strong holds on life, I shouldn't be sorry to leave this world. I like it less and less: the makeup of the girls; Vietnam; the laziness and economic folly of this country."[51] On his return from South America, he prepared a travelogue as usual, published under the title *Between Maule and Amazon* (1967); yet on the day he submitted the manuscript to Oxford University Press he confessed to his friend that he and Veronica were "wrecks from being 'state guests' in those hectic American countries."[52]

Things did not improve in 1967 when he went to lecture for a last time in the United States. By 1967 American military operations in Vietnam had reached serious proportions, and Toynbee's emphatic disapproval of that venture became a matter of controversy. Negotiations for Toynbee's return to the University of Pennsylvania fell through, and when one of his admirers arranged for him to visit Stanford instead, a professor in the business school angrily protested to the university's president, writing: "He is a congenital wind bag, and now a senile wind bag. Of course he cannot help that. But he could stay home.[53] When Toynbee saw a copy of this letter he responded: "If you and the President feel that it is in the University's best interest to call off my visit, do not hesitate to tell me so."[54] And even when officially assured of his welcome, he undertook to be careful of what he said. "If I am quizzed about Viet Nam, I shall try to combine frankness with answering in a way that will not cause the University avoidable embarrassment."[55]

The visit, begun under this sort of cloud, became bleaker when he learned in April that Rosalind had lung cancer, although she was then expected to live for several more months.[56] His own health and Veronica's also proved troubling. "Death stares at me in a formidable way now," he wrote to Columba, "for I have just heard from Philip that Rosalind—about whom we have all been rather concerned for several months—has been diagnosed as having cancer of the lungs, and that the doctors do not expect her to live for more than three to five months longer. I shall see her, still alive, when we get back. . . . She is at home at Low Holm, Cumwhitton, Cumberland, and was very happy to come home from hospital. Of all living people, Richard Stafford is, I suppose, the closest to her, and is perhaps the only person who has come up to her expectations. He is like an adopted son, and has done for her, I should say, the equivalent of what Veronica has done for me."[57]

Nine days later a cablegram from his sons arrived, saying "Mummy has died." The shock was severe. "The telephoned telegram hit me like a bullet, and, all yesterday, I felt as if I had received a physical wound. Not to have seen Rosalind again; not to have been with her at the moment of death; not to have been—or have the right to be—the person who was doing for her what was being done by Richard and his sisters: this hurt a lot; but what hurt more was the flood of the return of the life we had shared for thirty years, and contrition—welling up, as strong as ever after a quarter of a century—for my failure to save our marriage from eventually breaking down." And he added as postscript: "I remember, at the time of the break, you rightly reproved me for having, as I had confessed to you, felt towards Rosalind more as if she were a goddess than as if she were the human being that she was. I must still have some of this feeling left, because I can't imagine how Death can have had the audacity to take her."[58]

Tragedy mixed with farce for Toynbee during this visit, for the American Friends Service Committee, which had agreed to pay him to lecture in a number of American cities, decided to withdraw its sponsorship. As a result of this and other disappointments, he failed to accumulate anything like his accustomed lecture fees.[59] Toynbee therefore returned from America in June 1967 with a keen sense of financial and personal failure. His visit to Japan at the end of the year may have done something to restore his equilibrium. At the least, it helped to kindle his subsequent fame in that country, as we have seen. And as his eightieth birthday approached, friends and admirers in the United States set out to arrange a gala birthday party for him in New York. But the outburst of reconciliation and mutual esteem that might have blossomed on that occasion never occurred. Instead, the celebration had to be canceled when Toynbee suffered his coronary almost on the eve of his scheduled departure for the United States. The long love affair between Americans and Toynbee was therefore cheated of a grand finale. Instead, his last visit to the country, coming, as it did, in the midst of the war in Vietnam, left a sour taste in his mouth.

Despite the wearing out of his American connection, Toynbee's last years attracted fresh tributes at home and in Europe, as well as in Japan. The most surprising was his installation in succession to Winston Churchill as a Foreign Associate of the French Academy of Moral and Political Sciences. Toynbee had actually been elected to this dignity in 1965, but the exigencies of scheduling academy meetings and his various travels and ill health postponed the ceremonial consummation until 1 April 1968. As was customary, the president of the academy introduced the new member with a brief summation of his career, concluding: "You have, my dear fellow member, achieved the most lucid overview of our age, and you have persuaded us of the legitimacy of your dismay. It is for that reason that we have unanimously invited your presence among us. You maintain in this position a long tradition of Franco-British friendship to which we are much attached."[60]

Toynbee, as custom required, responded with a eulogy of his predecessor,

Winston Churchill. He ended his remarks by calling for a union of all of Europe, both East and West, as the only way to compete with the new superpowers of the postwar era. "The pressure which each of them exerts on Europe is heavy," he said, "and although the attitude of the Soviet Union may appear less menacing for Europe in 1968 than it was in the time of Stalin, the attitude of the United States has become more menacing. . . . Europe cannot permit itself the risk of becoming a second Viet Nam."[61]

Toynbee's health and vigor were by no means fully restored when he went to Paris for the ceremony of his installation as a member of the Academy. Indeed, as a general rule he gave up travel and lecturing after 1969. He made an exception of Greece, which he visited again as a guest of Constantine Doxiadis in 1969 and, for a final time, in 1972. Doxiadis, too, was growing old by this time and Greece was in the grip of an upstart military dictatorship (1967–1973), whose leaders disliked and distrusted Doxiadis. Accordingly, the Delos cruise of July 1972 was planned as the last of the series of similar cruises that had started ten years before. Toynbee, a veteran of the first such event, was assigned the role of senior soothsayer. In Athens, he spoke outdoors on (or near) the spot where St. Paul had once preached, declaring once again that "the state is not a god" but a "public utility" and should be treated as such. This was all but treasonous in the superheated patriotic atmosphere that then prevailed in Greece, but Toynbee was not molested. A week later, at Delos, he addressed Apollo in classical Greek, imploring the god to set western minds aright, so that they might give up excessive pursuit of material goods and, following the example of St. Francis, espouse Lady Poverty so as to keep the biosphere habitable.[62]

Though he thoroughly enjoyed this last trip to a country that had meant so much to him, the effort of travel strained both Toynbee and Veronica to the limit. "We were most tenderly looked after," he wrote to Columba, "but it was too much for us. Veronica got bronchitis on return, and, the Monday before last, I fell over backwards in the kitchen from sheer physical fatigue (X-ray shows that I haven't injured myself, but I stiffened into helplessness and am only just finding my feet again, thanks to skilled massage)."[63] After they had recovered, he explained: "Travelling we really have stopped—with enormous regret, but firmly. . . . The unfulfilled bits of my travel agenda sometimes trouble me, but I have been lucky in seeing a great deal of the World."[64]

Staying home had its rewards. Toynbee's foreign fame, his longevity, and his continued literary productivity did something to restore his reputation in England, especially after 1971 when an organization of bibliophiles known as the National Book League organized an exhibition in his honor, wittily styled "A Study of Toynbee." Toynbee cooperated by supplying samples of his youthful essays from school, the outline of the abortive philosophy of history he had sketched on the train in 1921 from which *A Study of History* eventually emerged, together with the notebooks from which he worked and many other items from his files. He also wrote captions for the exhibits, explaining the significance of each.

The display of his literary works was, indeed, impressive and the opening ceremonies, at which he spoke, attracted favorable attention in the press. Toynbee was gratified. "Yes, I was very grateful to the *Times,*" he wrote, "and it was a happy occasion—but an exhausting one. . . . The exhibition was splendidly set out by the National Book League, the O.U.P. and my marvellous secretary, Louise Orr. Veronica and I fished out some amazing early stuff, e.g., an essay on twentieth-century Byzantine history in a drawing book of 1903—sixty-nine years before the proofs of *Constantine Porfyrogenitus,* which are now coming in."[65]

Twice the Queen of England invited him to dine at Windsor Castle, and both times he declined, pleading ill health.[66] The excuse was genuine, but Toynbee may also have felt that the Crown owned him more than an invitation to dinner. At any rate, three years before he declined the queen's invitations, he confessed to Father Columba: "Two days ago I had a twinge when an historian—not me—was given the O.M., Veronica Wedgwood. She must be about 20 years younger than me, and, I suppose, is not of my calibre. But 'refraining from grasping' has fortunately unclutched me now from coveting the O.M., which, for some reason, I have perhaps been wanting, to force the English reluctantly to give me the recognition I have had in other parts of the world."[67] After such disappointments, the exhibition "A Study of Toynbee" was therefore doubly welcome. Unofficial, literary circles in England did thereby accord him something like "the recognition I have had in other parts of the world."

By 1972, increasing debility made life in a rented London flat burdensome. Veronica had difficulty finding suitable household help, and exposure to unpredictable rent increases roused Toynbee's irrational fear of finding himself impoverished. Consequently, in 1972, when Lawrence invited them to build a cottage on his property, next door to Ganthorpe, where they could live in peace for the rest of their days, Toynbee accepted the offer gratefully. It meant return to the place he had loved above all others, and where he had composed the first six volumes of his great book.

It was also, of course, a place replete with bittersweet memories of Rosalind and their life together. Lawrence had moved to Ganthorpe in 1965, giving up his teaching career in Oxford to become a country squire. Four years later, when Toynbee was recovering from his coronary attack, he visited Ganthorpe for a few weeks in the summer of 1969. It was the first time he had been there since the break with Rosalind; now that she was dead his return constituted, for Toynbee, a sort of reconciliation with Rosalind's formidable shade. "I am slightly like a revenant here," he wrote to Philip, "but being a welcome revenant is a rather amusing unusual experience—like reading one's own obituary, published by mistake."[68] A few days later he explained to Columba: "The Buddhist supplement (an essential one) to the Christian love is refraining from 'tanha' (from 'grasping'). Writing that poem helped me to unclutch my fingers from Ganthorpe, and I now take pure pleasure in Bun and Jean being here and making such good use of it."[69]

But Toynbee's reconciliation with the dead Rosalind and his coming closer to Lawrence involved an unexpected cost. For many years, Philip had aligned himself with his father, whereas Lawrence had, in general, submitted to Rosa-

Philip Toynbee
Lawrence Toynbee photo

Lawrence Toynbee in his studio
Northern Echo

lind's imperious charm and maintained only a distant cordiality with his father. But now that she was dead, Philip suddenly saw his father gravitating
toward Lawrence and feared that Toynbee might, in effect, discard him, just
as Rosalind had done years before in his childhood.

Philip's personal affairs were in a parlous condition anyhow, for he had
decided to act on newly acquired ecological convictions by establishing a community of like-minded spirits who would live simply, supplying their wants by
labor on the land while avoiding environmentally destructive practices. He
therefore set out to remodel Barn House, where he had been living for several
years, to accommodate the expected recruits to his community. But costs es-
calated unexpectedly, and in December 1969 Philip's bank refused to honor
one of his checks. He turned to his father for help. Toynbee transferred the
necessary funds, but asked to be repaid. "You took me by surprise, and, for
an old coronarian, being asked out of the blue for an instant £2000—'or else'—
is not good medicine. It gave me a shock that I am now digesting, but I lost
my sleep and my power of concentration." [70]

Worse was to follow, for as Toynbee's plans for building a cottage on Lawrence's property moved toward fulfillment, Philip's difficulties increased. Communal living turned into something of a disaster, and Philip responded by making new demands. He resented it when Richard Stafford sold the farm at Low
Holm where he had lived with Rosalind. "He has a perfect right to do so, of
course," he wrote to his father, "but Mummy was very definite that it was to
come to me on the assumption that he dies before I do. . . . This seems to me
another example of some animals being more equal than others." [71]

A few months later, family relations exploded. Some time after their mother's death, in bibulous conversation with Philip, Lawrence had agreed in principle to an equalization of their inheritances from Rosalind, but when Philip
asked Lawrence to do something about it by giving him an interest-free loan
of £30,000 for his Barn House community, Lawrence refused on the ground
that what Philip planned to do with the money was absurd and unworkable.
Philip then turned to his father, and Toynbee promised to transfer £10,000 from
his account with Oxford University Press to Philip, but only after he had himself moved into the new cottage at Ganthorpe and knew exactly what that was
going to cost.

This was not entirely unreasonable, since what had been first estimated to
cost £10,000 actually cost £18,000 to build, and Toynbee had begun to draw
more heavily on his accumulated royalty accounts with the Press to make up
for vanished lecture fees. Still, as he explained to Lawrence in 1972, "Last
year . . . the amount in that fund had slightly risen, not fallen, owing to new
royalties." [72] He could therefore have given Philip what he wanted right away,
but Toynbee's attitudes toward money reflected deep childhood conditioning
and were not really rational. Accordingly, he reproached Philip for spendthrift
ways while promising him gifts—eventually.

Philip was driven almost frantic by what was happening to him. The collapse
of his experiment in communal living was bad enough; on top of that he felt
deprived not only of his just share of Rosalind's properties but of his father's

affections as well. "I still feel strong resentment against Bun," he wrote to Toynbee, "But do believe me when I tell you that however hard I examine myself, I can find no trace of the assumption that I *ought* to have money just because I want it. What I saw was a rich brother and a rich father; a relatively poor brother/son who was genuinely trying to do something morally respectable but who could get no help from his family when he most needed it." [73] Toynbee responded: "I have been pretty severely shell shocked by what happened between you and me at Barn House, and by the follow up correspondence. It was unexpected, and it came on top of the strain of preparation for our move. At moments I have been afraid that I was going to collapse. I have always had that fear at the back of my mind since my father's breakdown, from which he did not recover, and this has aggravated my inherited anxiety about money. Veronica is much more generous minded, though her family was quite as hard up as mine." [74]

The quarrel with Philip mingled with a different kind of shock to Toynbee, for in June 1973 he learned that Richard Stafford, whom he had supposed to have been like a son to Rosalind, had in fact been her sexual partner. This information hit him hard, perhaps as hard as the news of her death had five years previously. A letter from Veronica to Philip's wife, Sally, offers the only direct account of his reactions. "After we came back from Ganthorpe," she wrote, "where Arnold heard for the first time about relations between Richard and Rosalind (not in the least Jean's [i.e., Lawrence Toynbee's wife's] fault; Arnold suddenly asked her about it) I waited to write until he was 'over the hump' about it. I do think he is now, and Philip's letters have helped him a lot. With Arnold, a complication was that it brought up again a most traumatic experience which he had a few months before Bun was born, when he had to accompany Lady Mary . . . to a remote village in France where Agnes was dying; she had gone there with a young Greek man, with whom she was living, and was overtaken by acute appendicitis. Medical help came too late and she died in great pain. The thing that has now struck Arnold to the heart is that Rosalind was quite implacable in her attitude to Agnes, and Lady Mary nearly broke with her because she would not express sympathy. Then some thirty years later, Rosalind . . . was also overtaken by acute appendicitis, and would also have died if Richard had not had the sense to summon adequate medical help in time. I must say one does wonder what Rosalind thought about that!

"Arnold's first reaction, as everyone's was, was how could Rosalind have condemned Celia [one of Lawrence's daughters who for a while had lived with a man whom she later married], but soon this memory of Agnes' death swamped even that from his mind. But, as I say, I think he has got over the worst now. It has been a terrible shock for Philip and Bun, as well as for Arnold, since all three were so much under Rosalind's spell, though I doubt if either of her sons put her quite as much on a pedestal to be worshipped as Arnold did." [75]

Philip was indeed able to comfort his father in this distress. "I too felt extreme animosity towards Mummy," he wrote, "when Richard blurted the whole thing out to us at Low Holm: most obviously of all, I longed to have her back so that I could confront her with this. She had marvellous qualities which were recognized by all sorts of different people who knew and loved her. But there must have been something terribly wrong with her deep inside her own mind and heart. Something which inevitably destroyed all her relationships when they made any real demands on her.''[76] A week earlier he had told his father: ''My own feeling is that Mummy was capable of a greater—indeed a more ludicrous—degree of self-deception than anyone I have known. . . . What was lacking in her was the ability to look at herself from the outside; the ability even to conceive the fact that she might be wrong. And yet . . . she had courage (far more than I have); she had a delightful gaiety when things were going well for her; she was capable of great, though sporadic, love; she had tremendous creative energy—though I think it was misdirected when she tried to write books.''[77] Philip was a sensitive observer, with a talent for writing, and this assessment of Rosalind's character is the most adequate available account of her extraordinary personality.[78]

Lawrence, too, was troubled by the family upheaval, and wrote to his father about his ''growing disillusionment with Mummy. While she was alive she was so vital and such fun to be with—and of course I was quite infatuated by her. Now that one sees her more in perspective, the view is a great deal less pleasant and I do find this traumatic. To have lost one's faith both in Catholicism and her in the same period is a bit depressing.''[79]

Like Lawrence, Toynbee found himself ''resentful and disillusioned towards Rosalind,'' but also blamed himself for blaming her, ''She did spellbind me,'' he confided to Columba, ''and Philip and Lawrence too. The reaction is proportionate but one ought not to feel resentment towards the poor dead, since they can do nothing more in this world to set right what they have done wrong.''[80] Yet he could not really forgive, however much he tried, declaring in another letter: ''R's implacability towards Agnes and Celia is what distresses me. I have no personal feelings. I divorced her, and thereby left her free.''[81]

Toynbee achieved a partial reconciliation with Philip, but his pending move to the cottage meant that he was about to come firmly into Lawrence's orbit, and the two brothers remained at loggerheads. This distressed their father, who explained in the letter quoted above: ''Unhappily, Philip had broken with Lawrence over the unequal division of the inheritance. Philip .   . has a genuine grievance, but to be obsessed by wrongs suffered by oneself is unprofitable and unhappy. Veronica has done her best, via Philip's wife, Sally, to make peace between the two families. V. and I are determined to remain on good terms with both.''

By March 1974 the cottage was nearly ready for their occupancy. Toynbee and Veronica arrived at Ganthorpe at the end of that month, together with their household goods—a large library of books chiefly—in order to supervise the final stages of work and to begin moving in. Everything took longer than expected, and it was late in June before they actually transferred to their new abode. ''We are in the cottage,'' Toynbee wrote to Columba on 27 June 1974, ''we slept here for the first time last Friday. We are utterly exhausted, but we shall gradually recover, and I think we are here for the rest of our lives.''[82]

The exertion of rearranging his books was a big chore for Toynbee to undertake, but he insisted on doing it himself. An extra strain came when he was invited to take part in a special ceremony at Winchester, his old school, on 29 June 1974. The ceremony whereby an honored visitor was formally received "Ad portas" by the masters and other members of the school was a rare and extraordinary affair, and Toynbee counted it as the most precious of all his honors. Yet it came to him at an awkward time. "I shall be welcomed in Latin and shall reply in it—great fun and a rare honour. But I wish there had been a longer interval between this and the move, for I shall be intimidated by the Warden and the Head and Second Masters, though they are babes compared to me—inside gate, I shall feel like a junior school boy again." [83]

Toynbee traveled by car from Ganthorpe to Winchester—a longish drive for a single day—to be received Ad Portas. He described the affair to his friend Father Columba as follows: "It was a formidable ceremony: three sides to 'Chamber Court' lined with 600 boys; a pack of V.I.P.s behind me; the head of the school facing me. He made a short speech to me in Latin, and I answered in Latin. I spoke for a quarter of an hour (I had said it to myself every night in bed for several weeks, so I had learnt it by heart). I used my speech to commemorate my school-fellows who were killed in 1915–1916—recalling them in the place where I had first come to know them. Veronica and I have not quite recovered from the fatigue. It was a rash expedition, but I could not have borne refusing to make it, when invited." [84]

Three weeks later, Toynbee suffered a stroke, on the night of 2–3 August 1974. "It happened after a quite normal day," Veronica wrote, "but I suppose was probably to be explained by the unusual strain of the Ad Portas ceremony at Winchester following so soon after the long strain of the prolonged move into our new house." [85] At first Veronica hoped that he would recover, at least partially; but in fact he was never again able to speak or walk and proved to be occasionally unruly as a patient. Veronica believed that he understood at least part of what was said to him and that he recognized visitors, but even if that was so, he could no longer communicate with those around him and spent the last months of his life helplessly bedridden in a nursing home in York. It was a sad ending for so energetic a man, and painful for Veronica, who was destined to outlive him for five years. She attended him daily until his death, more than a year later, on 22 October 1975.

On 17 December 1975 friends and family arranged a Service of Remembrance and Thanksgiving in his honor at St. James' Piccadilly, close by Chatham House, where he had worked for so many years. Philip read the Scriptural lesson; Lawrence drafted a statement for the officiating clergyman to read; and Toynbee's cousin Noble Frankland gave a formal address.

Lawrence wrote: "We are gathered here today to remember before God the life and character of Arnold Joseph Toynbee. We remember him as a great scholar, historian and writer, whose influence throughout the world has been immense. His deep love of humanity manifested itself in many ways, but especially in his search for a peaceful life for all mankind. But we remember him particularly for himself: for his kindness, his gentleness, and above all his charity."

Noble Frankland compared him with his namesake. "Let me remind you of the words which he used himself to describe his uncle, the first Arnold Toyn-bee. 'The vision that I caught,' he says, 'was one of simplicity, sincerity, disinterestedness, and ardour; and these qualities combined to strike an unmistakable note of greatness.' Such words convey to my mind the vision of our Arnold Toynbee, who, unlike his uncle, was to live into old age, to labour almost to the end, and to achieve a rich consummation.''

Both these ecomiums seem apt and entirely appropriate to the life they celebrated. Among the many obituary notices of Toynbee's death the most perceptive came from a former associate at Chatham House, Martin Wight, who wrote: ''He suffered very deeply in his life, but he never lost his boyish simplicity. . . . He combined political insight with political unsophistication. . . . He greatest weakness was an inability to learn from criticism. . . . His absorption in historical knowledge was in the last resort poetical and mystical.''[86]

These carefully considered words, coming from a son, from a cousin, and from a colleague, serve fittingly and publicly to commemorate the remarkable life of Arnold Joseph Toynbee.

Conclusion

Like other interesting people, Toynbee was full of contradictions. In politics, he habitually looked ahead and jumped to conclusions that outran everyday fumbling reality. Historical analogy convinced him that the political system of his time could not last, and that some sort of world government would have to emerge very soon. He hoped for the League of Nations and again for the United Nations; but he believed world empire was far more likely. Yet his attitude toward the future empire of the earth that he foresaw was utterly ambiguous. He both welcomed the prospect, because it would end war, and deplored the violence needed to establish an imperial global government. When push came to shove, he was unwilling to pay the price and adhered instead to inherited Liberal values. Consequently, when first Germany, and then the United States, began to act in a way that seemed to foreshadow the establishment of a global hegemony, he hesitated, but in the end opposed them both.

In personal matters, too, Toynbee's behavior was a tissue of contradictions. He deplored greed and denounced the materialism of modern civilization, yet he was almost pathologically grasping in money matters. He deplored self-love, yet never wavered in pursuing his own self-appointed goals, regardless, or almost regardless, of those around him. He delighted in consorting with the great and powerful, yet always felt himself an outsider. Shy, and almost apologetic in his public manner, he nonetheless expected deference and often got it.

These and other foibles kept him human while his formidable memory and indefatigable capacity for desk work approached the superhuman. How important was his work? How should his accomplishment be appraised now that the controversy and publicity surrounding him have faded into the past?

His own recorded estimates of his career are characteristically modest. In 1965, asked how he would like to be remembered, he replied: "As someone who has tried to see it whole, and . . . not just in western terms."[1] Some years later he answered another journalist's question: "I would like to think I had done a useful job in persuading Western peoples to think of the world as a whole.''[2] In a more lapidary phrase he told another reporter: "I always had the desire to see the other side of the moon,"[3] meaning that he had learned, from childhood, to wonder about the peoples left out of conventional history— Persians, Carthaginians, Moslems, and the like.

Broadening of sympathy and extending knowledge beyond the limits set by other historians were, surely, central contributions Arnold Joseph Toynbee made to his chosen discipline. This was no small accomplishment for before Toynbee enlarged our perspective, history as studied in Western schools and taught in Western universities dealt with Europeans—ancient, medieval, and modern— and their descendants overseas, while other peoples entered on the scene only when discovered or civilized and conquered by Europeans. Everyone knew that India, China, and Islam also had lengthy histories; but these were special fields of inquiry, disregarded by historians, consigned instead to linguists and students of comparative religion.

A few bold spirits from outside academic life had tried to connect European and non-European history before Toynbee came along. H.G. Wells was the most important such intellectual adventurer, but his *Outline of History* (first published in 1920) took less than two years to write, and, as Wells himself pointed out, most of his information was culled from the *Encyclopedia Britannica*. Moreover, Wells' history was a history of progress, and the progress he valued arose in Europe for the most part, so that other peoples played only a marginal and subordinated role. For the centuries before 1500, this constituted a gross distortion, projecting back upon the past the pattern of world relationships that happened to prevail toward the close of the nineteenth century. For these reasons, Wells' amateur effort to repair the defects of the academic tradition of historical study, though very popular and widely read, did not do much to transcend older limits.

Spengler, another outsider to the academic establishment, was far more learned than Wells, and, as we saw in chapter 5, his vision of separate civilizations gave Toynbee an essential clue for structuring *A Study of History*. Moreover, Spengler treated Western and (some) non-Western civilizations as equivalent one to another, in a way Wells had failed to do. He, in effect, revived and updated an eighteenth century tradition that had viewed the sages of China and other wise men of the East as equals or superiors to their European counterparts.[4] But Toynbee's learning was broader and far more omnivorous than Spengler's; and his delight in detail, illumined by striking and often surprising comparisons across time and space, conformed to the tradition of historical writing far better than Spengler's oracular pronouncements. Toynbee therefore, more than any other single person, was able to introduce to a large portion of the world's reading public[5] the simple truth that Asians, Africans, Amerindians, and even specialized peoples like the Eskimo had a history that was independent of and analogous to the history of Europeans. The vision of a human past, cast, as he said, "not just in western terms" was, therefore, his great and central contribution to our tradition of learning, and ought to become his enduring claim to fame.

He was justly hailed for this feat in the 1930s; but after World War II, when his public fame peaked, undeniable defects in his civilizational pattern of rise and fall allowed most academic historians to reject Toynbee's vision of the whole along with details of his interpretation. Instead of accepting the challenge of trying to find a more accurate and adequate framework for all of human history, most historians moved in an opposite direction, developing increasingly specialized, arcane researches. In an age when interaction among the different branches of humankind has been getting more important with every passing year, the prevailing repudiation of global history in academic circles is absurd. What future citizens need to know, after all, is how other peoples and civilizations have interacted in times past if they are to navigate intelligently in the world of the future. By the 1980s, there were some signs that a rehabilitation of world history might yet occur in American and other classrooms, and if that happens, Toynbee's reputation will undoubtedly emerge from the eclipse into which it has fallen. I even dare to hope that this biography, by weaving his thought and times together, may start to redress the balance.

It may therefore be worthwhile to conclude with some remarks about Toynbee's place in intellectual history, addressing the theme on a grand scale, as befits his own habit of mind. First and foremost, Toynbee, like other historians, ought not to be dismissed simply because of factual inaccuracies—however real they were. Inaccuracy is inevitable, since even the most perfect text soon becomes outmoded and inaccurate through the advance of scholarly investigation. Moreover, to put factual accuracy above all else in judging a historian is an epistemologically naive view that begs the very difficult question of how words can or do relate to human acts and other "facts" of history while seeking to capture an ever-elusive Truth.

Instead, Toynbee should be judged as we judge other artists; that is, as a practitioner of a style of thought and sensibility that grows and changes, varying with time and place, and responding to all sorts of stimuli in quicksilver fashion. Personal and public experiences affect an artist's output, and the more poignant such experiences are, the more transparently they alter what he has to say through his art, as this biography shows to have been true in Toynbee's case.

By this measure, Toynbee is a real though not too promising candidate for the dignity of ranking as a major figure in European (and an emerging global) literary and historical culture. Though it may seem far-fetched at first, an appropriate comparison is with such figures as Milton and Dante—poets who, in their time, mastered most (or a great deal) of what was knowable, and distilled it into their writings. The difference between poetry and history is real enough: what happened does constrain historians' thoughts in a way that is not true for poets. Yet the role of imagination in weaving the facts of history together is nonetheless what gives a work of history its meaning and structure, for facts do not speak for themselves nor do they arrange themselves intelligibly. Thus Toynbee's history, like the work of every other historian, is poetic in essence and ought to be evaluated as such.

From this perspective, Toynbee's truly remarkable command of history and

European literature resembled the mastery of received learning that had characterized Dante and Milton, each in his time, though Toynbee's abysmal ignorance of natural science constituted a defect his poetic predecessors had not suffered from to anything like the same degree. Obviously, Toynbee's religious speculations, altered, attenuated, and yet continued the Christian heritage so central to Dante and Milton, and he shared Milton's sympathy for Satan—the archetypical underdog, rebel, and outsider.

But Toynbee's link with the two poets runs deeper than such obvious continuities. Human beings use words and other symbols to give meaning to the world. We routinely work wonders on the strength of agreed fictions that mobilize millions of otherwise isolated individuals for common undertakings. This capacity is at the very core of the human epos. What allows us to bracket Toynbee with Dante and Milton is that all three of them devoted heroic effort to the task of reaffirming and redefining the world's intelligibility in the light of everything they knew and believed to be true. Their respective understandings of things human and divine belong to and help to define the tradition of Western Christendom. As long as that tradition seems worth studying and inheriting, Toynbee should therefore rank as a twentieth century epigon to his poetic predecessors, for he, like them, possessed a powerful and creative mind that sought, restlessly and unremittingly, to make the world make sense.

Whether Toynbee, the epigon, will turn out to be a dwarf or a giant remains to be seen. That will depend on what historians and other intellectual heirs of our culture do with world history in time to come, for a writer's stature is defined less by intrinsic merit than by how others respond to what he or she has to say.

***

It is also worth comparing Toynbee with the two ancient Greek historians who did so much to shape his intellectual outlook and aspiration. Remarkable and truly Toynbeean comparisons fairly leap to mind when one looks for parallels. First let me propose that as Herodotus was to Homer, Toynbee was to Milton. The fit is uncannily exact. Both Herodotus and Toynbee wrote in prose instead of the verse their predecessors had relied upon, and their inquiries roved across all of the geographically knowable world. Both Herodotus and Toynbee attenuated without entirely rejecting the theological views of their poetic predecessors. And both Herodotus and Toynbee tended to get lost in delightful details, cramming so many asides and excurses into their texts as almost to obscure the organizing ideas behind their respective histories. Yet both of them did have organizing ideas and insights, and commanded an effective literary style with which to clothe their discoveries.

Comparison with Thucydides is less obvious, for the spare rigor of Thucydides' text stands poles apart from Toynbee's discursive superfluity. Yet there are two connections worth thinking about. First and most obvious, Toynbee borrowed the tragic form Thucydides gave to his monographic history, applying it first to the whole of Graeco-Roman civilization, and then to other civilizations everywhere. There is a second resemblance, if one believes, with Cornford, that Thucydides started to write his history using up-to-date, newfangled ideas and then changed his mind by reaffirming the validity of *Hubris* and *Ate* in order to account for Athens' tragic fall. Toynbee, too, started to write his great book with a set of ideas which derived partly from the ancients and partly from such up-to-date writers as Frazer, Bergson, Teggart, Freeman, and Spengler. Then, as we have seen, personal and public experience convinced him that the world could not be understood without taking account of a transcendental Spiritual Reality that he often called God, without ever quite assimilating that Reality to the omnipotent Deity of the Christian tradition. By summoning *Hubris* and *Ate,* Thucydides, too, had made room for supernatural forces in human affairs and did so by drawing on his religious heritage in quite the same attenuated way that Toynbee drew on his Christian heritage.

Such analogies are worth pondering. They put Toynbee in high company, matching him with some of the greatest writers of the Western tradition. An alternative future would convert Toynbee into a twentieth century bodhisattva for Soka Gakkai. Such canonization would presumably require the prior dissolution and abandonment of the Western tradition of learning—an ironical vindication of his vision of civilizational cycles. It seems improbable, but history abounds with improbabilities and this one should not be dismissed as impossible.

Most probable of all is to suppose that Toynbee's reputation will recover somewhat from the unjust dismissal he suffered at the hands of small-minded critics in the 1950s, but that the errors and defects of his history, and its inordinate length, will persuade future historians to treat him as a minor figure, full of quirks and far too lengthy to summarize conveniently. Much depends on how the discipline of history develops, and whether historians and other intellectuals persevere in trying to reduce humankind's manifold adventures on planet earth to an intelligible whole. If they do, Toynbee's pioneering role will surely be recognized and applauded.

Appendix

S. Fiona Morton, *A Bibliography of Arnold J. Toynbee* (Oxford University Press, 1980) lists 2974 separate publications that appeared between 1910 and 1979 either written by Toynbee or reviews of his publications. A second part of this book lists books and articles written by others about him and is almost as bulky. In preparing this biography I have consulted only a small part of what he wrote. Much was of fugitive interest; much was repetitious; much was difficult of access. Moreover, his important books are easy to recognize amidst the multitude of entries in Morton's bibliography.

It may help readers of this book to list them here for convenient reference in the order of their publication.

1915 *Nationality and the War*
1916 *The Treatment of the Armenians in the Ottoman Empire, 1915–1916*
1921 *The Tragedy of Greece*
1922 *The Western Question in Greece and Turkey: A Study in the Contact of Civilizations*
1925 *Survey of International Affairs, 1920–1923*
1926 *Survey of International Affairs, 1924*
1927 *Survey of International Affairs, 1925;* volume I *The Islamic World since the Peace Settlement*
1928 *Survey of International Affairs, 1926*
1929 *Survey of International Affairs, 1927*
1929 *Survey of International Affairs, 1928*
1930 *Survey of International Affairs, 1929*
1931 *Survey of International Affairs, 1930*
1932 *Survey of International Affairs, 1931*
1933 *Survey of International Affairs, 1932*

1934 *British Commonwealth Relations*
1934 *A Study of History*, volumes I–III
1934 *Survey of International Affairs, 1933*
1935 *Survey of International Affairs, 1934*
1936 *Survey of International Affairs, 1935*
1937 *Survey of International Affairs, 1936*
1938 *Survey of International Affairs, 1937*
1939 *A Study of History*, volumes IV–VI
1940 *Christianity and Civilization*
1941 *Survey of International Affairs, 1938*, volume I
1946 *A Study of History*. Abridgement of volumes I–VI by D.C. Somervell
1948 *Civilization on Trial*
1953 *The World and the West*
1954 *A Study of History*, volumes VII–X
1956 *An Historian's Approach to Religion*
1957 *A Study of History*. Abridgement of volumes VII–X by D.C. Somervell
1961 *A Study of History*, volume XII *Reconsiderations*
1962 *America and the World Revolution and Other Lectures*
1965 *Hannibal's Legacy: the Hannibalic War's Effects on Roman life*
1967 *Acquaintances*
1968 *Man's Concern with Death*
1969 *Experiences*
1969 *Some Problems of Greek History*
1970 *Cities on the Move*
1971 *Surviving the Future*
1973 *Constantine Porphyrogenitus and His World*
1976 *Mankind and Mother Earth*
1976 *The Toynbee–Ikeda Dialogue: Man Himself Must Choose*

Notes

1. Great Expectations

1. Gertrude Toynbee, ed., *Reminiscences and Letters of Joseph and Arnold Toynbee* (London, n.d.), pp. 57, 72, 80.
2. Arnold J. Toynbee, *Acquaintances* (London, 1967), p. 3; Charles L. Mowat, *The Charity Organisation Society, 1869–1919: Its Ideas and Work* (London, 1961), p. 105.
3. Arnold Toynbee, *Lectures on the Industrial Revolution in England: Popular Addresses, Notes and Other Fragments, together with a short Memoir by Benjamin Jowett, Master of Balliol* (London, 1884), p. xviii.
4. Gertrude Toynbee, ed., *Reminiscences and Letters of Joseph and Arnold Toynbee*, letter, 25 March 1883.
5. Arnold Toynbee, *Lectures on the Industrial Revolution*, p. 25.
6. Gertrude Toynbee, *Reminiscences and Letters of Joseph and Arnold Toynbee*, letter, 18 April 1880.
7. Mowat, *The Charity Organisation Society*, p. 105.
8. Bodleian Library, Toynbee Papers, letter, Edith Toynbee to AJT, 22 October 1907, writing about a gift Paget had made to her son: "I dare say you will be tempted to think you'd rather not have the money at all from him, and I know very well from much experience that it is very hard to take money presents simply and gracefully. . . . But that is really a form of pride, and you will be able to feel more warmly about it if you remember that it is really a great help to Father, and feel grateful for his sake, though he would not want you to say anything like that to Uncle Paget."
9. A.J. Toynbee, *Acquaintances*, pp. 3–4.
10. Cf. the delightful sketch of Uncle Harry in A.J. Toynbee, *Acquaintances*, pp. 1–20.
11. Bodleian Library, Toynbee Papers, National Education TV Broadcast, "History and the Historian," 1967.
12. A.J. Toynbee, *Civilization on Trial* (New York, 1948), p. 3–4.
13. (London, 1891), p. 27.

14. Bodleian Library, Toynbee Papers, Juvenilia.
15. Margaret Toynbee, personal interview, 15 February 1986.
16. Bodleian Library, Toynbee Papers, letter, AJT to Edith Toynbee, 20 January 1906.
17. Mowat, *The Charity Organisation Society,* p. 147. This admirable book provides the basis for my remarks about the society and its limitations.
18. Arnold J. Toynbee, *Experiences* (New York and London, 1969), pp. 4–5.
19. Bodleian Library, Toynbee Papers, letters, Edith Toynbee to AJT, 18 May 1899 and 17 June 1899.
20. Bodleian Library, Toynbee Papers, letters, Edith Toynbee to AJT, 25 September 1899 and 6 October 1899.
21. Bodleian Library, Toynbee Papers, letter, Edith Toynbee to AJT, 3 February 1900.
22. Bodleian Library, Toynbee Papers, Juvenilia.
23. Bodleian Library, Toynbee Papers, letter, Harry Toynbee to AJT, 14 July 1901.
24. Bodleian Library, Toynbee Papers, letter, Edith Toynbee to AJT, 15 July 1901.
25. Bodleian Library, Toynbee Papers, letter, AJT to David Robertson, 4 November 1961; A.J. Toynbee, *Experiences,* pp. 7–9.
26. Margaret Toynbee, personal interview, 15 February 1986. Arnold's dormouse died while he was away at Winchester taking his examination for the second time, but the news was withheld until later lest it disrupt his concentration.
27. Bodleian Library, Toynbee Papers, letter, AJT to Edith Toynbee, 28 May 1907; cf. the essay devoted to Rendall in A.J. Toynbee, *Acquaintances,* pp. 37–42.
28. Bodleian Library, Toynbee Papers, postcard and letter, AJT to his parents, 3 July 1906 and 30 July 1906.
29. Bodleian Library, Toynbee Papers, engraved invitation to the award ceremony, 31 July 1907, lists prizes and winners. A cutting from *The Times* is also included listing the winners.
30. Bodleian Library, Toynbee Papers, letter, AJT to his parents, no date.
31. Bodleian Library, Toynbee Papers, postcard, AJT to his parents, dated only "Friday evening."
32. Bodleian Library, Toynbee Papers, letters, AJT to his parents, 4 November 1906; AJT to Edith Toynbee, 25 July 1907.
33. Bodleian Library, Toynbee Papers, letter, AJT to his parents, 31 March 1907.
34. A.J. Toynbee, *Experiences,* p. 295.
35. Bodleian Library, Toynbee Papers, Juvenilia, "The Roman Empire under the Macedonian Dynasty," read to the XVI Club at Winchester in 1907.
36. Bodleian Library, Toynbee Papers, Juvenilia.
37. Bodleian Library, Toynbee Papers, letter, AJT to Edith Toynbee, 7 July 1907 contains a lengthy and learned discourse on New Testament criticism.
38. Bodleian Library, Toynbee Papers, letter, AJT to Edith Toynbee, 24 February 1907.
39. Bodleian Library, Toynbee Papers, letter, AJT to Edith Toynbee, no date. The first pages of this letter are missing, but the fact that it goes on to tell of finishing *Faust* "yesterday. It is almost too dreadful to translate aloud" shows that the letter dates from his school years, probably 1906.

2. Balliol and the Breakup of Toynbee's Parental Home

1. Bodleian Library, Toynbee Papers, letter, AJT to his parents, 30 July 1906.
2. J.W. Mackail, *James Leigh Strachan-Davidson, Master of Balliol* (Oxford, 1925), p. 99.
3. Drusilla Scott, *A.D. Lindsay: A Biography* (Oxford, 1971), pp. 47–48.

4. Mackail, *Strachan-Davidson*, p. 54.
5. *Balliol College Register* (Oxford, 1953). Incidentally, in spite of remarks Toynbee regularly made in later life about the toll of lives among his friends in World War I, only seven of Toynbee's classmates at Balliol died between 1914 and 1918. Men with Balliol connections presumably drifted into staff positions very quickly, minimizing casualties. Those several years junior to Toynbee did suffer heavy losses, however. The highest ratio came in 1915, when sixteen of twenty-nine matriculants were killed.
6. Bodleian Library, Toynbee Papers, letter, AJT to his parents, 6 December 1906.
7. Bodleian Library, Toynbee Papers, letter, AJT to Edith Toynbee, 20 October 1907.
8. Bodleian Library, Toynbee Papers, letter, Edith Toynbee to AJT, 22 October 1907.
9. Bodleian Library, Toynbee Papers, letter, AJT to R.S. Darbishire, 9 April 1911.
10. Bodleian Library, Toynbee Papers, letter, AJT to R.S. Darbishire, 28 December 1910.
11. Bodleian Library, Toynbee Papers, letter, AJT to R.S. Darbishire, 27 January 1911.
12. Bodleian Library, Toynbee Papers, letter, AJT to R.S. Darbishire, 6 November 1912.
13. Gilbert Murray, *An Unfinished Autobiography* (London, 1960), p. 15.
14. Gilbert Murray delivered the Presidential address for the Society for Psychical Research on 9 July 1915, reporting some of his results. See *Proceedings of the Society for Psychical Research*, 29 (1918), 46–63. See also A.W. Verrall, "Report on a Series of Experiments in 'Guessing,' " *Proceedings of the Society for Psychical Research*, 29 (1918), 64–110.
15. Bodleian Library, Toynbee Papers, letters, AJT to R.S. Darbishire, no date [but, by deduction, early March 1910], and AJT to R.S. Darbishire, 12 March 1910.
16. Bodleian Library, Toynbee Papers, Juvenilia.
17. Bodleian Library, Toynbee Papers, Juvenilia.
18. Bodleian Library, Toynbee Papers, letter, AJT to R.S. Darbishire, not dated, but sometime in 1910.
19. Bodleian Library, Toynbee Papers, letter, AJT to R.S. Darbishire, 5 June 1910.
20. This book, based initially on lectures delivered at Columbia University, New York in 1912, was subsequently reissued, with an additional chapter, as *Five Stages in Greek Religion* (New York and London, 1925).
21. Bodleian Library, Toynbee Papers, letter, AJT to Edward R. Morgan, 22 April 1911.
22. Bodleian Library, Toynbee Papers, letter, AJT to R.S. Darbishire, 19 April 1946. Hasluck was a classical scholar who became specially interested in Christian–Moslem relations and interpenetration of the rival faiths that had occurred among peasants of the remoter Balkans. He thus became a pioneer European ethnologist.
23. Bodleian Library, Toynbee Papers, letters, AJT to R.S. Darbishire, 9 November 1909 and 30 March 1910.
24. Bodleian Library, Toynbee Papers, Juvenilia.
25. Arnold J. Toynbee, *Experiences* (New York and London, 1969), pp. 94–95.
26. Bodleian Library, Toynbee Papers, letter, AJT to Edith Toynbee, 30 June [1907].
27. Three tutorial essays prepared for Strachan-Davidson survive on such themes as "Lex Acilia de Repetundis" and "The status of Greece under Roman rule before the erection by Augustus of the province of Achaea." Bodleian Library, Toynbee Papers, Juvenilia. They are lengthy, learned, exact.
28. *Classical Review*, 24 (1910), 236–238.
29. Bodleian Library, Toynbee Papers, letter, AJT to R.S. Darbishire, 22 February

1910. In old age Toynbee resumed while altering this assignment by writing *Constantine Porphyrogenitus and His World* (London and Oxford, 1973).

30. Bodleian Library, Toynbee Papers, letter, AJT to R.S. Darbishire, 30 January 1910.

31. Bodleian Library, Toynbee Papers, letter, AJT to R.S. Darbishire, 21 May 1911.

32. Bodleian Library, Toynbee Papers, letter, AJT to R.S. Darbishire, 22 June 1911.

33. Bodleian Library, Toynbee Papers, Juvenilia.

34. Bodleian Library, Toynbee Papers, letter, AJT to Edith Toynbee, 27 March 1912.

35. Bodleian Library, Toynbee Papers, letter, AJT to Edith Toynbee, 19 January 1912.

36. Bodleian Library, Toynbee Papers, letter, AJT to Edith Toynbee, 26 January 1912.

37. Bodleian Library, Toynbee Papers, letter, AJT to Edith Toynbee, 3 March 1912.

38. Bodleian Library, Toynbee Papers, letter, AJT to R.S. Darbishire, 24 April 1910.

39. Bodleian Library, Toynbee Papers, letter, AJT to R.S. Darbishire, 30 January 1910.

40. Bodleian Library, Toynbee Papers, letter, AJT to R.S. Darbishire, no date, but reference to a trip to Ireland shows it was written in July 1910.

41. Bodleian Library, Toynbee Papers, letter, AJT to R.S. Darbishire, 12 June 1910.

42. Bodleian Library, Toynbee Papers, letter, AJT to R.S. Darbishire, 17 September 1911.

43. Bodleian Library, Toynbee Papers, letter, Margaret Toynbee to AJT, 23 February 1939. "We" referred to Margaret and her elder sister, Jocelyn.

44. Surviving documents do not make family financial arrangements clear. But letters Toynbee wrote to his mother from Greece in 1912 about how to reduce the costs of Harry's care make the circumstances fairly obvious. See especially Bodleian Library, Toynbee Papers, letter, AJT to Edith Toynbee, 3 February 1912.

45. Arnold J. Toynbee, *Acquaintances* (London, 1967), pp. 18–19.

46. Bodleian Library, Toynbee Papers, letter, AJT to R.S. Darbishire, 19 January 1913. Toynbee described the trauma, years afterward, as "the time, when I was 19, when my Father went out of his mind and my Mother became entirely absorbed in him, so that I then virtually lost my Mother (with whom I had had a particularly close relation till then)." Letter, AJT to Father Columba, 19 May 1944, in Christian B. Peper, ed., *An Historian's Conscience: The Correspondence between Arnold J. Toynbee and Columba Carey-Elwes, Monk of Ampleforth* (Boston, 1986), p. 162.

47. Private possession, letter, AJT to R.S. Darbishire, 3 October 1946.

48. Bodleian Library, Toynbee Papers, letter, AJT to Edith Toynbee, no date. The Dell was the name of his Aunt Grace's house.

49. Bodleian Library, Toynbee Papers, letter, AJT to R.S. Darbishire, 2 August [1910].

50. Bodleian Library, Toynbee Papers, letter, Mary Murray to AJT, 15 September 1910.

51. Bodleian Library, Toynbee Papers, letter, Margaret Toynbee to AJT, 23 February 1939.

52. Bodleian Library, Toynbee Papers, letter, AJT to R.S. Darbishire, 19 January 1913.

53. Bodleian Library, Toynbee Papers, letter, AJT to R.S. Darbishire, 19 January 1913.

54. Bodleian Library, Toynbee Papers, letter, AJT to R.S. Darbishire, 28 December 1910.

55. Bodleian Library, Toynbee Papers, letters, AJT to R.S. Darbishire, 30 March 1910, 24 April 1910, and 28 December 1910.

56. Bodleian Library, Toynbee Papers, letters, AJT to R.S. Darbishire, 6 November 1912 and 8 May 1910.

57. Bodleian Library, Toynbee Papers, letter, AJT to R.S. Darbishire, 6 August 1910.

58. Bodleian Library, Toynbee Papers, letter, AJT to R.S. Darbishire, 24 August 1911.

59. Bodleian Library, Toynbee Papers, postcard, Cyril Bailey to AJT, 22 March 1911.

60. Bodleian Library, Toynbee Papers, letter, A.D. Lindsay to AJT, 21 March [1911].

61. Bodleian Library, Toynbee Papers, letter, AJT to R.S. Darbishire, 4 November 1910.
62. Bodleian Library, Toynbee Papers, letter, AJT to R.S. Darbishire, 7 May 1911.

3. Grand Tour, Donhood, and Marriage

1. Bodleian Library, Toynbee Papers, letter, AJT to Edith Toynbee, 27 December 1911.
2. Bodleian Library, Toynbee Papers, letter, AJT to Edith Toynbee, 9 October 1911.
3. Bodleian Library, Toynbee Papers, letter, AJT to Edith Toynbee, 6 December 1911.
4. Bodleian Library, Toynbee Papers, letter, AJT to R.S. Darbishire, 25 September 1912.
5. Arnold J. Toynbee, *Experiences* (New York and London, 1969), pp. 18–40.
6. Bodleian Library, Toynbee Papers, letter, AJT to Edith Toynbee, 22 September 1911.
7. Bodleian Library, Toynbee Papers, letter, AJT to Edith Toynbee, 27 December 1911.
8. Bodleian Library, Toynbee Papers, letter, AJT to Edith Toynbee, 20 October 1911.
9. Bodleian Library, Toynbee Papers, letter, AJT to Edith Toynbee, 2 January 1912.
10. Bodleian Library, Toynbee Papers, letter, AJT to Edith Toynbee, 3 March 1912.
11. Bodleian Library, Toynbee Papers, letter, AJT to Edith Toynbee, 19 January 1912.
12. See, for example, *Experiences,* p. 38.
13. He wrote: "Today I ate bread for the first time in six weeks." Bodleian Library, Toynbee Papers, letter, AJT to R.S. Darbishire, 25 September 1912.
14. Bodleian Library, Toynbee Papers, letter, AJT to Edith Toynbee, 11 December 1911.
15. Bodleian Library, Toynbee Papers, letter, AJT to Edith Toynbee, 19 November 1911.
16. Bodleian Library, Toynbee Papers, letter, AJT to Edith Toynbee, 24 July 1912.
17. p. 35.
18. Bodleian Library, Toynbee Papers, letter, AJT to R.S. Darbishire, 6 November 1912.
19. Bodleian Library, Toynbee Papers, letter, AJT to Edith Toynbee, 2 January 1912.
20. Bodleian Library, Toynbee Papers, letter, AJT to Edith Toynbee, 5 November 1911.
21. Bodleian Library, Toynbee Papers, letter, AJT to Edith Toynbee, 22 June 1912.
22. Bodleian Library, Toynbee Papers, letter, AJT to Edith Toynbee, 26 June 1912.
23. Arnold J. Toynbee, *A Study of History* (London, 1954), X, 134–135.
24. Ibid., X, 136–137.
25. Ibid., X, 130–131.
26. Ibid., X, 139.
27. Ibid., X, 107–111.
28. Notes for these lectures are preserved in the Bodleian Library, Toynbee Papers.
29. A topographical diary survives in the Bodleian collection, but it breaks off with his arrival in Greece. Either Toynbee stopped making daily records of what he saw and surveyed, or, more likely, he abandoned the diary format for a more systematic record which does not appear among his papers.
30. Bodleian Library, Toynbee Papers, letter, AJT to R.S. Darbishire, 25 September 1912.
31. Bodleian Library, Toynbee Papers, letter, AJT to R.S. Darbishire, 6 November 1912.

32. Bodleian Library, Toynbee Papers, letter, AJT to R.S. Darbishire, 15 December 1912.
33. Bodleian Library, Toynbee Papers, letter, A.D. Lindsay to AJT, no date, but its content makes it clear that this must have been written during the spring examination period of either 1913 or 1914.
34. Bodleian Library, Toynbee Papers, letter, AJT to R.S. Darbishire, 15 February 1913.
35. Arnold J. Toynbee, "The Growth of Sparta," *Journal of Hellenic Studies,* 33 (1913), 246–275.
36. Bodleian Library, Toynbee Papers, letter, Gilbert Murray to AJT, 20 July 1914.
37. Bodleian Library, Toynbee Papers, letter, J.B. Bury to AJT, 1 July 1913.
38. Bodleian Library, Toynbee Papers, letter, Walter Leaf to AJT, 23 November 1913.
39. Private possession, letter, AJT to R.S. Darbishire, 11 May 1913.
40. Bodleian Library, Toynbee Papers, letter, AJT to R.S. Darbishire, 6 November 1912.
41. Bodleian Library, Toynbee Papers, letter, AJT to R.S. Darbishire, 19 January 1913. "Albanophone Hellenes" were people who spoke Albanian but worshiped in Greek Orthodox churches. The Greek delegates argued, of course, that such people were Greeks, not Albanians, and that the region of what the Powers eventually (1914) decided should become southern Albania, where they constituted a majority, ought, accordingly, to be incorporated into Greece.
42. Bodleian Library, Toynbee Papers, letter, AJT to Edith Toynbee, 26 September 1911.
43. Bodleian Library, Toynbee Papers, letter, AJT to Edith Toynbee, 3 March 1912.
44. Private possession, letter, AJT to R.S. Darbishire. This letter lacks a first page and is therefore undated, but reference to "this summer" makes it sure that it was written in the autumn of 1912.
45. Bodleian Library, Toynbee Papers, letter, AJT to R.S. Darbishire, 25 September 1912.
46. Bodleian Library, Toynbee Papers, letter, AJT to R.S. Darbishire, 19 August 1912.
47. Bodleian Library, Toynbee Papers, letter, AJT to R.S. Darbishire, 19 January 1913.
48. Private possession, letter, AJT to R.S. Darbishire, no date, but from the autumn of 1912.
49. Private possession, letter, AJT to R.S. Darbishire, 9 March 1913.
50. Private possession, letter, AJT to R.S. Darbishire, 11 May 1913.
51. Bodleian Library, Toynbee Papers.
52. Private possession.
53. Bodleian Library, Gilbert Murray Papers, Carton 231, letter, Rosalind Murray to Jean Smith, 15 August 1958.
54. The Howards attained prominence in Tudor times, and one of the family commanded the English attack on the Spanish Armada in 1588. The Earls of Carlisle were descended from a junior branch of the family, rising to fame and fortune after 1688. Castle Howard was built early in the eighteenth century, by the same architect who built Blenheim Palace for the Duke of Marlborough, and it is almost on the same scale. The Stanleys, however, came to England with William the Conqueror, a fact that the Countess of Carlisle and Rosalind, her granddaughter, never forgot; but, like the Howards, the Stanleys also became part of the Whig aristocratic establishment after 1688.
55. Two books, both written by members of the immediate family, provide the basis

for my remarks about Rosalind's grandmother. Dorothy Henley, *Rosalind Howard, Countess of Carlisle* (London, 1959) and Charles Roberts, *The Radical Countess: The History of the Life of Rosland, Countess of Carlisle* (London, 1962). Dorothy Henley was a daughter, Charles Roberts a son-in-law, and both accounts exhibit some reticences about a truly astounding personality.

56. Bodleian Library, Toynbee Papers, letter, AJT to Francis West, 26 October 1972.
57. Bodleian Library, Gilbert Murray Papers, Carton 476, letter, Gilbert Murray to Lady Carlisle, 26 March 1892.
58. Bodleian Library, Gilbert Murray Papers, Carton 476, letters, Gilbert Murray to Lady Carlisle, 27 March 1893 and 30 May 1894.
59. Bodleian Library, Gilbert Murray Papers, Carton 476, letter, Gilbert Murray to Lady Carlisle, 24 November 1897.
60. Bodleian Library, Gilbert Murray Papers, Carton 550, letter, Gilbert Murray to Lady Carlisle, 25 December 1906.
61. Bodleian Library, Gilbert Murray Papers, Carton 550, letter, Gilbert Murray to Lady Carlisle, 20 September 1903.
62. Bodleian Library, Toynbee Papers, letter, G.B. Shaw to Gilbert Murray, 4 March 1911. From a copy, sent to Toynbee in August 1968 by Sidney P. Albert of Pasadena, California, who had found it among Shaw's letters.
63. Bodleian Library, Gilbert Murray Papers, Carton 567, letter, John Galsworthy to Gilbert Murray, 5 January 1910. Rosalind's preferred title, "A First Encounter," was ruled out by Lady Mary as "vulgar." Letter, Gilbert Murray to Rosalind Murray, 2 March 1910.
64. Bodleian Library, Gilbert Murray Papers, Carton 567, letter, Gilbert Murray to Rosalind Murray, 23 January 1910. The Prince Kropotkin in question may have been the son of the famous revolutionary of that name, who wintered in Italy in 1909, presumably with his family.
65. In 1968 Toynbee found a MS sonnet among Rosalind's papers "unsigned but I am pretty sure, by Rupert Brooke. (Rosalind made friends with him about 1912; I think this gave her parents some anxiety.) . . . I believe Rosalind would have shrunk from having this made public, though it is no more than a straightforward expression of admiration, not going to the length of declaring love." Bodleian Library, Toynbee Papers, letter, AJT to Jean Smith, 9 February 1968. I found no trace of the sonnet in question, which may well have been destroyed along with Rosalind's other private papers.
66. Bodleian Library, Gilbert Murray Papers, Carton 567, letter, Gilbert Murray to Rosalind Murray, 16 October 1911.
67. Bodleian Library, Toynbee Papers, letter, AJT to Francis West, 2 November 1972.
68. Bodleian Library, Gilbert Murray Papers, Carton 567, letter, Gilbert Murray to Rosalind Murray, 29 June 1912.
69. Bodleian Library, Gilbert Murray Papers, Carton 567, letter, Gilbert Murray to Rosalind Murray, 14 December 1912.
70. Bodleian Library, Gilbert Murray Papers, Carton 567, letter, Gilbert Murray to Rosalind Murray, 18 January 1913.
71. In his youth, Murray had been accused of "teetotalism, enthusiasm and toynbeeism." Gilbert Murray, *An Unfinished Autobiography* (London, 1960), p. 105.
72. Bodleian Library, Toynbee Papers.
73. Bodleian Library, Murray Papers, Carton 567, letter, Gilbert Murray to Penelope Wheeler, 8 June 1913.
74. Rosalind Murray, *Unstable Ways* (London, 1914), pp. 255, 257, 262, 284.

75. Toynbee recalled, long afterward: "After Rosalind and I became engaged, he [Gilbert Murray] questioned me, I remember, on my religious beliefs, and showed satisfaction at finding they were, more or less, Stoical, in the classical sense." Bodleian Library, Toynbee Papers, letter, AJT to Douglas Woodruff, 5 February 1972.
76. Duncan Wilson, *Gilbert Murray, OM 1866–1957* (Oxford, 1987), p. 212.
77. Bodleian Library, Toynbee Papers, letter, Rosalind Murray to AJT, 5 June 1913.
78. Bodleian Library, Toynbee Papers, letter, J.L. Strachan-Davidson to AJT, 16 August 1913.
79. Bodleian Library, Toynbee Papers, letter, Charlotte M. Toynbee to AJT, 29 August 1913.
80. Bodleian Library, Toynbee Papers, letter, Rosalind Murray to AJT, 27 August 1913.
81. Bodleian Library, Toynbee Papers, letter, Rosalind Murray to AJT, 23 August 1913.
82. Bodleian Library, Toynbee Papers, letter, Rosalind Murray to AJT, 8 September 1913.
83. Bodleian Library, Gilbert Murray Papers, Carton 567, letters, Gilbert Murray to Rosalind Toynbee, 26 September 1913 and 29 September 1913.
84. Bodleian Library, Toynbee Papers, letter, Rosalind Murray to AJT, 27 August 1913.
85. Bodleian Library, Gilbert Murray Papers, Carton 567, letter, Gilbert Murray to Rosalind Toynbee, 29 September 1913.
86. Bodleian Library, Toynbee Papers, letter, AJT to Mary Murray, 25 June 1914.
87. Bodleian Library, Toynbee Papers, letters, AJT to Mary Murray, 4 July 1914 and 1 August 1914.
88. This list appears on a loose sheet of paper, and apparently describes the establishment as managed by whoever it was that lived in the house before the Toynbees moved in. Bodleian Library, Toynbee Papers, loose sheet.

4. The Great War and the Peace Conference

1. Bodleian Library, Toynbee Papers, letter, Edith Toynbee to AJT, 26 August 1914.
2. Bodleian Library, Toynbee Papers, letter, Edith Toynbee to AJT, 15 October 1915.
3. Bodleian Library, Toynbee Papers.
4. Bodleian Library, Gilbert Murray Papers, Carton 165, sheets 161–162, letter, Rosalind Toynbee to Bertrand Russell, 5 December 1917. Russell had just published a book attacking the morality and justice of the war, and Rosalind wrote to express her agreement with what he had said.
5. Bodleian Library, Toynbee Papers, letter, AJT to Edith Toynbee, 18 October 1914.
6. Bodleian Library, Toynbee Papers, letter, AJT to Edith Toynbee, 14 October 1915.
7. Bodleian Library, Toynbee Papers, letter, AJT to Gilbert Murray, 15 October 1915.
8. Bodleian Library, Toynbee Papers, letter, H.T. Gillett, M.D. to AJT, 13 October 1915.
9. Bodleian Library, Toynbee Papers, letter, AJT to Edith Toynbee, 12 December 1915.
10. Bodleian Library, Toynbee Papers, letter, Alexander Lindsay to AJT, 27 September [1915?]. The year is not indicated on this letter, but it must have been written either in 1914 or in 1915 before the two men broke off relations.
11. Bodleian Library, Toynbee Papers, letter, Edith Toynbee to AJT, 15 October 1915.
12. The preface, dated 15 February 1915, indicates when he completed writing. Manufacture was also rapid—taking only six weeks.
13. Arnold Toynbee, *Nationality and the War* (London, 1915), pp. 333–334.

14. Ibid., p. 29.
15. Ibid., p. 494.
16. Ibid., pp. 510, vii.
17. Bodleian Library, Toynbee Papers, letter, Charlotte M. Toynbee to AJT, 12 November 1915. The account of this episode in *Acquaintances* (London, 1967, p. 33) makes it clear that Toynbee did find it troublesome to be named after his famous uncle. He waited until all his aunts and uncles had died, then reverted to publishing books as Arnold Toynbee, without the middle initial. But he had to wait until he was sixty-seven years old to claim the preferred form of his name for himself!
18. Bodleian Library, Toynbee Papers, letter, AJT to Edith Toynbee, 28 April 1915.
19. Bodleian Library, Toynbee Papers, letter, AJT to Edith Toynbee, 28 April 1915. Toynbee's initial assignment was to serve as assistant to Sir Gilbert Parker, whom he described as a "figurehead (the man who signs the correspondence)." Later changes in personnel seem to have relegated Parker to obscurity; at any rate, he played no significant part in Toynbee's subsequent career with the government.
20. Bodleian Library, Toynbee Papers, letter, AJT to Gilbert Murray, 31 May 1915.
21. Bodleian Library, Toynbee Papers, letter, AJT to Gilbert Murray, 8 July 1915.
22. Bodleian Library, Toynbee Papers, letter, AJT to Gilbert Murray, 25 October 1915.
23. Bodleian Library, Toynbee Papers, letter, AJT to Gilbert Murray, 21 September 1916.
24. James Bryce (1838–1922) was a distinguished historian, sometime professor of law at Oxford, Liberal politician, and by 1915 an elder statesman, with special ties to the United States, where he had been British ambassador from 1907 to 1913. Though Toynbee said privately at the time that Bryce was as "obstinate as he is distinguished," nevertheless the older man became something of a role model for him, combining, as he did, an active career in public affairs with distinction as a historian. This Toynbee made plain in the chapter devoted to Bryce in his book *Acquaintances,* pp. 149–160.
25. (London, 1916).
26. Bodleian Library, Toynbee Papers, letter, AJT to R.S. Darbishire, 16 September 1917.
27. Sir James Headlam-Morley, *A Memoir of the Paris Peace Conference, 1919* (London, 1972), pp. xx–xxi.
28. Bodleian Library, Toynbee Papers, letter, AJT to R.S. Darbishire, 25 December 1917.
29. Public Record Office, Foreign Office, Political Intelligence Department, 371.4353 Peace Series. I am indebted to Christopher Collins of Oxford, who kindly allowed me access to Xerox copies of his transcriptions of Toynbee's minutes and other writings from 1917–1919 that are on deposit in the Public Records Office.
30. Ibid.
31. Public Record Office, Foreign Office, Political Intelligence Department, 681-697, Minute by Headlam-Morley, 26 November 1919, with additional comments by Hardinge and Curzon.
32. Bodleian Library, Toynbee Papers, letter, AJT to Edith Toynbee, 12 December 1917. Edward Henry Carson (1854–1935) was the Ulster leader who opposed Irish Home Rule in 1914 with quasi-revolutionary actions and later became a member of the War Cabinet. Lord Robert Cecil of Chelwood (1864–1958), scion of a famous aristocratic family, was Assistant Secretary of State for Foreign Affairs in 1916. Alfred Milner (1854–1925), a Balliol man and onetime follower of Arnold Toynbee, was another member of the inner War Cabinet. John Buchan (1875–1940), the

novelist, was Director of Information. Toynbee gives no hint as to which he found "good," though one may guess he meant Lord Cecil, an old-fashioned aristocrat, who later took a leading role in League of Nations activities in close association with Gilbert Murray.

33. Bodleian Library, Toynbee Papers, letter, AJT to Edith Toynbee, 14 November 1918.

34. Bodleian Library, Gilbert Murray Papers, Carton 35, sheets 60 and 61, letters, AJT to Gilbert Murray, both dated 12 November 1917.

35. Public Records Office, Foreign Office, Political Intelligence Department, 371.4363, General McDonough to Tyrrell, 13 July 1918 with Toynbee Minute, 19 July 1918, and further exchanges involving Headlam-Morley, Tyrrell, Toynbee, and McDonough through 24 July 1918.

36. Bodleian Library, Toynbee Papers, letter, AJT to Edith Toynbee, no date, but probably written in July 1918 at Castle Howard, from which Toynbee's riposte to General McDonough was sent in the same month. This letter may, indeed, reflect his irritation at the accusation the DMI had made against him.

37. *The Diary of Virginia Woolf, vol. I   1915–1919,* edited by Anne Oliver Bell (London, 1977), p. 108.

38. Bodleian Library, Toynbee Papers, letter, AJT to Edith Toynbee, 12 November 1916. For his views on a German Mitteleuropa, see letter, AJT to Edith Toynbee, 19 November 1916.

39. Bodleian Library, Toynbee Papers, letter, AJT to Edith Toynbee, a loose sheet, with no date, but stored between letters dated 23 January 1917 and 1 April 1917, so presumably referring to the February revolution.

40. Bodleian Library, Toynbee Papers, letter, Rosalind Toynbee to AJT, 11 March 1919, written immediately after her return to England.

41. Bodleian Library, Toynbee Papers, letter, AJT to Edith Toynbee, 26 January 1919. T.E. Lawrence (1888–1935), known as "Lawrence of Arabia," was a British war hero and advocated Arab independence at the Peace Conference. Jan Christian Smuts (1870–1950) fought against the British in the Boer War, played a key role in the founding of the Union of South Africa, and in 1917 became Minister for Air in the British Cabinet. At the Peace Conference he distinguished himself as a prominent supporter of the League of Nations and advocate of a moderate peace with Germany.

42. Bodleian Library, Toynbee Papers, letter, AJT to Edith Toynbee, 13 February 1919. Georges Clemenceau (1841–1929) was Premier of France and official host of the Peace Conference.

43. Bodleian Library, Toynbee Papers, letter, AJT to Edith Toynbee, 24 March 1919.

44. Public Record Office, Peace Conference, Smyrna Files, 1919, Doc. F. 6102, AJT Minute dated 3 April 1919.

45. Harold Nicolson (1886–1968) was a Balliol man, like Toynbee, but unlike him was a full-fledged member of the Foreign Service and the son of a British diplomat who rose to head the permanent staff of the Foreign Office. In later life he served in Parliament and wrote no fewer than thirty-five books.

46. Public Records Office, Peace Conference, Future Frontiers of Turkey, Delegation Minute, Doc. F.7335, dated 15 April 1919.

47. Harold Nicolson, *Peacemaking, 1919* (rev. ed. London, 1943) entry under 14 April 1919. On 16 April he referred to the proposal as "Toynbee's scheme," as undoubtedly it was.

48. Bodleian Library, Toynbee Papers, letter, Rosalind Toynbee to AJT, 17 May 1916. Rosalind was staying with her mother in Oxford at the time.

49. Bodleian Library, Toynbee Papers, letter, AJT to Mary Murray, 28 October 1915.

50. Bodleian Library, Toynbee Papers, letter, AJT to Mary Murray, 2 October 1919, refers to "a threat of a miscarriage. . . . It isn't at all certain, though we mustn't hope too much, as it is the second time." Two weeks later he reported to his mother-in-law "Rosalind had an operation." A plausible interpretation of another letter, AJT to Edith Toynbee, 20 October 1918, offers the best surviving evidence for the date of her first miscarriage.

51. Bodleian Library, Gilbert Murray Papers, Carton 550, sheet 11, letter, Mary Murray to the Countess of Carlisle, 15 March 1917.

52. Bodleian Library, Gilbert Murray Papers, Carton 550, sheet 79, letter, Mary Murray to the Countess of Carlisle, 6 May 1918.

53. Bodleian Library, Gilbert Murray Papers, Carton 568, sheet 47, 19 February 1919.

54. Bodleian Library, Toynbee Papers, letter, Edith Toynbee to AJT, New Years' Eve, 1915.

55. (London, 1961), p. 33 and passim. Philip also attributed three miscarriages to his fictional mother. This may well reflect accurate family reminiscence, but no trace of a third miscarriage for Rosalind survives in available letters.

56. Bodleian Library, Toynbee Papers, letter, Rosalind Toynbee to AJT, 16 March 1919.

57. Bodleian Library, Toynbee Papers, letter, Rosalind Toynbee to AJT, 16 January 1921.

58. Bodleian Library, Toynbee Papers, letter, Rosalind Toynbee to AJT, 31 July 1917.

59. Bodleian Library, Toynbee Papers, letter, AJT to Mary Murray, 18 July 1917.

60. Bodleian Library, Toynbee Papers, letter, AJT to Edith Toynbee, 28 April 1915.

61. "Is it necessary for you to do anything at all save what you yourself wish to do? Gibbon did nothing, Lord Acton did nothing. You imagined it folly to travel with a toothbrush only in Greece, but the toothbrush was essential and the haversack was full of superfluities. Your essential, I take it, is lively interest in learning everything. . . . Your superfluities [are] politics with journalism, or the manufacture of Firsts [in Oxford examinations]." Bodleian Library, Toynbee Papers, letter, R.S. Darbishire to AJT, 22 April 1915.

62. Bodleian Library, Toynbee Papers, letter, Lady Carlisle to Rosalind Toynbee, 21 November 1915.

63. Bodleian Library, Toynbee Papers, letter, Lady Carlisle to AJT, 25 November 1915. In a note attached to this letter, dated 26 February 1969, Toynbee explained the background, as follows: "Rosalind took a horror of living at Palace Green; Lady Mary [Murray] said I must take a furnished house. I did. I was now extremely hard pressed financially and I became so worried that Rosalind suggested I should ask Lady Carlisle for some income after all. This explains her letter of November 25, 1915."

64. Bodleian Library, Toynbee Papers, letter, Alexander Lindsay to AJT, 11 December [1915].

65. Bodleian Library, Toynbee Papers, letter, Alexander Lindsay to AJT, 25 December 1915.

66. Bodleian Library, Gilbert Murray Papers, Carton 568, sheet 29, letter, Gilbert Murray to Rosalind Toynbee, 18 December 1918.

67. Bodleian Library, Toynbee Papers, letter, Charlotte M. Toynbee to AJT, 20 December 1915.

68. Bodleian Library, Toynbee Papers, letter, Lady Carlisle to AJT, 14 January 1917.
69. Bodleian Library, Toynbee Papers, letter, Rosalind Toynbee to AJT, 10 August 1917.
70. Bodleian Library, Toynbee Papers, letter, Rosalind Toynbee to AJT, 24 January 1918.
71. A few of these essays were republished as *The New Europe: Some Essays in Reconstruction* (London, 1915).
72. S. Fiona Morton, ed., *A Bibliography of Arnold J. Toynbee* (Oxford, 1980) gives references, year by year; but even the editor's devoted labors could not trace many of Toynbee's casual writings, as she points out in the introduction.
73. Bodleian Library, Toynbee Papers, letter, J.A.C. Tilly to AJT, 5 July 1918. This letter, on official stationery, and written on behalf of the Foreign Secretary, Mr. Balfour, was more than a year late, inasmuch as Toynbee's service in the PID began in May 1917. Presumably routine business in the Foreign Office simply fell behind.
74. Bodleian Library, Toynbee Papers, letter, AJT to Gilbert Murray, 1 August 1918.
75. Bodleian Library, Toynbee Papers, letter, AJT to Gilbert Murray, 26 March 1919.
76. Bodleian Library, Toynbee Papers, letter, Rosalind Toynbee to AJT, 9 January 1919.
77. Bodleian Library, Toynbee Papers, letter, Ronald M. Burrows to AJT, 29 May 1919. Richard Clogg, *Politics and the Academy: Arnold Toynbee and the Koraes Chair* (London, 1986) provides a rich background of detail about the establishment of the Chair and Toynbee's later quarrel with the donors.
78. Bodleian Library, Toynbee Papers, letter, AJT to R.S. Darbishire, 21 July 1919.
79. Bodleian Library, Toynbee Papers, letter, Rosalind Toynbee to AJT, 11 January 1919.
80. Bodleian Library, Toynbee Papers, letter, Rosalind Toynbee to AJT, 8 April 1919.
81. Bodleian Library, Toynbee Papers, letter, Lady Carlisle to AJT, 2 May 1919.
82. Bodleian Library, Toynbee Papers, letter, AJT to Lady Carlisle, 3 May 1919.
83. Bodleian Library, Toynbee Papers, 6 May 1919.
84. Bodleian Library, Toynbee Papers, 23 May 1919.
85. Bodleian Library, Toynbee Papers, letter, Rosalind Toynbee to Lady Carlisle, 31 May 1919.
86. Bodleian Library, Toynbee Papers, letter, Rosalind Toynbee to Lady Carlisle, 13 June 1919.
87. Bodleian Library, Toynbee Papers, letter, AJT to Edith Toynbee, 23 March 1919.
88. Bodleian Library, Toynbee Papers, letter, Edith Toynbee to AJT, 24 March 1919.
89. Arnold J. Toynbee, *A Study of History* (London, 1954), X, 139–140.

5. The Koraes Professorship

1. Bodleian Library, Gilbert Murray Papers, Carton 550, sheet 93, letter, Mary Murray to Countess of Carlisle, 25 November 1918.
2. Bodleian Library, Gilbert Murray Papers, Carton 550, sheet 115, letter, Mary Murray to Countess of Carlisle, 4 May 1919.
3. A.J. Toynbee, *The Place of Medieval and Modern Greece in History. Inaugural Lecture of the Koraes Chair of Modern Greek and Byzantine Language, Literature and History* (London, 1919).

4. Ibid., where Gennadius' speech appears as a foreword almost as long as the lecture itself.

5. Some of these lectures are preserved in Toynbee's papers. They show a remarkable grasp of the whole sweep of Greek history and a sensitivity to patterns of human geography that resembles the mastery of Mediterranean landscapes later exhibited by Fernand Braudel.

6. Richard Clogg, *Politics and the Academy: Arnold Toynbee and the Koraes Chair* (London, 1986), pp. 60, 62.

7. Bodleian Library, Toynbee Papers, letter, AJT to R.S. Darbishire, 5 May 1918. The echo of his student essay contrasting "Mechanism" with "Life" is obvious in Toynbee's choice of metaphor. As we shall see, Toynbee's most ambitious and systematic thought remained under the sway of this sort of vague Bergsonianism (mediated and no doubt colored by Lindsay) until after 1920.

8. Arnold J. Toynbee, "My View of History" in *Civilization on Trial* (New York, 1948), pp. 7–8.

9. Bodleian Library, Toynbee Papers, letter, AJT to R.S. Darbishire, 5 May 1918.

10. Bodleian Library, Toynbee Papers, letter, AJT to Edith Toynbee, not dated, but written at Castle Howard, probably in 1918.

11. Bodleian Library, Toynbee Papers, letter, AJT to R.S. Darbishire, 13 April 1919.

12. Arnold Toynbee, *Experiences* (New York and London, 1969), p. 101.

13. Bodleian Library, Toynbee Papers, Juvenilia.

14. Ibid.

15. Ibid.

16. Ibid.

17. A.J. Toynbee, *The Tragedy of Greece. A lecture delivered for the Professor of Greek to Candidates for Honours in Literae Humaniores at Oxford in May 1920* (Oxford, 1921).

18. Bodleian Library, Toynbee Papers, letter, Gilbert Murray to AJT, 15 November 1920.

19. Bodleian Library, Toynbee Papers, undated clipping from *Contemporary Review*.

20. Thucydides, III, 82. Crawley translation.

21. A.J. Toynbee, *The Tragedy of Greece,* pp. 4, 51.

22. (London, 1907).

23. X, 230. Cf. in the same volume, pp. 124–125, where Cornford is quoted approvingly in connection with a discussion of sin and punishment, freedom and necessity in history.

24. In 1952 I had occasion to ask Toynbee about his opinion of Cornford's interpretation of Thucydides. He did not answer directly, preferring to praise Cornford's mind and character in general, while refraining from direct expression of disagreement. This was entirely characteristic, for Toynbee always strove to avoid explicit articulation of his differences with others, whether on theoretical or on practical matters.

25. Arnold J. Toynbee, *Civilization on Trial,* pp. 9–10. Cf. Bodleian Library, Toynbee Papers, letter, AJT to H. Stuart Hughes, 4 June 1951, which, however, adds nothing to the published account.

26. It is not possible to tell whether Toynbee read Spengler before, during, or after starting this essay. It seems likely that he took up Spengler (and Teggart) only after running into difficulties with what he was writing, since the surviving manuscript shows no trace of their ideas.

27. These pages, together with all the other manuscript materials related to *A Study of*

History, are in Nihon University Library, Tokyo. I consulted Xeroxed sheets thanks to the courtesy of the librarian and other authorities of Nihon University, who kindly copied them for my use.

Toynbee gathered the manuscript pages together and gave them the title they bear on the occasion of a special exhibition in his honor held in London in 1971 by the National Book League. Since what became *A Study of History* had not yet been conceived in 1920, the title he chose is rather misleading. I prefer to call it *The Mystery of Man,* echoing Sophocles' phrase.

28. Arnold Toynbee, *Experiences,* p. 101.
29. *Mystery of Man,* p. 3, revised version, Nihon University Library.
30. Ibid., pp. 3–4.
31. Ibid., p. 18.
32. Bodleian Library, Toynbee Papers, article in *Jerusalem Post,* 29 October 1954.
33. Bodleian Library, Toynbee Papers, letter, AJT to Margaret T. Hodgen, 22 June 1965. Toynbee's recollection cannot be entirely correct, for in 1920 Teggart had only two books in print, *Prolegomena to History, The Relations of History to Literature, Philosophy and Science* (1918) and *The Processes of History* (1918). His *Theory of History* (1925) and *Rome and China* (1929) appeared only after Toynbee had made up his mind on key issues for his own great work. Yet in listing Teggart among those who gave him intuitions and ideas for *A Study of History* (X, 232), Toynbee quotes from Teggart's 1925 work. What he got from Teggart in 1920 cannot have been what he picked up after 1925. In retrospect, Toynbee's memory obviously elided things he had learned at different times.
34. Frederick J. Teggart, *The Processes of History* (New Haven, 1917), p. 4.
35. Ibid., p. 37.
36. Ibid., p. 48.
37. Ibid., pp. 111, 151.
38. Bodleian Library, Toynbee Papers, letter, AJT to Gilbert Murray, 9 March 1920.
39. Bodleian Library, Toynbee Papers, text of motion passed by the senate of the University of London, 21 October 1920.
40. Bodleian Library, Toynbee Papers, account sheet, undated, but filed with other papers connected with this trip.
41. Bodleian Library, Toynbee Papers, letter, AJT for family circulation, 4 February 1921.
42. Bodleian Library, Toynbee Papers, letter, AJT to Edith Toynbee, 2 April 1921.
43. Bodleian Library, Toynbee Papers, letter, AJT for family circulation, 7 February 1921.
44. Bodleian Library, Toynbee Papers, letter, AJT for family circulation, 19 February 1921.
45. Arnold J. Toynbee, *The Western Question in Greece and Turkey: A Study in the Contact of Civilisations* (London, 1922), p. 233.
46. Bodleian Library, Toynbee Papers, letter, Rosalind Toynbee to Mary Murray, 28 May 1921.
47. Bodleian Library, Toynbee Papers, letter, Rosalind Toynbee to Mary Murray, 28 May 1921.
48. Bodleian Library, Toynbee Papers, letter, Rosalind Toynbee to Mary Murray, 18 June 1921.
49. Bodleian Library, Toynbee Papers, letter, AJT to Edith Toynbee, 12 June 1921.
50. Bodleian Library, Toynbee Papers, letter, C.P. Scott to AJT, 1 June 1921 [?]. This

letter and its sequel, cited below, are both actually dated 1922, but that seems almost surely an error, since Toynbee's revelations about Greek atrocities came in 1921.

51. Bodleian Library, Toynbee Papers, letter, C.P. Scott to AJT, 8 June 1921 [?].
52. Bodleian Library, Toynbee Papers, letter, Eric Forbes Adams to AJT, 17 January [1922].
53. Bodleian Library, Toynbee Papers, letter, Ernest Barker to AJT, 7 April 1922.
54. Bodleian Library, Toynbee Papers, letter, AJT to Edith Toynbee, 11 May 1921.
55. The manuscript went to press in March 1922, less than six months after his return.
56. Arnold J. Toynbee, *The Western Question*, pp. 267, 362–363.
57. A.J. Toynbee, *The Tragedy of Greece*, p. 4.
58. Arnold J. Toynbee, *Experiences*, p. 101.
59. Nihon University Library kindly provided me with a Xerox copy of this document. Its heading was created in 1971 for display at the National Bank League Exhibition in Toynbee's honor; and the document has nine, not a dozen, main headings, none of which corresponds exactly to any of the thirteen main headings of the mature work.
60. Nihon University Library, Outline Section headed "Institutions as Machines," pp. 5–6, 7.
61. Bodleian Library, Toynbee Papers, letter, Rosalind Toynbee to AJT, 22 August 1921.
62. Bodleian Library, Toynbee Papers, letter, Rosalind Toynbee to AJT, 22 August 1921.
63. Bodleian Library, Gilbert Murray Papers, Carton 568, sheet 61, letter, Gilbert Murray to Rosalind Toynbee, 7 January 1922.
64. Bodleian Library, Gilbert Murray Papers, Carton 568, sheet 65, letter, Gilbert Murray to Rosalind Toynbee, 3 March 1922.
65. Bodleian Library, Toynbee Papers, letter, Rosalind Toynbee to AJT, 18 August 1925.
66. Like other sensitive family matters, nearly everything related to this episode has been removed from Gilbert Murray's papers. Letter, Veronica Toynbee to Mrs. Philip Toynbee, 4 July 1973, in private possession, says something about the af-' fair. Cf. also Christian B. Peper, ed., *An Historian's Conscience. The Correspondence of Arnold J. Toynbee and Columba Cary-Elwes, Monk of Ampleforth* (Boston, 1986), p. 571. Rosalind's attitude presumably reflected anger at her sister's disregard of rules to which she had herself submitted when she married Toynbee. Rosalind's restlessness in the bonds of matrimony remained liverly, as we shall soon see.
67. Bodleian Library, Toynbee Papers, letter, Gilbert Murray to Rosalind Toynbee, 20 August 1922.
68. Bodleian Library, Gilbert Murray Papers, Carton 568, sheet 59, letter, Gilbert Murray to Rosalind Toynbee, 11 February 1921.
69. Bodleian Library, Gilbert Murray Papers, Carton 568, sheet 66, letter, Gilbert Murray to Rosalind Toynbee, 5 April 1923.
70. Bodleian Library, Gilbert Murray Papers, Carton 568, sheet 98, letter, Gilbert Murray to Rosalind Toynbee, 21 November 1926.
71. *The Diary of Virginia Woolf, vol. 1, 1915–1919,* edited by Anne Oliver Bell (London, 1977). Entry for 18 December 1921.
72. Rosalind Murray, *The Happy Tree* (London, 1926), p. 330.

73. Almost thirty years later, Gilbert Murray concluded: "I think The Happy Tree is really the best." Gilbert Murray papers, Carton 568, sheet 183, letter, Gilbert Murray to Rosalind Toynbee, 1 November 1954.

74. Cf. Philip Toynbee, *Pantaloon* (London, 1961), p. 25: "The anatomy of mothers provides for two sons only." Other passages are equally revelatory, for instance, on pp. 97, 232, and, indeed, throughout the poem.

75. Jessica Mitford, *Faces of Philip: A Memoir of Philip Toynbee* (New York, 1984) offers a particular and quite limited view of his character.

76. Philip Toynbee, *Pantaloon* provides the best evidence for these assertsions about family relations in the early 1920s.

77. Ibid., p. 140.

78. In 1924 Toynbee's income from journalism was about £220 a year, more than a third of his university salary. Cf. Bodleian Library, Toynbee Papers, letter, AJT to Gilbert Murray, 3 February 1924.

79. Bodleian Library, Toynbee Papers, letter, AJT to Marry Murray, 31 October 1923.

80. Bodleian Library, Toynbee Papers, letter, AJT to Mary Murray, 16 October 1919.

81. Bodleian Library, Toynbee Papers, letter, Jocelyn Toynbee to AJT, 17 January 1923.

82. Oral information from Margaret Toynbee, 15 February 1986.

83. Bodleian Library, Toynbee Papers, letter, AJT to Mary Murray, 15 January 1926.

84. Clogg, *Politics and the Academy*, pp. 65–67.

85. Bodleian Library, Toynbee Papers, letter, AJT to Ernest Barker, 6 May 1920.

86. Bodleian Library, Toynbee Papers, letter, C.P. Scott to AJT, 16 August 1922.

87. Bodleian Library, Toynbee Papers, letter, AJT to Rosalind Toynbee, 13 April 1923.

88. Bodleian Library, Toynbee Papers, letter, Eric Forbes Adams to AJT, 11 September [1923]. Curzon was Foreign Secretary at the time.

89. Clogg, *Politics and the Academy*, p. 74.

90. Sir Cooper Perry, Vice Chancellor of the University of London, was emphatic in attacking Toynbee in private correspondence with the Subscribers' Committee but urged the committee to eschew public confrontation since it is "much more dignified to treat Toynbee with the contempt he deserves" by disregarding him. Gennadion Library, Athens, Ioannes Gennadius Papers, File 11, letter, Sir Cooper Perry to Ioannes Gennadius, 13 February 1924. Other letters in this file show that Ernest Barker, too, backed away from his initial endorsement of Toynbee's behavior, at least in dealings with the Greeks. See, for example, N. Eumorfopoulos to I. Gennadius, 25 November 1923, reporting a conversation with Barker about how to redefine the terms of the Koraes Professorship.

91. Bodleian Library, Toynbee Papers, letter, R.W. Seton-Watson to AJT, 4 January 1924.

92. Bodleian Library, Toynbee Papers, draft dated 5 February 1924.

93. Bodleian Library, Toynbee Papers, letter, Mary Murray to AJT, 31 October 1923.

94. Bodleian Library, Toynbee Papers, letter, AJT to Mary Murray, 1 November 1923.

95. Bodleian Library, Toynbee Papers, letter, AJT to Mary Murray, 20 January 1925.

96. "Don't talk of America yet. . . . I don't want Yanks as grandsons." Bodleian Library, Toynbee Papers, letter, Edith Toynbee to AJT, 14 June 1924.

6. Chatham House and Ganthorpe

1. Bodleian Library, Toynbee Papers, letter, J.W. Headlam-Morley to AJT, 16 January 1924.

2. Bodleian Library, Toynbee Papers, letter, J.W. Headlam-Morley to AJT, 21 November 1922, expressed gratitude on behalf of himself and Lord Curzon.

3. So named from the fact that it had been the private dwelling of William Pitt, Earl of Chatham, when he was Prime Minister. Gladstone had also lived there.

4. Cf. Stephen King-Hall, *Chatham House: A Brief Account of the Origins, Purposes and Methods of the Royal Institute of International Affairs* (London, 1937), pp. 11–22. This book was prepared as part of a fund drive in the late 1930s and is correspondingly celebratory in tone, but it remains the only published account of the institute's early development.

5. Toynbee contributed a substantial essay to volume 6, entitled "The Non-Arab Territories of the Ottoman Empire since the Armistice of the 30th October, 1918."

6. Bodleian Library, Toynbee Papers, letter, AJT to Gilbert Murray, 3 February 1924.

7. Bodleian Library, Toynbee Papers, letters, AJT to J.W. Headlam-Morley, 2 February 1924, and J.W. Headlam-Morley to AJT, 12 February 1924. The institute also guaranteed £300 for secretarial assistance.

8. Bodleian Library, Toynbee Papers, letter, AJT to Gilbert Murray, 15 February 1924.

9. Bodleian Library, Toynbee Papers, Memorandum from the Publications Sub-Committee of the B.I.I.A., dated 26 February 1924.

10. Bodleian Library, Toynbee Papers, letter, AJT to Veronica Boulter.

11. *The Times*, 20 July 1925.

12. *Survey of International Affairs, 1920–1923*, p. viii.

13. Bodleian Library, Toynbee Papers, letter, AJT to Veronica Boulter, 11 December 1924.

14. Aside from her novels, the best evidence of this is the fact that Rosalind was almost surely responsible for renaming Toynbee's philosophy of history "The Nonsense Book." It was referred to by this name from the mid-1920s within the bosom of the family, and Toynbee himself often used the phrase. Cf. the testimony of Lawrence Toynbee: "My parents were not . . . 'well suited.' . . . She tended to belittle what he did—to make me feel he was slightly ridiculous. . . . I deeply regret that I was in a way prevented from appreciating him sooner." Christian B. Peper, ed., *An Historian's Conscience: The Correspondence between Arnold J. Toynbee and Columba Carey-Elwes, Monk of Ampleforth* (Boston, 1986), p. xvii.

15. Bodleian Library, Toynbee Papers, letters, AJT to Veronica Boulter, 11 August 1926 and 4 September 1926.

16. *Survey of International Affairs, 1920–1923*, pp. v–vi.

17. Bodleian Library, Toynbee Papers, letter, AJT to Gilbert Murray, 4 December 1924.

18. Bodleian Library, Toynbee Papers, letter, AJT to Mary Murray, 22 January 1925.

19. Bodleian Library, Toynbee Papers, letter, AJT to Gilbert Murray, 9 March 1925.

20. Bodleian Library, Toynbee Papers.

21. Bodleian Library, Toynbee Papers, letter, AJT to Gilbert Murray, 31 March 1925.

22. Bodleian Library, Toynbee Papers, letter, AJT to Gilbert Murray, 8 April 1925.

23. Where, of course, his clash with the donors of the Koraes Chair had provoked sharp controversy just eighteen months before.

24. Bodleian Library, Toynbee Papers, letter, AJT to Edith Toynbee, 12 August 1925.

25. Bodleian Library, Toynbee Papers, letter, AJT to Gilbert Murray, 12 August 1925.

26. Bodleian Library, Toynbee Papers, letter, AJT to Veronica Boulter, 10 September 1925.

27. Her best novel, *The Happy Tree* (London, 1926), was published by Chatto and Windus with some kind of financial guarantee from Toynbee in an edition of 600 copies, but only after it had been rejected several times by other publishers. Bod-

leian Library, Toynbee Papers, letter, AJT to Gilbert Murray, 14 April 1926, describes the deal Toynbee negotiated on Rosalind's behalf, though without explaining the extent of the financial guarantee he gave the publisher. Reviews were disappointing. Cf. Bodleian Library, Toynbee Papers, letter, AJT to Gilbert Murray, 24 October 1926.

28. Bodleian Library, Toynbee Papers, letter, Rosalind Toynbee to AJT, 20 July 1925.
29. Bodleian Library, Toynbee Papers, letter, Rosalind Toynbee to AJT, 22 August 1925.
30. Bodleian Library, Toynbee Papers, letter, Rosalind Toynbee to AJT, 28 August 1925. Boothby was another Howard–Murray establishment, located in Cumberland. Rosalind planned to go there after wearing out her welcome at Castle Howard, where, as we saw in Chapter 4, her thinly disguised claim to succeed her grandmother as mistress of the whole establishment made relations with her Uncle Geoffry's wife, who actually held that position, uneasy at best.
31. Bodleian Library, Toynbee Papers, letter, Rosalind Toynbee to AJT, 9 June 1925.
32. Bodleian Library, Toynbee Papers, letter, Rosalind Toynbee to AJT, 26 September 1925.
33. Bodleian Library, Toynbee Papers, letter, Rosalind Toynbee to AJT, 3 August 1925.
34. Bodleian Library, Toynbee Papers, letter, Rosalind Toynbee to AJT, 20 July 1925.
35. Bodleian Library, Toynbee Papers, fragments of inaugural lecture, Stevenson Chair of International History. I have not seen the full text of this inaugural lecture, which is at Nihon University Library, Tokyo. With some emendation, it later became the opening pages of *A Study of History*.
36. Bodleian Library, Toynbee Papers, memorandum on the Stevenson Chair, 23 June 1953. Toynbee wrote this memorandum in response to accusations that he had failed to fulfill the duties of a Director of Studies properly, preferring to write books of his own and leaving the administration of Institute research projects to Miss Anne Cleeve, Secretary to the Research Board, while scamping his duties at the University of London entirely. The question arose when his retirement neared and the duties of a successor to the Stevenson Chair had to be defined.
37. Rushton Coulborn (d. 1968) was one such student who, in 1927 or 1928, became a research assistant for Toynbee, helping him with preparations for the *Study*. Coulborn worked in the Toynbee home and found lunchtime bickering between Toynbee and Rosalind quite embarrassing, as he told his wife long afterward. I owe this information to a letter from Dr. Imogen Seger-Coulborn to W.H. McNeill, dated 15 February 1987.
38. Bodleian Library, Toynbee Papers, memorandum on the Stevenson Chair, 23 June 1953.
39. Bodleian Library, Toynbee Papers, letter, AJT to Veronica Boulter, 24 July 1928.
40. Bodleian Library, Toynbee Papers, letter, AJT to Gilbert Murray, 9 August 1928.
41. This, with all the other materials pertaining to *A Study of History*, is in Nihon University Library, Tokyo. I have not been able to see it, but have examined a copy of a twenty-one page table of contents. This shows that all the main headings of the actual work were in place by 1929, though, as we shall see, Toynbee moved some of the subject matter from one part of the book to another when it came time to write it out.
42. The *Economist* paid Toynbee £200 a year. The new contract prescribed that the institute would withhold the same sum from Toynbee's nominal salary and allow Toynbee to use that amount to pay for any outside help he might need in writing

the surveys. In practice, therefore, his salary became £1000, supplemented by a travel allowance as before, and by the sum of £200 which he could use at will to hire others to write parts of the survey volumes.

43. Memoranda by AJT, dated 23 June 1927 and 28 December 1928, defining his wishes and what he agreed to, and addressed to Sir Daniel Stevenson, the University of London, and the Royal Institute, are preserved in Bodleian Library, Toynbee Papers, together with a calculation he prepared for himself in 1927 about how to budget more time for his great book by diminishing his hours on the survey and hiring extra assistance.

44. He did collaborate in a way with Kenneth Kirkwood, a former teacher at Smyrna, on a book for Ernest Benn's series *Nations of the Modern World* entitled *Turkey* (1926). Toynbee undertook this job while still Koraes Professor as a way of making extra money, but he gave it up when he started working for the institute. The publisher then asked Kirkwood to complete the manuscript. As he explained to Toynbee, "As team mates we are unpractised and rather unequal in competancy. . . . We needs must drive in tandem, you in lead, I following. All I hope is that I may pull my weight." Bodleian Library, Toynbee Papers, letter, Kenneth Kirkwood to AJT, 30 January 1926. The two did not actually meet, since Kirkwood after leaving Smyrna got a job in Canada. Their collaboration therefore merely required Toynbee to read and make a few suggestions for changes in what Kirkwood wrote.

45. So did others. At the ceremony christening the Institute Royal, on 31 May 1926, the Prince of Wales declared: "The history we most need for practical politics is the history of the last few years. For public men who have always to work against time, it is most difficult to frame accurate and connected accounts of recent events. This institute is now giving them that history through the medium of the annual volumes of 'The Survey of International Affairs.' " *The Times,* 1 June 1926.

46. Everything else for 1925 was entrusted to C.A. Macartney and others, *Survey of International Affairs, 1925* (London, 1928). Veronica Boulter wrote the section of this volume dealing with southeastern Europe and also compiled a *Chronology of International Events and Treaties, 1 January 1920–31 December 1924,* which was published as a separate volume. This marked the first time she allowed her name to be publicly associated with the surveys.

47. *Survey of International Affairs 1925* (London, 1927), p. 1.

48. "If you looked in the right places, you could doubtless find some old fashioned Islamic Fundamentalists still lingering on," he wrote in 1929. "You would also find that their influence was negligible." *A Journey to China; Or, Things which are Seen* (London, 1931), p. 117.

49. Bodleian Library, Toynbee Papers, letter, AJT to Veronica Boutler, 20 September 1927.

50. Bodleian Library, Toynbee Papers, letter, AJT to Veronica Boulter, 24 June 1931.

51. Bodleian Library, Toynbee Papers, letter, AJT to Veronica Boulter, 26 June 1931.

52. Bodleian Library, Toynbee Papers, letter, AJT to Veronica Boulter, 2 July 1931.

53. Bodleian Library, Toynbee Papers, letter, Rosalind Toynbee to AJT, 31 July 1927. Rosalind admitted in this letter: "It is rather depressing to find how exhausting their [the two elder boys'] arrival is."

54. Bodleian Library, Toynbee Papers, letter, AJT to Gilbert Murray, 22 June 1928. One may guess that Toynbee greatly wished that his son had gained a scholarship so as not to have to depend on Lady Mary's generosity.

55. Bodleian Library, Toynbee Papers, letter, AJT to R.S. Darbishire, 21 October 1928.
56. Bodleian Library, Toynbee Papers, letter, Rosalind Toynbee to AJT, 19 January 1928.
57. Bodleian Library, Toynbee Papers, letter, AJT to Mary Murray, 2 May 1929. In practice, Toynbee's driving was clumsy and erratic and he gladly let Rosalind take the wheel throughout the trip—and subsequently. He never really learned to drive, relying on others to chauffeur him about.
58. Bodleian Library, Toynbee Papers, letter, to Rosalind Toynbee, with signature page missing, dated Sofia, 16 September 1928.
59. Bodleian Library, Toynbee Papers, informal notes of the trip, probably prepared by Rosalind but unsigned. Cf. A.J. Toynbee, *A Journey to China*, pp. 27–31. This book consists of collected articles he sent back to various newspapers and weekly journals during his trip.
60. Bodleian Library, Toynbee Papers, letter, Rosalind Toynbee to AJT, 27 September 1929.
61. Bodleian Library, Toynbee Papers, letter, Rosalind Toynbee to AJT, 27 September 1929, written en route from Chalons-sur-Marne.
62. Bodleian Library, Toynbee Papers, letter, Rosalind Toynbee to AJT, 10 November 1929.
63. Bodleian Library, Toynbee Papers, letter, Rosalind Toynbee to AJT, 10 November 1929.
64. Bodleian Library, Toynbee Papers, letter, Rosalind Toynbee to AJT, 6 December 1929.
65. Bodleian Library, Toynbee Papers, letter, AJT to Edith Toynbee, 24 July 1929.
66. Bodleian Library, Toynbee Papers, letter, AJT to R.S. Darbishire, 26 May 1919. Maida Vale is a district of London.
67. Bodleian Library, Toynbee Papers, memorandum, dated 22 June 1929.
68. Bodleian Library, Toynbee Papers, letter, AJT to Veronica Boulter, 6 August 1929.
69. *A Journey to China*, pp. 220–226.
70. Bodleian Library, Toynbee Papers, letter, AJT to Veronica Boulter, 11 September 1929.
71. *A Journey to China*, p. 250.
72. Bodleian Library, Toynbee Papers, letter, AJT to Veronica Boulter, 14 November 1929.
73. Cf. *A Journey to China*, pp. 116–119, 148, 256 for explicit statements that India, China, and the whole Islamic world were "going West" like Turkey.
74. Bodleian Library, Toynbee Papers, letter, AJT to Gilbert Murray, 22 October 1929.
75. Bodleian Library, Toynbee Papers, letter, AJT to Gilbert Murray, 10 November 1929.
76. Cf. *A Journey to China*, p. 289: "Was it possible that the Japanese might be successfully solving the problem of how to secure the material advantages of industrialism without suffering the spiritual impoverishment which the West suffered from it?"
77. Eileen Power (1889–1940) was an accomplished medievalist. Her most famous book, *Medieval People* (London, 1924), is a pioneer example of social history.
78. John K. Fairbank, later professor at Harvard, met Eileen Power in China in 1929 and remembered her as "a remarkable phenomenon. A professor at London, she had a Meryl Streep kind of startling good looks, except perhaps a firmer jaw. But her primary characteristic was that she had an LSE mind (I mean this as high praise in the 1920s). She was incisive and intellectually well organized and rather made a

point of being well dressed.'' Letter, John K. Fairbank to W.H. McNeill, 17 January 1986.

79. Exact time and place is unclear, but probably this crisis boiled up in Shanghai between 29 December 1929 and 2 January 1930.

80. On 2–3 January 1930 according to the itinerary published in the last pages of *A Journey to China*.

81. Bodleian Library, Toynbee Papers, letter, AJT to Eileen Power, no date. This is presumably a draft of the letter actually sent. It is in pencil, written on cheap paper. Why Toynbee saved it is hard to imagine, unless he felt that even his clumsy peccadilloes deserved to be fully recorded.

82. Bodleian Library, Toynbee Papers, letter, Eileen Power to AJT, 23 February 1930. The letter was written from New York, where she was teaching at Barnard College for a term.

83. Translated for me by Elizabeth Meyer of Yale University.

84. He had probably broken off marital relations with his wife sometime before Lawrence's birth in 1922. The unexpected way in which Philip had been conceived in 1915 meant that Toynbee could only trust total abstinence to prevent unwanted births, and the financial and personal strains of family life undoubtedly made them both wish to call a halt to Rosalind's repeated pregnancies.

85. The phrase comes from his book *Experiences* (New York and London, 1969), p. 127.

86. Acts 17:23.

87. *Experiences,* p. 176.

88. Toynbee had been in that city on 17–18 November 1929, according to the itinerary published in the last pages of *A Journey to China*.

89. Bodleian Library, Toynbee Papers, letter, Rosalind Toynbee to AJT, 11 December 1929. I have altered punctuation in Rosalind's letters, for she used dashes so liberally that, as written, each paragraph consists of one long sentence.

90. Toynbee's delayed return resulted from the fact that the ordinary route from China across Siberia was blocked by political conflict between the Soviet Union and China over control of the Chinese Eastern Railway. Instead of traveling through Manchuria, therefore, Toynbee had to return to Japan, take a ship from Kobe to Vladivostok, and travel from there to Moscow, Berlin, and so home.

91. Bodleian Library, Toynbee Papers, letter, Rosalind Toynbee to AJT, 14 December 1929.

92. *A Journey to China,* pp. 299–318.

93. Bodleian Library, Toynbee Papers, letter, Rosalind Toynbee to AJT, 10 April 1930. "Last Christmas" was, of course, when Toynbee's abortive affair with Eileen Power reached its climax.

94. Bodleian Library, Toynbee Papers, letter, AJT to Mary Murray, 15 April 1930.

7. Triumph and Defeat

1. The exact date is obscure. In her autobiographical account, published by John A. O'Brien, ed., *The Road to Damascus,* 2 vols. (London, 1949), I, 143–146, she says she accepted the Roman Catholic faith in 1932; but if so, she must have kept it secret until 1933, to judge by letters from her family, quoted below. Margaret Toynbee remembers the date as 1933.

2. Bodleian Library, Toynbee Papers, letter, AJT to Veronica Boulter, 5 August 1931.

3. Bodleian Library, Toynbee Papers, letter, Rosalind Toynbee to AJT, 31 December 1931.
4. "How long it will take you to finish your Great Work is something that you probably will find out only when you get going. (I reckoned to write the whole of mine in two summer long vacations after I had got it in the shape of notes, and it has taken me twenty-five years.)" Bodleian Library, Toynbee Papers, letter, AJT to William H. McNeill, 18 June 1954.
5. Bodleian Library, Toynbee Papers, letter, AJT to Veronica Boulter, 24 July 1930.
6. Bodleian Library, Toynbee Papers, letters, AJT to Veronica Boulter, 11 August 1930, 28 August 1930, and 13 September 1930.
7. Bodleian Library, Toynbee Papers, letter, AJT to Veronica Boulter, 25 October 1930. The Book of Genesis was, of course, the section of *A Study of History* titled "II The Geneses of Civilizations," which, with its annexes and an introductory part I, comprises the first volume of the published work.
8. Bodleian Library, Toynbee Papers, letter, AJT to Veronica Boulter, Good Friday, 1931.
9. Veronica Toynbee, typescript, "Relations between Arnold and Veronica Toynbee, 1924–1975."
10. Ibid.
11. Bodleian Library, Toynbee Papers, letter, AJT to Gilbert Murray, 17 July 1931.
12. Royal Institute of International Affairs, Archives, 4 Toynbee Carton 1, letter, AJT to Anne Cleeve, 18 July 1931.
13. Arnold J. Toynbee, assisted by V.M. Boulter, *Survey of International Affairs, 1931* (London, 1932), pp. 1, 7, 14, 25. Toynbee had in mind French and American tariff policy in accusing them of being mainly responsible for the depression. Quincy Wright, an American professor of international relations, wrote to congratulate Toynbee "on the way in which you have converted the events of this 'annus terribilis' into a unity." Bodleian Library, Toynbee Papers, letter, Quincy Wright to AJT, 29 December 1932.
14. *Survey of International Affairs, 1931* (London, 1932), p. 432.
15. Bodleian Library, Toynbee Papers, letter, AJT to Gilbert Murray, 16 February 1932.
16. Bodleian Library, Toynbee Papers, letter, AJT to Gilbert Murray, 18 February 1932.
17. Bodleian Library, Toynbee Papers, letter, AJT to Gilbert Murray, 22 February 1932.
18. Bodleian Library, Toynbee Papers, letter, AJT to Gilbert Murray, 7 July 1932.
19. See, for instance, *Survey of International Affairs, 1933* (London, 1934), pp. 484–518.
20. Bodleian Library, Toynbee Papers, letter, AJT to Veronica Boulter, 3 January 1933.
21. Arnold J. Toynbee, ed., *British Commonwealth Relations* (London, 1934).
22. Bodleian Library, Toynbee Papers, letter, AJT to G.M. Gathorne-Hardy, 1 January 1934.
23. Bodleian Library, Toynbee Papers, letter, AJT to Veronica Boulter, 28 October 1933.
24. Bodleian Library, Toynbee Papers, letter, AJT to Veronica Boulter, 20 October 1933.
25. Bodleian Library, Toynbee Papers, letter, AJT to Veronica Boulter, 4 November 1933.
26. Bodleian Library, Toynbee Papers, letter, AJT to Veronica Boulter, 28 November 1933.
27. Bodleian Library, Toynbee Papers, letter, Lawrence Toynbee to AJT, 14 November 1933.

28. Bodleian Library, Toynbee Papers, letter, AJT to Veronica Boulter, 3 December 1933.
29. Bodleian Library, Toynbee Papers, letter, AJT to Veronica Boulter, 26 September 1933.
30. Bodleian Library, Toynbee Papers, letter, Rosalind Toynbee to AJT, 3 November 1933.
31. Bodleian Library, Toynbee Papers, letter, Rosalind Toynbee to AJT, 21 November 1933.
32. Bodleian Library, Toynbee Papers, letter, Rosalind Toynbee to AJT, 28 November 1933.
33. O'Brien, *The Road to Damascus,* I, 143–146. This book assembled autobiographical accounts of the conversion of a number of distinguished British and American literary figures, including Evelyn Waugh, Claire Booth Luce, and others less famous. Among them was Toynbee's elder sister, Jocelyn, who became Catholic in 1929 when, like Cardinal Newman, she concluded that Anglican ordination was canonically invalid. Jocelyn's account, II, 195–201, is severely impersonal, in striking contrast to the story Rosalind chose to tell.
34. Gilbert Murray, *Four Stages of Greek Religion* (New York, 1912). Chapter 3 is so titled.
35. Bodleian Library, Toynbee Papers, letter, Rosalind Toynbee to AJT, 25 March 1929.
36. Bodleian Library, Toynbee Papers, letter, AJT to Mary Murray, 18 April 1930.
37. O'Brien, *The Road to Damascus,* I, 148.
38. Rosalind Murray, *The Good Pagan's Failure* (London, 1939), p. 141 and passim.
39. A generational change at Castle Howard also altered Rosalind's relation to that ancestral pile. In 1932 Geoffry's wife died suddenly and unexpectedly, bringing her long friction with Rosalind to an end. Gilbert Murray commented to his wife: "Looked in at Castle Howard this afternoon. It is curious and very pathetic to see the effect of the poor mistress's absence. The children gayer and more talkative, the guests more at their ease—especially of course Rosalind who gets on famously with Geoffry." Bodleian Library, Gilbert Murray Papers, Carton 472, letter, Gilbert Murray to Mary Murray, 17 September 1932. In 1935 Geoffry died, and the new heir, a boy, had no affinity for Rosalind, so her franchise at Castle Howard was restricted to a mere three years.
40. Philip Toynbee, *Part of a Journey: An Autobiographical Journal, 1977–1979* (London, 1981), p. 9.
41. In addition to *The Good Pagan's Failure,* Rosalind published four tracts: *Time and the Timeless* (London, 1942); *The Life of Faith* (London, 1943); *The Forsaken Fountain* (London, 1948); and *The Further Journey: In My End is My Beginning* (London, 1953). Her career as an apologist resembled her career as a novelist in the sense that *The Good Pagan's Failure* attracted much more attention than what followed.
42. Bodleian Library, Toynbee Papers, letter, Marry Murray to AJT, 3 August 1933.
43. Bodleian Library, Toynbee Papers, letter, AJT to Mary Murray, 8 August 1933.
44. Bodleian Library, Toynbee Papers, letter, AJT to Mary Murray, 22 September 1930.
45. Bodleian Library, Toynbee Papers, letter, Rosalind Toynbee to AJT, 12 August 1932.
46. Bodleian Library, Toynbee Papers, letter, AJT to Edith Toynbee, 30 November 1932.
47. Bodleian Library, Toynbee Papers, letter, Rosalind Toynbee to AJT, 8 September

1933. Cf. Philip's report of his first "puppy love" in Philip Toynbee, *Friends Apart: A Memoir of Esmond Romilly and Jasper Ridley in the Thirties* (London, 1954), pp. 37, 46. The Bonham Carters were a prominent Liberal family.

48. Bodleian Library, Toynbee Papers, letter, Edith Toynbee to AJT, 27 September 1933.

49. Bodleian Library, Toynbee Papers, letter, AJT to Edith Toynbee, 9 October 1933.

50. Bodleian Library, Toynbee Papers, letter, Rosalind Toynbee to AJT, 6 September 1933.

51. Bodleian Library, Toynbee Papers, letter, Rosalind Toynbee to AJT, 27 September 1933.

52. Bodleian Library, Toynbee Papers, letter, Rosalind Toynbee to AJT, 28 October 1933.

53. Bodleian Library, Toynbee Papers, letter, Rosalind Toynbee to AJT, 3 November 1933.

54. Bodleian Library, Toynbee Papers, letter, Philip Toynbee to Rosalind Toynbee, 8 October 1933.

55. Bodleian Library, Toynbee Papers, letter, Antony Toynbee to Rosalind Toynbee, no date.

56. Bodleian Library, Toynbee Papers, letter, Rosalind Toynbee to AJT, 13 September 1933.

57. Bodleian Library, Toynbee Papers, letter, AJT to Veronica Boulter, 24 July 1930.

58. *A Study of History,* II, 361.

59. There was, of course, a sense in which the Arab conquerors did indeed profit from the longstanding cultural divergence between Hellenized upper classes and semitic-speaking lower classes of Syria and Mesopotamia. But the new Moslem civilization that swiftly took shape in the wake of the initial conquest added Greek and Persian elements to the original Arab–Moslem core, and it seems entirely unreasonable to dub that amalgam "Syriac" or equate it with the cultural landscape of Achaemenid times, as Toynbee does.

60. Bodleian Library, Toynbee Papers, letter, AJT to Humphrey Milford, 31 December 1931. This is the first time this modest camouflage title appears in Toynbee's papers, and may have been invented for the occasion.

61. Bodleian Library, Toynbee Papers, letter, Humphrey Milford to AJT, 28 January 1932.

62. "I have been remarkably lucky to have got Milford to believe in the book, for I think this kind of thing is still very much out of fashion." Bodleian Library, Toynbee Papers, letter, AJT to Veronica Boulter, 21 August 1933.

63. On 18 December 1931, before Toynbee even approached Humphrey Milford, E.V. Rieu invited Toynbee to submit his "Philosophy of History" to Methuen; when he learned that Oxford would have it, he wrote: "I won't pretend that I am not disappointed that we are not to publish so important a work as yours, in my opinion, will prove." Bodleian Library, Toynbee Papers, letter, E.V. Rieu to AJT, 1 March 1932.

64. Bodleian Library, Toynbee Papers, letter, AJT to Veronica Boulter, 28 August 1930.

65. Bodleian Library, Toynbee Papers.

66. Bodleian Library, Toynbee Papers, letter, AJT to Gilbert Murray, 27 December 1930. Toynbee did indeed send subsequent batches of his book to Murray for criticism; two years later, Murray told his wife, "I am reading the nonsense book all the time. It is very brilliant and amazingly learned." Bodleian Library, Gilbert

Murray Papers, Carton 472, letter, Gilbert Murray to Mary Murray, 11 September 1932.

67. Toynbee told Humphrey Milford in December 1931 that Zimmern, Hammond, Baynes, and Paton had read the manuscript; the rest on my list are attested independently in his papers, but it is probable that others were also consulted. Moreover, the roster of critics consulted ahead of time for the second set of volumes expanded substantially, with the result that Toynbee engaged nearly all of the leading historians of Britain in connection with one or another segment of the work.

With very few exceptions, their responses are not among the Toynbee papers in the Bodleian Library, having presumably accompanied the manuscript of *A Study of History*, along with Toynbee's notes and other miscellaneous papers connected with the work to Nihon University Library, Japan. I was unable to consult these papers, since the authorities at Nihon University Library will not open the collection to the public until they have constructed a suitable index in Japanese for the entire collection. Nevertheless, it is clear that by the time the first volumes of *A Study of History* came out, English literary and historical circles were primed and prepared to admire his work.

68. Arnold J. Toynbee, *Civilization on Trial* (New York, 1948), p. 10.

69. *Times Literary Supplement*, 4 October 1934; *New Statesman and Nation*, 18 August 1934; *Manchester Guardian*, 26 June 1934; *The Observer*, 24 June 1934. Toynbee wrote Murray in acknowledgment: "You were the first person to read the first batch of the MS and to give me some confidence that it wasn't nonsense after all. You may not realise how much that encouraged me in going on; so I have two reviews to thank you for really: the private one of a year or two ago, as well as the public one yesterday." Bodleian Library, Toynbee Papers, letter, AJT to Gilbert Murray, 24 June 1934.

70. *The Spectator*, 6 July 1934. Privately, Toynbee's old mentor, A.D. Lindsay, was much sharper, protesting: "A person of your pretentions to scholarship really ought not to write such nonsense as you do about Calvinism." Bodleian Library, Toynbee Papers, letter, A.D. Lindsay to AJT, 23 July [1934].

71. Bodleian Library, Toynbee Papers, letter, AJT to Veronica Boulter, 14 January 1935: "Royalties from the nonsense book down to the end of December came to £333! Will you let me celebrate my three quarters of it in a way I should specially like: I mean by making you a present of the trip . . . whatever your plans may be?"

72. "P 4, line 1: It should be 'Das' not 'Die' Römische Staatsrecht," she told him. Bodleian Library, Toynbee Papers, letter, Edith Toynbee to AJT, 6 November 1934. And in another letter: "I have such faith in your wonderful memory that I was very chary of thinking I had found a tiny slip in the book." But she had, and rather gleefully pointed out how Toynbee had confused Rowland Hunt, Rowland Hill, and one Henry Fawcett, who, though blinded in his youth, became postmaster general in 1880. Bodleian Library, Toynbee Papers, letter, Edith Toynbee to AJT, Michaelmas Day, 1934.

73. Bodleian Library, Toynbee Papers, letter, Alfred Zimmern to AJT, 29 June 1934.

74. Bodleian Library, Toynbee Papers, letter, D.C. Somervell to AJT, 11 September 1934.

75. S. Fiona Morton, ed., *A Bibliography of Arnold J. Toynbee* (Oxford, 1980), pp. 4–5 lists a total of twenty-nine reviews of the first three volumes, of which thirteen appeared in Britain, ten in the United States and the rest in professional journals of Germany, France, Spain, Norway, and Japan.

76. Toynbee used the astronomical metaphor to describe the difference between Gilbert Murray's view of "the Jerusalem–Athens–Rome centered view of history" and his own not long after World War II. Bodleian Library, Toynbee Papers, letter, AJT to Gilbert Murray, 21 November 1953.

77. Toynbee must have remained unaware of Herder's great book, *Ideen zur Philosophie der Geschichte der Menschheit* (1784–1791). At any rate, Herder's name does not appear in the magnificent index to *A Study of History,* despite the many points on which Toynbee might have been expected to recognize anticipations of his own outlook in what Herder has to say.

78. Oswald Spengler was rather more oracular than Toynbee in tone, but his range was more restricted, since he dealt only with a few civilizations of Eurasia.

79. This grant, made in 1932, was earmarked for Chatham House researches and not for Toynbee's personal work per se. He prepared a long list of possible topics, but in practice the funds seem to have been mostly used to support his own work. Cf. Bodleian Library, Toynbee Papers, "Plan for Chatham House Research" by AJT, dated February 1932.

80. Bodleian Library, Toynbee Papers, letter, AJT to Gilbert Murray, 21 October 1935.

81. Bodleian Library, Toynbee Papers, letter, AJT to Veronica Boulter, 21 August 1936.

82. Bodleian Library, Toynbee Papers, letter, AJT to Veronica Boulter, 9 September 1935.

83. Bodleian Library, Toynbee Papers, letter, AJT to Veronica Boulter, 15 September 1935.

84. Bodleian Library, Toynbee Papers, letter, AJT to Veronica Boulter, 21 October 1936.

85. Bodleian Library, Toynbee Papers, letter, AJT to Veronica Boulter, 5 November 1936.

86. Bodleian Library, Toynbee Papers, letter, AJT to Veronica Boulter, 13 January 1937.

87. Bodleian Library, Toynbee Papers, letter, AJT to Veronica Boulter, 31 December 1937.

88. Bodleian Library, Toynbee Papers, letter, AJT to Edith Toynbee, 12 June 1934. Oswald Mosely was the leading British fascist and Philip managed to get himself roughed up at the meeting. Cf. Philip Toynbee, *Friends Apart* for a perceptive account of his own behavior and that of Esmond Romilly, who led where Philip followed in assaulting the English public school system and proclaiming the Communist cause.

Toynbee's letter to his mother was occasioned by the fact that he had been scheduled to accompany her on a visit to his father. He explained, "I can't unfortunately go to Northampton because I must keep my hands free till I have made some arrangement for [Philip]." This must have been a cruel blow to his mother, whose solitary visits to her husband were faithful, painful, and almost never supported by the company of her son.

89. Bodleian Library, Toynbee Papers, letter, AJT to Veronica Boulter, 4 September 1934. For a sympathetic account of Philip's youthful antics, see Jessica Mitford, *Faces of Philip: A Memoir of Philip Toynbee* (New York, 1984), pp. 11–38.

90. Philip Toynbee, "Arnold Toynbee—A Eulogy," *PHP: A Forum for a Better World* (May 1976), 6–8. *PHP* is a magazine published in English in Japan.

91. Private possession, letter, Rosalind Toynbee to Mary Murray, 4 August 1938. As

before, I have substituted my guesses at appropriate punctuation for the dashes on which Rosalind almost exclusively relied.

92. Bodleian Library, Toynbee Papers, letter, AJT to Gilbert Murray, 21 October 1935.
93. Bodleian Library, Toynbee Papers, letter, AJT to Veronica Boulter, 9 October 1935.
94. Royal Institute of International Affairs, Archives, 4 Toynbee Carton 1, letter, AJT to Ivison Macadam, 15 September 1935.
95. Bodleian Library, Toynbee Papers, letter, AJT to Gilbert Murray, 23 December 1935.
96. Bodleian Library, Toynbee Papers, letter, AJT to Veronica Boulter, 17 April 1936.
97. Bodleian Library, Toynbee Papers, letter, AJT to Grant Robertson, 29 January 1937.
98. Bodleian Library, Toynbee Papers, letter, AJT to Veronica Boulter, 17 July 1936.
99. In this instance, Toynbee modified the annual framework of the surveys by extending the narrative of the war and its diplomacy into 1936.
100. Royal Institute of International Affairs, Archives, 4 Toynbee, unsigned memorandum addressed to Toynbee, dated 17 August 1936.
101. Bodleian Library, Toynbee Papers, letter, AJT to Veronica Boulter, 24 December 1936.
102. Bodleian Library, Toynbee Papers, letter, AJT to Veronica Boulter, 8 September 1938.
103. Some six weeks after Hitler came to power, Toynbee proposed to Gilbert Murray that the League of Nations Union in Great Britain, of which his father-in-law was the leading spirit, ought to propose German–Polish boundary revisions, but Murray refused. Bodleian Library, Toynbee Papers, letter, AJT to Gilbert Murray, 20 February 1933.
104. Bodleian Library, Toynbee Papers, letter, AJT to Lord Lothian, 21 June 1934. It was on this occasion that Philip's escapade of running away from Rugby called Toynbee and Rosalind back home unexpectedly.
105. Bodleian Library, Toynbee Papers, typescript of speech.
106. Bodleian Library, Toynbee Papers, letter, Tracy Philipps to AJT, 29 March 1936. Toynbee also received a letter of congratulation from T.F. Breen of the British embassy in Berlin, dated 28 February 1936. On the other hand, a spokesman for Chatham House found it needful to declare that Toynbee's remarks in Berlin represented "his own personal views" and not those of Chatham House. These two documents may be found in the Royal Institute of International Affairs, Archives, 4 Toynbee.
107. For a summary of Hitler's remarks on this occasion see Arnold J. Toynbee and V.M. Boulter, *Survey of International Affairs, 1936* (London, 1937), p. 257.
108. Toynbee gave a somewhat different, though not necessarily contradictory, account of how he was chosen for this encounter in *Acquaintances* (London, 1967), pp. 283–285.
109. Toynbee did not initiate this action on his own. Instead, it arose from a dinner party at the Astors' on 8 March 1936, in course of which Toynbee told about his interview with Hitler. Thomas Jones, one-time secretary to the Cabinet and confidant of Baldwin, was among those present, and he decided that the Prime Minister ought to hear what Toynbee had to say. He recorded the matter in his diary as follows: "Toynbee has just returned from a visit to Germany. . . . He had an interview with Hitler which lasted one and three quarters hours. He is convinced

of his sincerity in desiring peace in Europe and a close friendship with England. . . . I have asked Toynbee to put his impressions down and shall have them typed here and handed to S.B. [Baldwin] and Eden first thing in the morning.

"What I am trying to do is to secure that S.B. should have his mind made up on the big major issue of accepting Hitler at his face value and trying him out fairly now that the last trace of humiliation has been removed." Thomas Jones, *A Diary with Letters, 1931–1950* (London, 1954), p. 181.

110. On that day, Baldwin remarked in Parliament: "In Europe we have no more desire than to keep calm, to keep our heads, and to continue to try to bring France and Germany together in a friendship with ourselves." *Survey of International Affairs, 1936,* p. 276.

111. *Acquaintances,* pp. 278–283 provides an account of the interview emphasizing the anti-Russian tone of Hitler's remarks to the exclusion of everything else. In this retrospective account, Toynbee chose to say nothing about his personal effort to influence British policy through the memorandum of 8 March 1936. Instead, he implied that Hitler, though very clever, had not convinced him of Nazi Germany's pacific intentions. The contemporary testimony of the memorandum shows otherwise.

112. This and preceding quotations come from Bodleian Library, Toynbee Papers, memorandum addressed to the Foreign Office, 8 March 1936.

113. Bodleian Library, Toynbee Papers, letter, AJT to A. Dufour von Ference, 20 March 1936.

114. Bodleian Library, Toynbee Papers, letter, AJT to Quincy Wright, 9 July 1936.

115. Royal Institute of International Affairs, Archives, 4 Toynbee Carton, 1, letter, AJT to Margaret Cleeve, 10 September 1938.

116. Bodleian Library, Toynbee Papers, letter, AJT to Veronica Boulter, 1 October 1938.

117. Bodleian Library, Gilbert Murray Papers, Carton 85, sheets 154–161.

118. Bodleian Library, Toynbee Papers, AJT paper, "After Munich," dated 18 November 1938, with undated note from G.M. Gathorne-Hardy attached.

119. Bodleian Library, Toynbee Papers, letter, AJT to Veronica Boulter, 8 September 1938.

120. Bodleian Library, Toynbee Papers, letter, AJT to S. Tetley, 27 September 1939.

121. Bodleian Library, Toynbee Papers, letter, AJT to Veronica Boulter, 29 August 1938.

122. Bodleian Library, Toynbee Papers, letter, Margaret Toynbee to AJT, 23 February 1939.

123. Bodleian Library, Toynbee Papers, letter, Jocelyn Toynbee to AJT, 18 February 1939.

124. Bodleian Library, Toynbee Papers, letter, Rosalind Toynbee to AJT, 4 January 1936.

125. Bodleian Library, Toynbee Papers, letter, AJT to Veronica Boulter, 11 May 1936.

126. Letter, Vaud-Scott to Anthony Toynbee, dated 4 July 1938, in the possession of Mrs. Philip Toynbee.

127. Bodleian Library, Toynbee Papers, letter, AJT to Veronica Toynbee, 15 July 1938.

128. Arnold J. Toynbee, *Experiences* (New York and London, 1969), p. 176.

129. Philip began his literary career with a novel, published in 1936. He worked for a while in Birmingham as a journalist, but had resigned from that job at the time of Tony's suicide—to his parents' distress and annoyance. He, too, had embarked on tumultuous affairs with several young women; but his contemplation of suicide

was dramatic posturing, like much of his youthful behavior. The tone of his response to Tony's death is perhaps suggested by a letter from his close friend Jasper Ridley, who remarked: "I would regard it as extreme discourtesy if you were to kill yourself immediately after seeing me." Philip Toynbee, *Friends Apart: A Memoir of Esmond Romilly and Jasper Ridley in the Thirties* (rev. ed. London, 1980), p. 143. Mrs. Philip Toynbee allowed me to see extensive extracts from Philip's diary from the time, which make clear his deep attachment to Tony, as well as his mercurial and ambivalent relations with his parents and with the various young women whom he was pursuing.

130. Bodleian Library, Toynbee Papers, letter, AJT to Veronica Boulter, 18 March 1939.

131. Bodleian Library, Toynbee Papers, letter, AJT to Veronica Boulter, 25 April 1939.

132. "And so it came to pass that the Gospel of a Jewish Messiah who was God himself incarnate was preached by Galileans and taken to heart by Gentiles." *Study of History,* IV, 263.

133. "A plurality of parochial totalitarian stakes will assuredly give place, sooner or later, to a single oecumenical totalitarian state in which the forces of Democracy and Industrialism will at any rate secure, at last, their natural world-wide field of operations." Ibid., IV, 179.

134. *Times Literary Supplement,* 19 August 1939; *Manchester Guardian,* 1 December 1939; Bodleian Library, Toynbee Papers, letter, R.C. Collingwood to AJT, 12 October 1939.

135. Bodleian Library, Toynbee Papers, letter, Gilbert Murray to AJT, 18 April 1935.

136. Bodleian Library, Toynbee Papers, letter, Rosalind Toynbee to AJT, 9 September 1939.

8. World War II

1. Bodleian Library, Toynbee Papers, letter, AJT to G.M. Gathorne-Hardy, 31 March 1938.

2. For the initial 80–20% budgetary arrangement, see Bodleian Library, Toynbee Papers, letter, AJT to E.H. Carr, 13 December 1939. Chatham House Archives, Carton 2/1/7b, contains a few miscellaneous papers from the period at Oxford. A proposed budget for 1942–1943 shows that the initial financial arrangement was modified as the service took on new tasks. According to the draft budget for 1942–1943, the Foreign Research and Press Service planned to expend £67,568, of which £7600 came from Oxford University and £3271 from the Royal Institute of International Affairs. Toynbee's salary was to be £1325; Veronica's, £600; Bridget Reddin's, £208.

3. Bodleian Library, Toynbee Papers, letter, AJT to David Davies, 15 December 1940.

4. Christian B. Peper, ed., *An Historian's Conscience: The Correspondence between Arnold J. Toynbee and Columba Carey-Elwes, Monk of Ampleforth* (Boston, 1986), pp. 80–81, letter, AJT to Columba, 5 December 1940.

5. *The Times,* letters to the editor, 8 November 1939 and 9 December 1939. Toynbee's principal assailant was E.H. Carr, who headed a competing Foreign Publicity Division based at London University. Cf. Bodleian Library, Toynbee Papers, letter, E.H. Carr to AJT, 11 December 1939 and Toynbee's somewhat acerbic reply, 13 December 1939.

6. R.I.I.A., Archives, 4 Toynbee Carton 1 for weekly meetings. For personnel total on 23 January 1943, R.I.I.A., Archives, Carton 2/1/7b.

7. Veronica Toynbee, typescript, "Relations between Arnold and Veronica Toynbee, 1924–1975," pp. 8–9.

8. Bodleian Library, Toynbee Papers, letter, AJT to Veronica Boulter, 5 June 1941.

9. Bodleian Library, Toynbee Papers, letter, AJT to Veronica Boulter, 14 January 1942.

10. Bodleian Library, Toynbee Papers, letter, W. Paton to AJT, 14 December 1939. This letter describes a visit to England by a Norwegian clergyman, Bishop Berggrav, bringing what he said were German peace terms.

11. Bodleian Library, Toynbee Papers, Report of Peace Aims meeting, Balliol College, 2–3 October 1941. Alfred Zimmern, A.D. Lindsay, and the Archbishop of York were among those gathered with Toynbee to hear what Van Dusen had to say about American attitudes toward a postwar settlement.

12. Bodleian Library, Toynbee Papers, Report of Peace Aims meeting, Balliol College, 6 July 1942. Dulles served as Secretary of State under President Eisenhower, 1953–1959. He was a prominent Presbyterian layman and New York lawyer in 1942.

13. Bodleian Library, Toynbee Papers, Foreign Office Note, 13 July 1942, signed by Francis Evans.

14. R.I.I.A., Archives, 4 Toynbee Carton 8, letter, Walter Mallory to AJT, 6 July 1942.

15. Bodleian Library, Toynbee Papers, letter, AJT to Veronica Boulter, 23 August 1942.

16. Bodleian Library, Toynbee Papers, letter, AJT to Veronica Boulter, 20 September 1942.

17. Bodleian Library, Toynbee Papers, letter, AJT to Veronica Boulter, 15 October 1942.

18. Bodleian Library, Toynbee Papers, letter, AJT to Veronica Boulter, 5 September 1942.

19. Bodleian Library, Toynbee Papers, letter, AJT to Veronica Boulter, 13 September 1942.

20. Bodleian Library, Toynbee Papers, memorandum re Toynbee visit to the United States, no date.

21. Bodleian Library, Toynbee Papers, letter, H. Van Dusen to W. Paton, 15 October 1942.

22. Bodleian Library, Toynbee Papers, Director of Foreign Research and Press Service, Confidential Report, dated October 1942.

23. Peper, *An Historian's Conscience*, pp. 171–173, letter, AJT to Columba, 20 September 1944.

24. Ibid., p. 16, letter, Columba to AJT, 23 February 1937.

25. Ibid., p. 43, letter, Columba to AJT, 29 November 1939.

26. Ibid., pp. 34, 35, letters, AJT to Columba, 24 July 1939 and 25 July 1939.

27. Ibid., p. 37, letter, AJT to Columba, 3 September 1939.

28. Ibid., p. 47, letter, Columba to AJT, 22 September 1939. The ancestral sin Columba had in mind was the English Reformation.

29. Bodleian Library, Toynbee Papers, letter, AJT to Veronica Boulter, 31 January 1941.

30. Bodleian Library, Toynbee Papers, text of sermon preached at University Church, Oxford, 1940, but without an exact date.

31. Republished in A.J. Toynbee, *Civilization on Trial* (New York, 1948), pp. 225–

252. When Columba read the lecture he found it "one of the most wonderful moments of my life. . . . Nothing grated; everything pure delight." Peper, *An Historian's Conscience,* p. 67, letter, Columba to AJT, 24 June 1940.

32. *Civilization on Trial,* pp. 235–236, 236, 239, 243.

33. Ibid., p. 249. He dismissed Protestantism as "premature" in discarding the institutional armor of Catholicism, p. 243.

34. The Gilbert Murray Papers in the Bodleian Library have been purged of everything related to Rosalind's break with Toynbee, though Gilbert Murray's habit of correspondence with his wife and other close associates can scarcely have avoided the topic of his daughter's divorce. Jean Smith, a belligerently pious convert to Catholicism, friend to Rosalind, and one-time secretary to Gilbert Murray, was one of three persons in charge of his papers after Murray's death, and she probably destroyed everything that struck her as embarrassing. This wholesale destruction was compounded by the fact that after Rosalind's death, her lover Richard and her son Philip decided to burn her diary, either in deference to what they thought to be her wishes or from shock at what they found in it.

35. Peper, *An Historian's Conscience,* p. 40, letter, AJT to Columba, 19 October 1939.

36. Ibid., p. 72, letter, AJT to Columba, 14 July 1940.

37. Ibid., p. 77, letter, AJT to Columba. The Sword of the Spirit was a Catholic organization to promote justice in British society. Cf. also Bodleian Library, Toynbee Papers, letter, AJT to David Davies, 15 December 1940: "She [Rosalind] likes the work and also has a slightly excessive taste for living in the Blitz."

38. Peper, *An Historian's Conscience,* p. 167, letter, AJT to Columba, 9 July 1944.

39. Rosalind Murray, *Unstable Ways* (London, 1914), p. 257.

40. Peper, *An Historian's Conscience,* p. 115, letter, AJT to Columba, 19 May 1942. Toynbee's collapse had occurred in December 1941 and brought Rosalind to the rescue. She first arranged to put him in a Catholic nursing home, and then allowed him to stay at her flat in London for a few days after his release. Bodleian Library, Toynbee Papers, letters, AJT to Veronica Boulter, 18 December 1941 and 30 December 1941.

41. Typescript, "Relations between Arnold and Veronica Toynbee, 1924–1975," p. 9. The unacceptable condition Rosalind set for continuation of their marriage was "to live apart but visit her by invitation." Cf. Peper, *An Historian's Conscience,* p. 142, letter, AJT to Columba, 29 July 1943.

42. Peper, *An Historian's Conscience,* p. 137, letter, AJT to Columba, 25 February 1943.

43. Toynbee probably described himself as a "man married to a goddess" in a lengthy quasi-confessional letter, which he sent to Columba in May 1943. It subsequently disappeared. Cf. Peper, *An Historian's Conscience,* pp. 135, 160–161. At any rate, after reading this lost letter, which Columba had sent to him, Gilbert Murray used the phrase a "man married to a goddess" to sum up what was wrong with the marriage. Ibid., pp. 140–141, letter, Gilbert Murray to Columba, 3 July 1943.

44. Typescript, "Relations between Arnold and Veronica Toynbee, 1924–1975," p. 11.

45. Peper, *An Historian's Conscience,* p. 159, letter, AJT to Columba, 12 March 1944.

46. Oral recollection, 15 February 1986.

47. Peper, *An Historian's Conscience,* p. 165, letter, Bridget Reddin to Columba, 3 July 1944.

48. Ibid., p. 137, letter, AJT to Columba, 18 June 1943.

49. Ibid., p. 132, letter, AJT to Columba, 25 February 1943.

50. Ibid., p. 144, letter, AJT to Columba.

51. Ibid., p. 157, letter, AJT to Columba, 23 February 1944.

52. Bodleian Library, Toynbee Papers, letter, AJT to Gilbert Murray, 20 October 1943.

53. Bodleian Library, Toynbee Papers, letter, AJT to Gilbert Murray, 26 October 1943.

54. The original version exists in typescript among Toynbee's papers at the Bodleian Library. The printed version (*Experiences* [New York and London, 1969], pp. 393–401), simply by dropping the two lines in which Rosalind's name is introduced, makes the "urgent muse," that is, his professional work, the object (along with Ganthorpe and its environs) that he has worshiped instead of God. The emendation may reflect an alteration in his judgment but is more likely due to a wish to avoid public exploration of the wound he suffered from Rosalind's rejection.

55. Peper, *An Historian's Conscience*, pp. 135, 160–161.

56. Ibid., p. 162, letter, AJT to Columba, 19 May 1944.

57. Ibid., p. 170, letter, AJT to Columba, 9 August 1944. Toynbee's phrase "after Rosalind had refused me a home" points to the way he equated his break with her to his earlier break with his mother, who also had "refused him a home" by precipitously abandoning the house he had grown up in. Toynbee remained almost pathologically afraid of finding himself suddenly homeless even after marrying Veronica.

58. Ibid., p. 136, letter, Rosalind Toynbee to Columba, 17 June 1943. The canon law of Rome had long held that valid marriage could not take place between a baptized person and one who was not baptized. After 1918, this was interpreted to apply only to cases of Roman Catholic baptism; but in 1913, when Rosalind and Toynbee married, a more sweeping interpretation prevailed according to which Toynbee's Anglican baptism sufficed to invalidate his marriage to an unchurched Rosalind!

59. Bodleian Library, Toynbee Papers, letter, AJT to Gilbert Murray, 24 September 1944.

60. Bodleian Library, Toynbee Papers, letter, AJT to Gilbert Murray, 2 October 1944.

61. Bodleian Library, Toynbee Papers, letter, AJT to Gilbert Murray, 12 October 1945.

62. Peper, *An Historian's Conscience*, p. 571, letter, AJT to Columba, 22 February 1974. It was on this occasion that Rosalind's diary was destroyed.

63. Rosalind Murray, *The Forsaken Fountain* (London, 1948), pp. 134–142 explains at some length how erotic love may serve as a way station on the path toward discovering the true love of God. Her argument perhaps reflects a version of her relation with Richard. The last religious tract she wrote, *The Further Journey: In My End Is My Beginning* (London, 1953), p. 7 further proclaims the need "on occasion to give onself away, to risk becoming a fool for Christ's sake." She remained emphatically Catholic throughout, and no doubt found a religious justification for her conduct, although her activity on behalf of Catholic causes faded once she set herself up with Richard on the farm in Cumberland.

64. Peper, *An Historian's Conscience*, p. 142, letter, AJT to Columba, 29 July 1943.

65. Typescript, "Relations between Arnold and Veronica Toynbee, 1924–1975," p. 10.

66. Ibid., p. 11.

67. Bodleian Library, Toynbee Papers, letter, AJT to Veronica Boulter, 5 April 1943.

68. Bodleian Library, Toynbee Papers, letter, AJT to Veronica Boulter, 7 August 1945.

69. Bodleian Library, Toynbee Papers, letter, AJT to Veronica Boulter, 11 August 1946.

70. "My chief asset was that I was among the admirers of the Study of History, and indeed of almost all his work" she wrote, in support of her claim that "he gradually

became happier with me than he had ever been with Rosalind since the early days of their marriage.'' Typescript, ''Relations between Arnold and Veronica Toynbee, 1924–1975,'' pp. 11–12.

71. Bodleian Library, Toynbee Papers, letter, AJT to Gilbert Murray, 27 March 1944.

72. Bodleian Library, Toynbee Papers, letter, AJT to Gilbert Murray, 28 June 1944.

73. Bodleian Library, Toynbee Papers, letter, AJT to Gilbert Murray, 27 October 1946.

74. Bodleian Library, Toynbee Papers, letter, AJT to Foreign Office, 28 May 1945. This seems a lame excuse, given American delight in British titles. Toynbee may simply have indulged a fit of pique at the offer of no more than the honor that was routinely assigned to senior officers of the foreign service.

75. Private possession, letter, AJT to Philip Toynbee, 8 December 1943.

76. Bodleian Library, Toynbee Papers, letter, Philip Toynbee to AJT, 1 January 1945. Lawrence does not believe that his mother ever blamed him for having suffered shell shock, but does remember her initial suspicions of his future wife. Later, however, Rosalind reconciled herself to her daughter-in-law, and the two became quite friendly. Oral communication, 18 December 1987.

77. ''I have just come back from a happy evening with Bun and Jean and little Rosalind.'' Bodleian Library, Toynbee Papers, letter, AJT to Veronica Boulter, 10 June 1946.

78. Bodleian Library, Toynbee Papers, letter, AJT to Veronica Boulter, 2 August 1945.

79. At first Gilbert Murray may have inclined to an opposite view. ''I believe that Arnold's acceptance of the attitude of 'man married to a goddess' is largely responsible for putting the relations of the two in a false position from the beginning,'' he wrote to Columba on 3 July 1943 when returning Toynbee's long confessional letter, since lost. Peper, *An Historian's Conscience*, p. 141.

80. Bodleian Library, Toynbee Papers, letter, AJT to Gilbert Murray, 25 September 1946.

81. Bodleian Library, Toynbee Papers, letter, AJT to Veronica Boulter, 27 April 1946.

82. Bodleian Library, Toynbee Papers, letter, AJT to Veronica Boulter, 3 May 1946.

83. Bodleian Library, Toynbee Papers, letter, AJT to Veronica Boulter.

84. Bodleian Library, Toynbee Papers, letter, AJT to Veronica Boulter, 26 April 1946.

85. Bodleian Library, Toynbee Papers, letter, AJT to Veronica Boulter, 7 May 1946.

86. Bodleian Library, Toynbee Papers, letter, AJT to Veronica Boulter, 17 July 1946.

87. Bodleian Library, Toynbee Papers, letter, AJT to Veronica Boulter, 6 May 1946.

88. Bodleian Library, Toynbee Papers, letter, AJT to Veronica Boulter, 19 August 1946.

89. Bodleian Library, Toynbee Papers, letter, AJT to Gilbert Murray, 25 September 1946.

90. Bodleian Library, Toynbee Papers, letter, AJT to E.R. Morgan, 11 August 1946.

91. Bodleian Library, Toynbee Papers. The document bears no date, but the decision was handed down on 13 December 1946 according to Peper, *An Historian's Conscience*, p. 137.

92. Private possession, letter, AJT to R.S. Darbishire, 3 October 1946.

93. Bodleian Library, Toynbee Papers, letter, AJT to Gilbert Murray, 18 May 1947.

94. Bodleian Library, Toynbee Papers. He was mistaken in saying that the packet contained ''all the letters I ever had from Rosalind.'' Several are located elsewhere among his papers.

9. Fame and Fortune

1. Unlike other Oxford colleges All Souls' had no undergraduates, and fellows therefore had no teaching duties.
2. Bodleian Library, Toynbee Papers, letter, AJT to Ivison Macadam, 4 April 1946.
3. Bodleian Library, Toynbee Papers, letter, AJT to Gilbert Murray, 27 October 1946.
4. Bodleian Library, Toynbee Papers, letter, AJT to Katherine McBride, 30 March 1953.
5. Bodleian Library, Toynbee Papers, letter, Joseph A. Willetts to Ivison Macadam, 25 March 1947.
6. Bodleian Library, Toynbee Papers, letter, AJT to Ivison Macadam, 31 March 1947.
7. Terms are recorded with other correspondence between AJT and the Rockefeller Foundation in Bodleian Library, Toynbee Papers. Subsequently, the Rockefeller Foundation granted an additional £60,000 for the wartime survey, according to a memorandum Toynbee wrote to Ivison Macadam on 17 May 1954.
8. In fact, Clark had started to plan a *New Cambridge Modern History,* which was eventually published, though only after innumerable editorial vicissitudes. The incoherence of its contents faithfully reflected the absence of consensus among historians as to what their subject encompassed—a great contrast to the wholeness, within its limited range, that had characterized the *Cambridge Modern History* as planned by Lord Acton at the beginning of the century.
9. Bodleian Library, Toynbee Papers, letter, Jocelyn Toynbee to AJT, 7 June 1947.
10. Bodleian Library, Toynbee Papers, letter, AJT to G.C.N. Clark, 18 June 1947.
11. In practice, Calvocoressi found it impossible to write a volume for 1947 in time to publish it in 1948. He was succeeded first by Geoffrey Barraclough and then by D.C. Watt, each of whom merely fell further and further behind. When, in 1977, what purported to be an "annual" series got to be thirteen years behind the events it chronicled, the enterprise was abandoned with D.C. Watt's belated survey for 1963!
12. Others, including one by George Kirk devoted to the Middle East and one by William H. McNeill devoted to relations among the Allies, had the virtue of a single point of view; but Kirk's animus against the Jews in Palestine led him to use language which had to be retracted when a reviewer called attention to it in 1952. This caused Toynbee and Chatham House very considerable embarrassment. Chatham House Archives, 4 Toynbee Carton 32 has full details.
13. Bodleian Library, Toynbee Papers, letter, D.C. Somervell to AJT, 12 September 1943.
14. Bodleian Library, Toynbee Papers, letter, AJT to D.C. Somervell, 22 September 1943.
15. Bodleian Library, Toynbee Papers, letter, D.C. Somervell to AJT, 24 September 1943.
16. Bodleian Library, Toynbee Papers, letter, AJT to D.C. Somervell, 26 October 1944.
17. Bodleian Library, Toynbee Papers, letter, AJT to D.C. Somervell, 1 December 1944.
18. Bodleian Library, Toynbee Papers, letter, D.C. Somervell to AJT, 7 December 1944. Somervell had actually started the project to try to interest his son, who had found the six volumes unduly formidable.
19. "The birth of which the angels then sang was not a rebirth of Hellas nor a new birth of other societies of the Hellenic species. It was the birth in the flesh of the

King of the Kingdom of God." *A Study of History,* Abridgement (New York, 1947), p. 532.

20. "The perpetual turning of the wheel [i.e., what Toynbee describes in the same passage as "the temporal history of man as this manifests itself in the geneses, growths, breakdowns and disintegrations of human societies"] is not a vain repetition if, at each revolution, it is carrying the vehicle that much nearer to its goal." Ibid., p. 556.

21. Ibid., p. 554.

22. Bodleian Library, Toynbee Papers, letter, H. Milford to Anne Cleeve, 23 October 1944. This royalty arrangement was, in fact, accepted.

23. Bodleian Library, Toynbee Papers, letter, AJT to D.C. Somervell, 8 October 1945.

24. Bodleian Library, Toynbee Papers, copy of speech made by Henry C. Walck of Oxford University Press at a publishers' award ceremony on 27 January 1948.

25. Bodleian Library, Toynbee Papers, letter, AJT to Gerard Hopkins, 18 June 1946. Toynbee wrote to Hopkins to protest the wording of an advertisement for the book. When told that he could not alter advertising copy in proof stage he supplied what he thought would be a proper sort of advertisement for future use, so as to "lie low about patterns" and emphasize the "factual side of the work."

26. Bodleian Library, Toynbee Papers, letter, AJT to Gilbert Murray, 8 March 1947.

27. Bodleian Library, Toynbee Papers, letter, AJT to Gilbert Murray, 19 April 1947.

28. John K. Jessup, ed., *The Ideas of Henry Luce* (New York, 1969), p. 10.

29. Ibid., pp. 70–71.

30. Letter, Sheldon Meyer to W.H. McNeill, 3 February 1987. Sales in London totaled 14,431 by 31 March 1948. In the United States, Book of the Month Club sales were also very substantial, but these figures appear to be irrecoverable and are not included in O.U.P. totals. Toynbee's U.S. royalties in the first year of publication amounted to £7926-1-9, according to a document in Chatham House Archives, 4 Toynbee Carton 3. This was almost four times his annual salary.

31. Bodleian Library, Toynbee Papers, letter, Virginia Carrick to AJT, 7 March 1947. Ms. Carrick was publicity director for Oxford University Press in New York, and was thus in a position to know some of the details. Cf. Whittaker Chambers, *Witness* (New York, 1952), p. 505, where he claims responsibility for *Time*'s cover story on Toynbee.

32. Bodleian Library, Toynbee Papers, letter, Virginia Carrick to AJT, 12 March 1947.

33. Bodleian Library, Toynbee Papers, letters, Virginia Carrick to AJT, 3 April 1947, and 8 April 1937.

34. By contributing to an obscure pamphlet, George Boas and Wheeler Harvey, eds., *Lattimore as Scholar: A Few Random Selections from the Responses of Widely Diverse Scholars* (Baltimore, 1953).

35. Bodleian Library, Toynbee Papers, letter, AJT to Gilbert Murray, 18 May 1952.

36. Bodleian Library, Toynbee Papers, letter, AJT to A.E. Raubitschek, 31 January 1950. Raubitschek was a professor of classics at Princeton and so knew that the battle of Zama ended the Second Punic War in 202 B.C. and that the battle of Actium established the Roman Empire in 31 B.C.

37. Bodleian Library, Toynbee Papers, letter, AJT to Frank H. Underhill, 15 November 1951. Underhill was a professor at the University of Toronto.

38. Christian B. Peper, ed., *An Historian's Conscience: The Correspondence between Arnold J. Toynbee and Columba Carey-Elwes, Monk of Ampleforth* (Boston, 1986), p. 244, letter, AJT to Columba, 28 April 1949.

39. Bodleian Library, Toynbee Papers, letter, AJT to Gilbert Murray, 21 October 1953.

40. Bodleian Library, Toynbee Papers, letter, AJT to A.E. Eurich, 15 August 1952.

41. Bodleian Library, Toynbee Papers, letter, AJT to Gilbert Murray, 29 September 1952.

42. Bodleian Library, Toynbee Papers, letter, AJT to Gilbert Murray, 18 May 1952.

43. Toynbee's ichthyology was clearly at fault, since catfish are freshwater herbivores. He presumably meant dogfish—a kind of small shark.

44. Bodleian Library, Toynbee Papers, speech to New York Herald Book and Author's Luncheon, 13 April 1949.

45. Bodleian Library, Toynbee Papers, letter, AJT to Gilbert Murray, 15 November 1952.

46. Bodleian Library, Toynbee Papers, letter, AJT to Gilbert Murray, 18 June 1954.

47. Peper, *An Historian's Conscience,* p. 265, letter, AJT to Columba, 15 August 1950.

48. Bodleian Library, Toynbee Papers, text of AJT talk on Voice of America. There is no date, but to judge from its position in the file with other broadcast talks it was presumably aired in 1950 or 1951.

49. Bodleian Library, Toynbee Papers, letter, AJT to R.L. Sylvester, 21 December 1954.

50. Peper, *An Historian's Conscience,* p. 199, letter, AJT to Columba, 31 August 1947.

51. Bodleian Library, Toynbee Papers, letter, AJT to Gilbert Murray, 25 December 1953.

52. Bodleian Library, Toynbee Papers, letter, AJT to Gilbert Murray, 25 December 1952.

53. A.J. Toynbee, "Poetical Truth and Scientific Truth in the Light of History," *International Journal of Psycho Analysis,* XXX (1949), 150.

54. X, 126.

55. Bodleian Library, Toynbee Papers, letter, AJT to David Davies, 21 August 1954.

56. Bodleian Library, Toynbee Papers, letter, AJT to David Davies, 25 August 1954.

57. Bodleian Library, Toynbee Papers, letter, AJT to Raymond Aron, 21 October 1954.

58. Bodleian Library, Toynbee Papers, letter, AJT to Robert Redfield, 1 July 1954.

59. Bodleian Library, Toynbee Papers, talk at Chatham House, "Politics and Opinion and Feeling in America in the Present Crisis," 15 February 1951. Weekly talks, of which this was one, were confidential with attendance limited to members of Chatham House.

60. Bodleian Library, Toynbee Papers, talk, "Britain's Future in the Light of Sweden's Past," delivered 17 March 1954.

61. Bodleian Library, Toynbee Papers, BBC broadcast, "The World and the West," 12 March 1953.

62. Bodleian Library, Toynbee Papers, letter, AJT to J.H. Willetts, 13 October 1948.

63. Peper, *An Historian's Conscience,* pp. 320–323. The Pope gave him a medal; he reciprocated with an offprint of an article he had written. Idem, pp. 323–334, letter, AJT to Columba, 14 October 1954.

64. Bodleian Library, Toynbee Papers, cutting from the *Moscow Literary Gazette.* The article appeared in English.

65. Compiled from S. Fiona Morton, ed., *A Bibliography of Arnold J. Toynbee* (Oxford, 1980).

66. The long gap between his completion of the text in 1951 and publication in 1954

was due to printers' delays as well as delays arising from Toynbee's own corrections, additions, and alterations.

67. *Civilization on Trial* (New York, 1948), pp. 39–41.

68. Published as *The World and the West* (London, 1953).

69. *"Counsels of Hope"*: *The Toynbee–Jerrold Controversy: Letters to the Editor of the Times Literary Supplement, with Leading Articles Reprinted* (London, 1954).

70. Bodleian Library, Toynbee Papers, letter, AJT to T.L.S. Editor, 16 April 1954.

71. Bodleian Library, Toynbee Papers, BBC Broadcast in Arabic, 12 March 1953. Toynbee did not speak Arabic; this text was presumably written to be translated and actually transmitted by someone else.

72. Bodleian Library, Toynbee Papers, letter, AJT to Stephen King-Hall, 22 September 1954.

73. Bodleian Library, Toynbee Papers, BBC broadcast, 4 January 1948.

74. Many of the contributors to Ashley Montague, ed., *Toynbee and History: Critical Essays and Reviews* (Boston, 1956) illustrate this academic habit.

75. Frederick L. Schuman, "The Paradoxes of Dr. Toynbee," *The Nation*, 6 November 1954, p. 405.

76. These figures are derived from Morton, *A Bibliography of Arnold J. Toynbee*. The languages in which these articles appeared indicates better than anything else the geographical distribution of Toynbee's reputation in the Western world, for fifty-five articles in English were matched against nine in German, five in Spanish, three in French, two in Russian, one in Dutch, and one in Portuguese.

77. Instead of revising his text in the light of comments made by readers to whom he submitted portions in advance, Toynbee decided to put remarks he took seriously into footnotes, leaving the text as it had stood originally.

78. Bodleian Library, Toynbee Papers, letter, AJT to Gilbert Murray, 26 August 1954.

79. Bodleian Library, Toynbee Papers, letter, AJT to John Crofton, 29 May 1968, refers to the "emotional and contorted style of these four volumes, which now makes me wince."

80. Bodleian Library, Toynbee Papers, letter, Margaret Toynbee to AJT, 15 December 1954 with postscript from Jocelyn; AJT draft reply, without date.

81. Bodleian Library, Toynbee Papers, Summary of Publicity, dated 1 February 1955.

82. *Time*, 17 March 1947, p. 71.

83. *A Study of History*, VII, 393.

84. Ibid., VII, 449.

85. Peper, *An Historian's Conscience*, p. 224, letter, AJT to Columba, 4 August 1948.

86. *A Study of History*, VII, 442.

87. Ibid., VII, 438.

88. Bodleian Library, Toynbee Papers, letter, AJT to David Davies, 6 August 1949.

89. Bodleian Library, Toynbee Papers, letter, AJT to Veronica Toynbee, 12 July 1947.

90. Bodleian Library, Toynbee Papers, letter, AJT to Veronica Toynbee, 13 July 1947.

91. Bodleian Library, Toynbee Papers, letter, Rosalind Murray to AJT, 2 July 1947.

92. Peper, *An Historian's Conscience*, p. 191, letter, AJT to Columba, 17 July 1947.

93. Ibid., pp. 193–194, reproduces the entire text. This poem, dated 16 July 1947, must have been written immediately after returning from Ampleforth.

94. Ibid., p. 249, letter, AJT to Columba, 21 September 1949.

95. Bodleian Library, Toynbee Papers, letter, Rosalind Murray to AJT, 11 November 1947.

96. Bodleian Library, Toynbee Papers, letter, Rosalind Murray to AJT, 29 March 1951.

97. Bodleian Library, Toynbee Papers, letter, AJT to Veronica Toynbee, 13 July 1947. Jean was Lawrence's wife; Rosalind, their eldest daughter.

98. Bodleian Library, Toynbee Papers, letter, AJT to Gilbert Murray, 25 December 1953.

99. Bodleian Library, Gilbert Murray Papers, Carton 567, letter, Gilbert Murray to Rosalind Murray, 26 September 1954.

100. Peper, *An Historian's Conscience,* p. 339, letter, AJT to Columba, 21 August 1954.

101. Bodleian Library, Toynbee Papers, letter, Philip Toynbee to AJT, 4 February 1949.

102. Peper, *An Historian's Conscience,* p. 280, letter, AJT to Columba, 8 February 1951.

103. Ibid., p. 343, letter, AJT to Columba, 16 January 1955.

104. In 1986 Polly Toynbee, Philip's second daughter, remembered her childish disappointment at going hungry from the table as well as the strain of forced conversation with her grandfather on these occasions.

105. "Arnold Toynbee—A Eulogy," *PHP: A Forum for a Better World* (May 1976), p. 8.

106. Chatham House Archives, 4 Toynbee Carton 8; Bodleian Library, Toynbee Papers, memorandum of his duties by AJT, dated 17 May 1954.

107. Bodleian Library, Toynbee Papers, letter, AJT to Ivison Macadam, 28 July 1953.

10. Toynbee as a World Figure

1. Bodleian Library, Toynbee Papers, letter, AJT to E.F. D'Arms, 3 June 1952.

2. Toynbee received a retainer of £500 a year for twelve to fifteen essays—including "reviews, articles from travels, and historical background of current events"— with £25 for any additional articles submitted and accepted, and reserving book rights for subsequent publication. Bodleian Library, Toynbee Papers, letter, K. Obank to AJT, 1 June 1954. Arnold J. Toynbee, *East to West: A Journey Round the World* (New York, 1958) republished a selection of seventy-three of Toynbee's *Observer* articles.

3. Bodleian Library, Toynbee Papers, letter, AJT to David Davies, 1 April 1956.

4. Bodleian Library, Toynbee Papers, letter, AJT to E.D. Myers, 18 March 1956.

5. Bodleian Library, Toynbee Papers, letter, AJT to Veronica Toynbee, 4 April 1956, and AJT to Gilbert Murray, 5 April 1956. Toynbee made light of his ailment in these letters and gave no details.

6. Bodleian Library, Toynbee Papers, letter, AJT to David Davies, 14 May 1957.

7. Arnold Toynbee and Edward D. Myers, *Historical Atlas and Gazetteer* (London, 1959) constituted volume XI of *A Study of History.*

8. Bodleian Library, Toynbee Papers, letter, AJT to Norah Williams, 5 November 1956. Norah Williams was his secretary at Chatham House.

9. Bodleian Library, Toynbee Papers, letter, AJT to Gilbert Murray, 27 November 1956.

10. Bodleian Library, Toynbee Papers, letter, AJT to David Davies, 29 May 1957.

11. Bodleian Library, Toynbee Papers, letter, AJT to David Davies, 12 June 1956. Gilbert Murray could boast the Order of Merit. This was the honor that Toynbee coveted. His chagrin was at least partly occasioned by the fact that he had failed to achieve the level of public recognition accorded to his onetime father-in-law.

12. Pp. 14–28.

13. Christian B. Peper, ed., *An Historian's Conscience: The Correspondence between Arnold J. Toynbee and Columba Carey-Elwes, Monk of Ampleforth* (Boston, 1986), p. 370, letter, Columba to AJT, 14 July 1957.
14. Bodleian Library, Toynbee Papers, letter, Stephen Spender to AJT, 16 May 1957.
15. *Study of History*, XII, 574.
16. Bodleian Library, Toynbee Papers, speech by AJT, 6 May 1964.
17. Bodleian Library, Toynbee Papers, letter, AJT to David Davies, 19 December 1956.
18. Bodleian Library, Toynbee Papers, speech for Japanese TV, 22 November 1956.
19. Bodleian Library, Toynbee Papers, letter, AJT to Gilbert Murray, 19 January 1955.
20. The annual stipend from Britain was initially set at £300 and from O.U.P. New York at £1650 ($5000). These sums remained unchanged for many years, and since in most years Toynbee's royalties exceeded the stipends paid out, the arrangement gave the Press possession and use of a substantial sum of accrued royalties down to and beyond his death in 1975. In 1959 undrawn royalties totaled £18,200 according to Toynbee's own calculation of his wealth. His investments totaled £20,000 and houses and land he valued at £18,000 at the same date. Bodleian Library, Toynbee Papers, letter, AJT to Rosalind Murray, 16 December 1959.
21. Bodleian Library, Toynbee Papers, contract with Aramco, dated 28 August 1957 offered £1000 for a paper on appropriate company policy toward Arabs. Toynbee produced an eighteen page memorandum on the subject and collected an additional £450 for delivering three lectures under the company's auspices in Dhahran, Saudi Arabia in July 1957.
22. Bodleian Library, Toynbee Papers contains a carton labeled "Journeys, 1955" with details of this tour, including a letter defining his fee at Minneapolis. Earlier, he gave lectures at three seminaries in and around New York, resulting in a published book, *Christianity among the Religions of the World* (New York, 1957). The lectureship required publication, but Oxford University Press did not want to publish these lectures, which, in fact, did little but reiterate what he had already said in the last volumes of the *Study*. When the three sponsoring seminaries arranged for Scribner's to publish the lectures, a nasty contretemps ensued, since Oxford University Press objected to the way their new contract with Toynbee had been infringed. This put him in fresh fear of the tax authorities—a fear perhaps accentuated by the amount of easy money he had raked in by lecturing.
23. Bodleian Library, Toynbee Papers, carton labeled "Journeys 1963–64" contains details, including a loose sheet with this sum entered on it.
24. *Democracy in the Atomic Age*, The Dyason Lectures for 1956 (Melbourne, 1957), p. 9; *The Present Day Experiment in Western Civilization*, The Beatty Memorial Lectures (London, 1962), pp. 51–74.
25. *Christianity among the Religions of the World*, p. 105.
26. *America and the World Revolution* (New York, 1962), p. 40 and passim.
27. *The Economy of the Western Hemisphere*, The Weatherhead Foundation Lectures delivered at the University of Puerto Rico (London, 1962), p. 33.
28. *Change and Habit: The Challenge of Our Time* (London, 1966). This book summarized his lectures at Denver and New College, Florida in 1964–1965.
29. Bodleian Library, Toynbee Papers, letter, AJT to David Davies, 7 February 1958. Sputnik was the first earth satellite, launched by the Russians in October 1957.
30. Bodleian Library, Toynbee Papers, letter, AJT to David Davies, 2 March 1958.
31. Bodleian Library, Toynbee Papers, text of AJT's remarks for CBS Symposium, 19 February 1961. Kissinger later became Secretary of State, Stevenson was Ambassa-

dor to the United Nations and a former candidate for the American presidency, and Strauss was head of the Atomic Energy Commission.

32. *Playboy Magazine,* April 1967, pp. 57–76, 166–169. Much of the interview was autobiographical reminiscence. Toynbee was presented as a crusty old character, with extreme opinions about practically everything.

33. These quotes come from a republication of the article by the Southern Student Organizing Committee as part of an anti-Vietnam War movement.

34. Bodleian Library, Toynbee Papers, letter, AJT to David Davies, 17 May 1958. Toynbee was probably influenced by his son Philip in this matter. Philip played a prominent role in Ban the Bomb agitation in Britain in the late 1950s in association with his distant cousin Lord Bertrand Russell.

35. Bodleian Library, Toynbee Papers, letter, AJT to Arthur Smith, 14 April 1964.

36. *McGill Daily,* 26 January 1961, p. 1. Toynbee first made this comparison in *A Study of History,* VIII, 290. In Montreal he was simply asked whether he still believed what he had written, to wit, that it was a "supreme tragedy" that Jews in 1948 had chosen "to imitate some of the evil deeds that the Nazis had committed."

37. The debate was recorded and copies of the recording sold widely in Canada and beyond. In addition, a transcript of the debate was published in Egypt in English; but these quotes come from Bodleian Library, Toynbee Papers.

38. Abba Eban, *The Toynbee Heresy* (New York, 1955) is a vigorous, shrill polemic.

39. "Pioneer Destiny of Judaism," *Issues,* 14 (Summer 1960), p. 14.

40. *Time,* 19 May 1961, p. 25.

41. James Parkes, "Toynbee and the Uniqueness of Jewry," reprinted in *Prelude to Dialogue: Jewish–Christian Relationships* (New York, 1969), p. 105. This essay originally appeared in the *Jewish Journal of Sociology* in 1962 and was reprinted several times in collections of essays and polemics against Toynbee.

42. This version of Niebuhr's remark is quoted from *The Baltimore Sun,* 21 February 1961, when Toynbee was about to appear before Jewish groups in that city.

43. Letter in *Jewish Frontier,* 10 January 1955.

44. *Acquaintances* (London, 1967) devotes two chapters to graceful sketches of these men.

45. Bodleian Library, Toynbee Papers, letter, AJT to Oskar K. Rabinowicz, 24 June 1968. Rabinowicz had just published a very well-documented article, "Toynbee's Pro-Zionism in World War I," *Herzl Yearbook,* VII (1968), 1–14.

46. Letter to the editor, *Jewish Frontier,* 10 January 1955.

47. Bodleian Library, Toynbee Papers, letter, AJT to George Metcalf, 21 February 1959. Metcalf was a Canadian student at Toronto.

48. Bodleian Library, Toynbee Papers, letter, AJT to Elizabeth Vice, 15 April 1971.

49. Bodleian Library, Toynbee Papers, letter, AJT to Major General Moustapha Tlass, 9 October 1973.

50. Bodleian Library, Toynbee Papers, letter, AJT to J.L. Buchanan, 11 September 1963.

51. *The Times,* 2 October 1971. Toynbee's letter provoked several hostile replies.

52. Bodleian Library, Toynbee Papers, letter, AJT to Elmer Berger, 29 May 1969.

53. Soon after the storm broke in 1961 Toynbee wrote to another correspondent: "At Hillel House, Montreal I had the strange sensation of having suddenly stepped out of Canada into Tel Aviv. But I do not want to say this in public." Bodleian Library, Toynbee Papers, letter, AJT to Harry Snellenburg, 3 April 1961.

54. Bodleian Library, Toynbee Papers, letter, AJT to David Davies, 11 June 1960.

55. Bodleian Library, Toynbee Papers, letter, AJT to David Davies, 18 May 1960.

56. Bodleian Library, Toynbee Papers, letter, AJT to David Davies, 12 April 1960.

57. Bodleian Library, Toynbee Papers, letter, AJT to David Davies, 27 February 1960.

58. *Between Oxus and Jumna: A Journey in India, Pakistan and Afghanistan* (New York and London, 1961), p. 180.

59. Ibid., pp. 183, 191. In fact, India and Pakistan fought an undeclared war over Kashmir in 1965 and Ayyub Khan was overthrown in 1969.

60. Cf. *Between Oxus and Jumna,* pp. 182–183 where Toynbee says: "Parliamentary democracy of the American or British type was, in fact, commended both by snobbery and by superstition. . . . it was a badge of respectability and distinction." But Toynbee did not adhere firmly to this devaluation of Indian electoral politics. Cf. *The Present Day Experiment in Western Civilization* (1962), p. 56, where he praises India's parliamentary democracy as a "credit to the Indian people."

61. *Between Oxus and Jumna,* p. 183.

62. Bodleian Library, Toynbee Papers, letter, Harold Beeley to AJT, 6 July 1961.

63. Bodleian Library, Toynbee Papers, carton labeled "Journeys 1957–59" contains a text and connected papers.

64. Bodleian Library, Toynbee Papers, letter, AJT to Countess Thurn-Valsassina, 30 October 1959. Edited versions of what was said at this conference were published as *L'histoire et ses interpretations: Entretiens autour de Arnold Toynbee* (Paris, 1961).

65. The organization still exists, publishes a journal, and meets annually, but it is now based entirely within the United States.

66. *Experiences* (New York and London, 1969), pp. 40–44 gives a summary of what Toynbee called his third Greek education, that is, what he learned from association with Doxiadis. In this scheme, his schooling at Winchester and Balliol constituted his first Greek education; the trip to Greece in 1911–1912 was his second.

67. Bodleian Library, Toynbee Papers, letter, AJT to Constantine Doxiadis, 12 April 1974.

68. Bodleian Library, Toynbee Papers, letter, AJT to David Davies, 20 January 1958.

69. *Study,* XII, 344–345, 443, 553–554.

70. Ibid., pp. 279, 563.

71. Ibid., p. 571.

72. Toynbee reiterated his view that Judaism betrayed its mission by clinging to the notion of a "Chosen People" and ended by summoning a new Jewish prophet to inspire "his fellow Jews at last to dedicate themselves to their universal mission wholeheartedly" (p. 517). These were, assuredly, fighting words.

73. Pieter Geyl, "Toynbee's Answer," *Mededelingen der Koninklijke Nederlandse Akademie van Wetenschappen,* Afd. Letterkunde, Nieuwe Reeks, Deel 24, No. 5 (Amsterdam, 1961), p. 26.

74. Bodleian Library, Toynbee Papers, letter, AJT to John Brown, 4 October 1963.

75. Bodleian Library, Toynbee Papers, letter, John Brown to AJT, 15 June 1962.

76. Bodleian Library, Toynbee Papers, letter, John Brown to AJT, 28 February 1969. Since custom decreed a spring and autumn publishing season this meant four books per annum, a figure Toynbee exceeded only in 1962 when five of his books, all based on lectures, came out.

77. Bodleian Library, Toynbee Papers, letter, AJT to A.H. McDonald, no date.

78. *TLS,* 2 December 1965.

79. Stuart Oost in *Classical Philology* (April 1967), p. 146. M.I. Finlay in *The New Statesman and Nation,* 24 December 1965, used far stronger language to convey

much the same meaning when he remarked that the book bored and sometimes repelled him.

80. Bodleian Library, Toynbee Papers, letter, AJT to Jean Smith, 14 January 1959. Jean Smith was Gilbert Murray's secretary and an incandescent convert to Catholicism. Toynbee was retorting to her expostulation rather than to Rosalind's, for Rosalind had rather grudgingly accepted his revised language about her father's religion a few days earlier in a letter dated 9 January 1959.

81. Bodleian Library, Toynbee Papers, letter, Rosalind Murray to AJT, 22 January 1963.

82. Bodleian Library, Toynbee Papers, sheet on which AJT totted up values for property divided between his two sons by the wills of Gilbert Murray, Rosalind, and himself. The sheet is undated but was clearly connected with his letter to Rosalind, dated 16 December 1959, in which he explained his decision to reallocate shares in his property on a two-thirds to one-third ratio between Philip and Lawrence.

83. Bodleian Library, Toynbee Papers, letter, Philip Toynbee to AJT, 8 January 1974.

84. Bodleian Library, Toynbee Papers, letter, Philip Toynbee to AJT, 10 April 1956.

85. Bodleian Library, Toynbee Papers, letter, Philip Toynbee to AJT, 13 May, no year. If Philip calculated accurately on 23 October 1972, when he wrote in another letter to his father "I am taking a sabbatical from *my* nonsense book; suddenly realised that I have been either planning or writing it for the last 19 years, with virtually no break," his decision to write in verse must have been taken in or soon after 1953.

86. Bodleian Library, Toynbee Papers, letter, Philip Toynbee to AJT, 11 October 1961.

87. Bodleian Library, Toynbee Papers, letter, AJT to Philip Toynbee, 24 September 1962.

88. Republished in *Experiences*, pp. 105–111.

89. Bodleian Library, Toynbee Papers, text of AJT's speech, 6 May 1964.

90. *The Times*, "Towards One World," 14 April 1964.

11. The Closing Decade of a Busy Life

1. Christian B. Peper, ed., *An Historian's Conscience: The Correspondence between Arnold J. Toynbee and Columba Carey-Elwes, Monk of Ampleforth* (Boston, 1986), pp. 464–465, letter, AJT to Columba, 5 February 1966.

2. Ibid., p. 474, letter, AJT to Columba, 21 December 1966.

3. *Toynbee on Toynbee: A Conversation between Arnold J. Toynbee and G.R. Urban* (New York, 1974).

4. *Cities on the Move* (London, 1970), p. 235.

5. Bodleian Library, Toynbee Papers, letter, Jan Sutton to AJT, 4 December 1967.

6. Toynbee's letters to Sutton and A. Shalit, the Israeli scholar who aroused his ire, are occasionally acrimonious, even spiteful. "I have put him in a fix—which is, of course, what I meant to do," Toynbee gloated after rewriting large parts of Shalit's text. Bodleian Library, Toynbee Papers, letter, AJT to Jan Sutton, 28 February 1968. Since Shalit's original text is unavailable it is impossible to judge the rights and wrongs of the quarrel. What becomes evident, however, is Toynbee's immense knowledge of political and military details of Roman–Jewish encounters. When challenged, Toynbee could be a formidable antagonist, as Shalit discovered.

7. *Constantine Porphyrogenitus and His World* (London, 1973), p. viii. Toynbee's study of army corps districts took up a problem he had first explored as a boy of

fourteen, when he drew a map of the Byzantine military "themes" while recovering from a bout of pneumonia.

8. Private possession, letter, AJT to Philip Toynbee, 26 August 1969.

9. Bodleian Library, Toynbee Papers, letter, Paul Alexander to AJT, 17 May 1974.

10. *Constantine Porphyrogenitus*, p. 5.

11. Ibid., pp. 24, 630.

12. Bodleian Library, Toynbee Papers, letter, AJT to Margaret Usborne, 20 July 1968.

13. Oxford University Press reserved the right to publish the volume; Thames and Hudson produced it; and a negotiated financial arrangement settled accounts between the two.

14. Bodleian Library, Toynbee Papers, letter, Jane Caplan to AJT, 29 August 1972.

15. It was not a commercial failure, however, being distributed in the United States by the Book of the Month Club, which according to Jane Caplan "accounted, no doubt, for the bulk of sales." Letter, Jane Caplan to W.H. McNeill, 16 November 1987.

16. *Mankind and Mother Earth* (New York and London, 1976), p. 412.

17. For instance, Toynbee explained the industrial revolution as a result of the breakdown of traditional restraints on human greed. Ibid., p. 564.

18. *Acquaintances* (London, 1967) and *Experiences* (New York and London, 1969) were the principal vehicles for Toynbee's reminiscences. Planned as a single volume, Oxford University Press editors decided to divide the manuscript in two and held back *Experiences* for more than a year, in accord with their policy of trying not to oversaturate the market with Toynbee's books.

19. Peper, *An Historian's Conscience*, p. 157, letter, AJT to Columba, 23 February 1944.

20. Bodleian Library, Toynbee Papers, letter, AJT to Richard Acland, 26 April 1973. Marcion was a second century heretic who held that the God of the Old Testament, as the creator of the material world, was different from and in eternal opposition to the transcendent, loving deity of the New Testament. Toynbee's ideas do not seem to resemble Marcion's in spite of what he said to Acland.

21. Peper, *An Historian's Conscience*, p. 537, letter, AJT to Columba, 25 January 1972.

22. Bodleian Library, Toynbee Papers, translated excerpt from *Asahi Shimbun*, 12 March 1966. *Asahi Shimbun* is a Tokyo newspaper.

23. Bodleian Library, Toynbee Papers, excerpts from the translation of an article by Fusao Hayashi, writing in *Asahi Shimbun*, 15 March 1966.

24. *Impressions of Japan* (Tokyo, 1968) pp. 13, 44, 61, 81. Quotes are from a pamphlet reprint published in Tokyo with annotation in Japanese, intended for use of persons learning English.

25. In 1967 Toynbee vetoed a proposal to set up a Toynbee Society in England, yet he approved of the use of his name in exactly the same way in Japan in 1968. "The use of my name would be bound to revive controversies about my work," he wrote. "I dislike controversy; and I do not want to be diverted by it again from more constructive uses of my time and energy." Bodleian Library, Toynbee Papers, letter, AJT to John Peart-Binns, 27 September 1967. Yet another letter from the same archive, AJT to Masunosuke Takashina, 1 May 1968, approved the use of his name. Perhaps Toynbee felt he was less controversial in Japan or trusted Professor Takashina, the first president of Japan's Toynbee Society, more than he did John Peart-Binns.

26. Letter, Keisuke Kawakubo to W.H. McNeill, 5 August 1986. Professor Kawakubo, who teaches English at Reitaku University, is a standing director of the Toynbee Society.

27. Letter, Professor Shigeru Nakayama to W.H. McNeill, 29 June 1985.

28. Bodleian Library, Toynbee Papers, letter, AJT to Masunosuke Takashina, 14 October 1971.

29. Bodleian Library, Toynbee Papers, letter, AJT to Masunosuke Takashina, 11 October 1973.

30. Preface to A.J. Toynbee and K. Wakaizumi, *What Will Be Man's Future?* [Oxford Progressive English Students' Texts] (Tokyo, 1972), p. v. This pamphlet, published in Japan for use by students of English, is an edited-down version of parts of a larger work, *Surviving the Future* (London, 1971), which digested and rearranged the original Toynbee–Wakaizumi interviews to appeal to an English-language readership. The textual imprecision arising from having different editors work over an original oral record for different audiences reproduces in high tech all the phenomena of oral literary tradition. Gaps between what was actually said, what the Japanese newspaper published, first in article and then in book form, and what is available in different English versions are probably significant, and quite beyond my powers of discovery.

31. Ibid., p. vi.

32. Bodleian Library, Toynbee Papers, letter, Kei Wakaizumi to AJT, 11 May 1970.

33. Bodleian Library, Toynbee Papers, letter, AJT to Kei Wakaizumi, 16 November 1970.

34. Bodleian Library, Toynbee Papers, letter, Kei Wakaizumi to AJT, 6 May 1971.

35. *Surviving the Future,* p. 59.

36. Ibid., p. 55.

37. Kei Wakaizumi, letter to *The Times* (London), 30 November 1975.

38. These figures come from a magazinelike publication of Soka Gakkai International, entitled *Waves of Peace Towards the New Century* (1987), p. 1. Some estimates of Soka Gakkai membership in Japan ran much higher—up to 16 million, according to *Time,* 1 August 1983, p. 60.

39. Bodleian Library, Toynbee Papers, letters, Kei Wakaizumi to AJT, 11 October 1971 and AJT to Daisaku Ikeda, 23 December 1971.

40. Toynbee asked that the "gift" be delivered in one lump sum and paid over at a date different from the date of the contract for his fees so as to guard "against any possible attempt of the UK tax authorities to contend that this gift is taxable." Bodleian Library, Toynbee Papers, letter, AJT to Daisaku Ikeda, 11 November 1971. Ikeda also brought some "humble gifts" of Japanese art when he arrived in London.

41. Bodleian Library, Toynbee Papers, letter, Daisaku Ikeda to AJT, 10 May 1972.

42. Daisaku Ikeda, *The Human Revolution* (New York, 1972), Preface.

43. "I find your association with President Ikeda and his vast organization, that is not without its questionable techniques of coerced conversions and its fund raising potentials, just a little out of character." Bodleian Library, Toynbee Papers, letter, John Grisdale to AJT, no date.

44. Bodleian Library, Toynbee Papers, letter, AJT to John Grisdale, 9 July 1974.

45. Bodleian Library, Toynbee Papers, letter, AJT to Rolf Italiander, 21 July 1972.

46. Arnold Toynbee and Daisaku Ikeda, *Choose Life: A Dialogue* (London, 1976), pp. 11–12.

47. Ibid., p. 284.

48. Ibid., p. 238.

49. Ibid., p. 106.

50. My remarks about Soka Gakkai and Daisaku Ikeda depend partly on a sampling of the sect's propaganda literature that Ikeda sent me, partly on a book by James White, *The Sokagakkai and Mass Society* (Stanford, 1970), and partly on a sampling of counterpropaganda sent to me by R.E. Schecter of the American Family Foundation. Conversation with my colleague at Chicago Professor Tetsuo Najita was also helpful.

51. Peper, *An Historian's Conscience*, p. 468, letter, AJT to Columba, 1 July 1966.

52. Ibid., p. 474, letter, AJT to Columba, 21 December 1966.

53. Bodleian Library, Toynbee Papers, letter, John Trexell to President Sterling, 30 December 1966.

54. Bodleian Library, Toynbee Papers, letter, AJT to Lyle Nelson, 11 January 1967.

55. Bodleian Library, Toynbee Papers, letter, AJT to Lyle Nelson, 27 January 1967.

56. Bodleian Library, Toynbee Papers, letter, Philip Toynbee to AJT, 15 April 1967.

57. Peper, *An Historian's Conscience*, pp. 477–478, letter, AJT to Columba, 2 May 1967.

58. Ibid., pp. 479–480, letter, AJT to Columba, 11 May 1967.

59. Bodleian Library, Toynbee Papers, letters, AJT to Stephen Cary, 3 February 1967 and AJT to E. Vaudrin, 19 January 1967. Lyle Nelson, the academic entrepreneur who arranged Toynbee's visit to Stanford, may have misled Toynbee and the American Friends Service Committee as to exact terms of agreement. But Toynbee was also at fault, inviting the break by greedily scheduling extra lectures for additional fees along the way while expecting the American Friends Service Committee to pay all his traveling expenses and those of Veronica. Like his financial arrangements with Ikeda in 1973, these letters betray an unbecoming and almost hysterical grasp after money that he did not really need—behavior starkly in contradiction to his public denunciations of greed and frequent praise of St. Francis' moral example. Toynbee's physical debilities apparently undermined his good manners and common sense in matters of money, allowing old anxieties to surface and dominate his behavior as never before.

60. Institut de France, Académie des Sciences Morales et Politiques, *Installation de M. Arnold Toynbee comme Associé Étranger* (Paris, 1968), p. 11. The translation is my own. The tone of the President's remarks about Toynbee seems distinctly cool, but any debates about his qualifications that took place among the members of the Academy before Toynbee's election remain secret. Toynbee's admirers in France were few, and his election to this dignity therefore remains surprising, all the more since the seat was not permanently assigned to an Englishman. In fact, Toynbee was succeeded in due course by a Swiss literary critic named Marcel Raymond.

61. Ibid., p. 35. My translation.

62. *The Times,* 10 July 1972 and 16 July 1972. An English version of Toynbee's invocation to Apollo is also to be found among his papers in the Bodleian Library.

63. Peper, *An Historian's Conscience*, pp. 545–546, letter, AJT to Columba, August 1972.

64. Bodleian Library, Toynbee Papers, letter, AJT to Robin Weeks, 13 August 1973.

65. Peper, *An Historian's Conscience*, pp. 539–540, letter, AJT to Columba, 14 March 1972.

66. Bodleian Library, Toynbee Papers, invitations for 12 April 1972 and 19 June 1973, with formal notes of declination.

67. Peper, *An Historian's Conscience*, p. 511, letter, AJT to Columba, 18 July 1969.

"Refraining from grasping" is a Buddhist expression. O.M. signifies Order of Merit, and was a higher honor than the C.H. (Companion of Honour) that Toynbee had received in 1956.

68. Private possession, letter, AJT to Philip Toynbee, 10 July 1969.

69. Peper, *An Historian's Conscience,* p. 511, letter, AJT to Columba, 18 July 1969. The poem in question was discussed in Chapter 8.

70. Private possession, letter, AJT to Philip Toynbee, 22 December 1969. For a telling sketch of the Barn House Community, and of how Philip failed to live up to his principles, running off to have fancy meals with old aristocratic friends, see Jessica Mitford, *Faces of Philip: A Memoir of Philip Toynbee* (New York, 1984), pp. 124–144. Philip Toynbee, *Part of a Journey: An Autobiographical Journal, 1977–1979* (London, 1981) also contains scattered references to his experiment in communal living (pp. 12, 365, and elsewhere).

71. Bodleian Library, Toynbee Papers, letter, Philip Toynbee to AJT, 4 June 1973. "Some animals being more equal than others" is a reference to George Orwell's *Animal Farm.*

72. Bodleian Library, Toynbee Papers, letter, AJT to Lawrence Toynbee, 15 November 1972.

73. Bodleian Library, Toynbee Papers, letter, Philip Toynbee to AJT, 8 January 1974.

74. Bodleian Library, Toynbee Papers, letter, AJT to Philip Toynbee, 15 January 1974.

75. Private possession, letter, Veronica Toynbee to Sally Toynbee, 4 July 1973. Agnes, Rosalind's younger sister, had died in 1922.

76. Bodleian Library, Toynbee Papers, letter, Philip Toynbee to AJT, 2 July 1973.

77. Bodleian Library, Toynbee Papers, letter, Philip Toynbee to AJT, 26 June 1973.

78. Philip's insight and sensitivity did not extend to his own behavior or allow him to recognize how much he resembled his mother in making financial demands on his brother and father, and, while pursuing an ideal of communal living, excepted himself from obedience to community rules.

79. Bodleian Library, Toynbee Papers, letter, Lawrence Toynbee to AJT, 26 June 1973.

80. Peper, *An Historian's Conscience,* p. 569, letter, AJT to Columba, 20 January 1974.

81. Ibid., p. 572, letter, AJT to Columba, 22 February 1974.

82. Ibid., p. 577, letter, AJT to Columba.

83. Ibid., p. 577, letter, AJT to Columba, 27 June 1974.

84. Ibid., pp. 579–580, letter, AJT to Columba, 11 July 1974. The text of Toynbee's remarks with a translation are reproduced on pp. 580–584.

85. Ibid., pp. 585–586, letter, Veronica Toynbee to Columba, 16 August 1974.

86. Martin Wight, "Arnold Toynbee: An Appreciation," *International Affairs,* January 1976, pp. 10–12.

Conclusion

1. Bodleian Library, Toynbee Papers, transcript of NBC program, "Issues and Answers," 10 January 1965.

2. Bodleian Library, Toynbee Papers, clipping from *Sunday Times,* 15 October 1972.

3. Bodleian Library, Toynbee Papers, *Daily Telegraph Magazine,* 17 April 1970, p. 17.

4. Montesquieu's *Lettres persanes* (1721) and Voltaire's *Essai sur les moeurs* (1756),

along with Herder's *Ideen zur Philosophie der Geschichte der Menschheit* (1784–1791), are the great monuments of this tradition.

5. Translations of Toynbee's works made his ideas widely accessible. Thus, for example, his greatest popular success, Somervell's abridgement of the first six volumes of *A Study of History,* appeared in Arabic, Danish, Dutch, Finnish, French, German, Gujarati, Hindi, Italian, Japanese, Norwegian, Portuguese, Serbo-Croat, Spanish, Swedish, and Urdu.

Index